The Royal Society and the Promotion of Science since 1960

The Royal Society is one of the world's oldest and most prestigious scientific bodies, but what has it done in recent decades? Increasingly marginalised by postwar developments and the reforms of civil science in the 1960s, the Society was at risk of resting on its laurels. Instead, it found ways of exploiting its unique networks of scientific talent to promote science. Creating opportunities for outstanding individuals to establish and advance research careers, influencing policy-making at national and international levels, and engaging with the public outside the world of professional science, the Society gave fresh expression to the values that had shaped its long history. Through unparalleled access to the Society's modern archives and other archival sources, interviews with key individuals and extensive inside knowledge, Peter Collins shows how the Society addressed the challenges posed by the astounding growth of science and by escalating interactions between science and daily life.

PETER COLLINS worked at the Royal Society from 1981 to 2013, responsible primarily for the science policy function and latterly for governance and for history of science. These roles included substantial engagement in international affairs and in often controversial public debates. As a long-term core member of senior staff, he was closely involved in development and delivery of the Society's strategy, and had a ringside seat at many key events in this period. In addition to many Royal Society reports, he has published on the history of the British Association and of the Royal Society, including a volume of conference proceedings on the Society in the twentieth century.

The Royal Society and the Promotion of Science since 1960

Peter Collins

CAMBRIDGE
UNIVERSITY PRESS

CAMBRIDGE
UNIVERSITY PRESS

University Printing House, Cambridge CB2 8BS, United Kingdom

One Liberty Plaza, 20th Floor, New York, NY 10006, USA

477 Williamstown Road, Port Melbourne, VIC 3207, Australia

314-321, 3rd Floor, Plot 3, Splendor Forum, Jasola District Centre, New Delhi - 110025, India

79 Anson Road, #06-04/06, Singapore 079906

Cambridge University Press is part of the University of Cambridge.

It furthers the University's mission by disseminating knowledge in the pursuit of education, learning and research at the highest international levels of excellence.

www.cambridge.org
Information on this title: www.cambridge.org/9781108705806

First published 2016
First paperback edition 2019

A catalogue record for this publication is available from the British Library

Library of Congress Cataloging in Publication data
Collins, Peter, 1952–
The Royal Society and the promotion of science since 1960 / Peter Collins
     pages   cm
Includes index.
ISBN 978-1-107-02926-2
1. Royal Society (Great Britain) – History.   2. Science – Great Britain – History – 20th century.   I. Title.
Q127.G4C65   2015
506'.041–dc23

2015029275

ISBN 978-1-107-02926-2 Hardback
ISBN 978-1-108-70580-6 Paperback

For Geralyn

Contents

Figures

Tables

Preface

The Royal Society is about science. Its Fellows, a group with a wide range of conflicting opinions on almost every subject, are united in their passion for science. The Society is embedded in the scientific life of the UK, and is recognised throughout the world where science flourishes. Its mission, in the words of its founders, is 'promoting by the authority of experiments the sciences of natural things and of useful arts, to the glory of God the Creator and the advantage of the human race'. But the Society does not directly employ scientific researchers. Rather, it uses its particular attributes and resources in other ways to promote the sciences of natural things. Exactly how it does this reflects its understanding of its own identity and of the opportunities available to it. That understanding and those opportunities developed markedly in the years after the Second World War, and particularly after 1960, against a background of astonishing growth in the scientific enterprise and unprecedentedly intense interactions between scientists and non-scientists.

My aim in this book is to analyse some key features of the Society's approach to promoting science during this period, and thus to uncover something of its identity. This can be only a partial undertaking, focused on the institutional life of the Society. Individuals who have encountered the Society may assess it in other terms, for example its positive or negative impact on their own careers, or its support or opposition for particular points of view, or what they see it as representing more generally. The Society can arouse strong feelings, and these feelings are germane to its identity. But they are difficult to disentangle on a sufficient scale from other factors operating at the personal level. My task is the simpler one of looking at the documentable behaviour of the Society as a corporate entity in the context of what was happening elsewhere in the world of science.

The Society's most defining characteristic is its concern with the highest standards in science. For example, it elects its Fellows in a hugely competitive process from among those it judges to have made the most substantial contributions to science. This is, of course, a

potentially hubristic undertaking. At various times the Society has been
slow to recognise not just particular individuals but also the achievements of women scientists, and of applied scientists, and of scientists
working outside the major academic centres or outside academe altogether. Nevertheless, by common consent, the Royal Society is unequivocally the elite body for natural science as a whole within its
geographical span.

There is an element of potential paradox here. Elitism, in the negative
sense of an arbitrary exercise of self-proclaimed authority, is a suffocating
force, destructive of scientific creativity. The Society's motto recognises
that, indeed warns against it. *Nullius in verba*, or 'Not committed to
swearing by the words of any master',[1] advocates freedom to reach
one's own conclusions, a rejection of human authority as a source of
truth in favour of evidence from critical observation and experiment.
The motto celebrates the fact that the talent needed to make scientific
advances can be found at almost any point in the spectrum of seniority. By
gathering into its fold some of those who have made the most substantial
contributions to scientific knowledge, and by securing enough public
assent to its judgement in this, the Society acquired authority in scientific
matters – despite its disavowal of the concept. It was accepted as specially
skilled in the business of sifting scientific evidence, elite in the positive
sense. But in order to flourish, or even survive, during the period covered
in this book, the Society had to exercise its authority with great care,
grounding it at all times in scrupulous attention to scientific evidence and
always alert to new developments that might challenge its established
positions. And it had to learn to look outward beyond itself and to use
its authority to work for the public good.

It all begins with the Fellowship. Election to the Fellowship brings a
tacit trade-off between the individual Fellow and the corporate Society:
the Society associates itself with the scientific achievements of the new
Fellow – which may or may not owe something to the Society's earlier
support – and the Fellow is affirmed by public recognition of his or her
merit as a scientist. There is mutual benefit, each bestowing kudos on the
other. The Society therefore subjects the election process to constant and
intimate scrutiny to ensure, within the subtle limitations of human behaviour, that it judges well whom to admit. At a strategic level, too, the
election process holds a key to the Society's identity, in that it can be used

---

[1] The full quotation, *Nullius addictus iurare in verba magistri*, is from Horace; the translation
is by the Society's President Andrew Huxley. The context is that of a newly freed slave who
no longer has to agree with everything his ex-master said. The common colloquial
translation, 'Take nobody's word for it', misses some of the richness of the original.
Andrew Huxley, 'Nullius in verba', *Nature*, 315 (23 May 1985), 272.

to express new priorities, new recognition of where and how the most substantial contributions to science are being made.

Beyond the individual achievements of its current Fellowship, the Society also carries considerable aura from the sheer fact of its longevity and from its association with many of the famous scientists one has heard of from earlier periods (as well as many now forgotten). Now well into its fourth century, the Society's unparalleled history adds to its authority, and to the need for care in how it is exercised.

The Royal Society was founded in 1660. Its early days have long attracted intense historical scrutiny. But its recent history has not. A two-day conference in April 2010 on the Society in the twentieth century was said, only partly in jest, to have tripled the amount of scholarship on the subject.[2] Longevity in human affairs may imply that something important is going on. The Society's motto was revolutionary when it was chosen in 1662. It is still revolutionary. An organisation trying to live up to such a motto, and on that account held in high esteem around the world, is a rich object of study.

In a slightly arbitrary way, this book focuses on the period 1960–2010, from the Society's 300th anniversary to its 350th. The time frame is not rigid. Some parts of my narrative start in 1945 or earlier, while others stop before 2010 or go beyond it almost to the present day. I have not gone back to the beginning of the century to pick up the story from where Marie Boas Hall left it,[3] nor have I set out to record all of the Society's postwar activities. Rather, I have concentrated on aspects of its work that usefully illustrate its identity. Some of the Society's core characteristics, such as its commitment to the highest standards in science, have been constant throughout the period of this book; others, most obviously its dealings with the public beyond the world of professional scientists, have developed radically. There is continuity and there is change, as with most human institutions, and both are important.

The Royal Society is very much alive, and the history of its recent past must therefore butt against its unfinished present. I have not dealt evenly with all parts of the period. The most recent years, in particular, are generally covered more sketchily than earlier years, not only for reasons of due confidentiality but also because of the difficulty in judging long-term significance. On the other hand, I have in several places, and especially in the final chapter, added comments, based on reflections on the

---

[2] Jeff Hughes, 'Introductory comments to final discussion session', *Notes Rec R Soc*, 64 (2010), S173.

[3] Marie Boas Hall, *All scientists now: the Royal Society in the nineteenth century* (CUP, 1984).

historical narrative, that may hold interest for those concerned with the current Society.

Contemporary history is not the only hazard of this undertaking. I worked for the Society from 1981 to 2013, responsible at various times for the policy advice function, for aspects of governance and for history of science. There is a fine tradition of people closely associated with the Society writing about its history,[4] so I am in good company, but I recognise that I cannot be wholly objective. On the other hand, there are some advantages in having inside knowledge. I have, in short, written the sort of history that someone in my position could write, in the hope that it will have value alongside the histories that may be written by other people with other prejudices.

This book is aimed at historians of science, at scientists and at others interested in the Society, including those concerned with national academies of science in other countries. It will also be of interest to those concerned with the postwar period more generally, since the Society's work touched many areas of public life. And, I hope, it will prove of use to those charged with guiding the Society through future phases of its existence. That range of audiences might ideally need a corresponding range of approaches, which is a further hazard: some may have wanted more about how the Society has affected particular areas of research, while others may have wanted more on the Society's dealings with other organisations, or more international comparisons, or more analysis of the accolade function, or a fuller historiographical commentary, and so on. I hope the current book will be found a useful starting point for such further studies.

The opening chapter, about the unusually controversial election of a new President in 1945, highlights some of the strategic challenges facing the Society at the end of the Second World War, in particular how much it was going to engage with public life and hence what sort of leadership it needed. This dilemma set stellar scientific achievement in contrast to political nous among the criteria for the incoming President. The issue of how much political nous mattered, and whether there was any scope (or need) for compromise on scientific achievement in order to secure it, would resonate increasingly in later elections. In describing how the 1945 election unfolded, the chapter uncovers some of the inner workings of the decision-making process at that time.

The second chapter examines the difficulties that the Society faced as, against its spirited opposition, the government took increasing control of

---

[4] For example, Thomas Sprat (1667), Thomas Birch (1756), C.R. Weld (1848), Archibald Geikie (1912), Henry Lyons (1940, 1944), Percy Andrade (1960), Harold Hartley (1960), John Rowlinson and Norman Robinson (1989).

civil science. The Trend review of civil science in the mid-1960s created new agencies and left the Society largely bereft of official function in running UK science. So, as an independent body outside the government system, and with relatively modest resources, the Society was challenged to reassess its particular strengths and weaknesses and to think imaginatively about just what it could do to promote science in these new circumstances. Under Howard Florey's leadership, the Society became more rather than less determined to be at the centre of affairs.

Subsequent chapters explore specific ways in which the Society responded to that challenge. Chapter 3 traces its financial and moral support for outstanding individual scientists at various stages of their careers, adding practical patronage of individual talent to its more nebulous accolade function. Such activity started before the Second World War but expanded massively afterwards, especially from the 1980s. In terms of fostering individual creativity and providing opportunity for the most promising researchers, the Society carved out an important and effective role for itself that played well to its instinctive strengths.

Chapter 4 deals with the Society's efforts to concern itself with the applications of science, the practical advantage of the human race. This was a more complicated niche, in which the Society struggled to define a clear role. Already starting to worry about the issue in the early 1960s, the Society was pushed into action by threats to create a separate elite body for engineering. It succeeded in heading off that dénouement for a while, and, when the Fellowship of Engineering was eventually established in 1976, insisted that the applications of science (including, but not limited to, engineering) remained very much within its purview. Its impact in this area has to date been at best ambivalent, but there is now a renewed commitment to the cause.

Chapter 5 analyses the Society's policy advice function and its efforts to defend the Science Base both to government and to public opinion. This is a long-running theme that has expanded very considerably in recent decades. The Society had clear and consistent views on how to manage the Science Base so as to promote fundamental science in particular, and it argued the case at every opportunity. It also addressed more publicly controversial aspects of research such as genetic engineering, animal experiments, in vitro fertilisation and cloning. The Society's direct involvement with the wider public, both in terms of promoting familiarity with scientific processes and findings and in terms of engagement on areas of scientific and social controversy, is the subject of Chapter 6. The Society made it respectable for successful research scientists to devote effort to improving public understanding of science. It also got deeply involved in public debates about such matters as acid rain, depleted uranium, BSE, genetic modification, nanotechnology and climate change. In so doing it gained valuable experience in interacting with policy processes, and it helped to shape the making of public policy.

The next four chapters consider the Society's impact at the international level. Chapter 7 deals with the Society's role in helping the British Government recognise that civil science, and not just military science, was relevant to international relations in postwar conditions. The Society's global reputation and its independence from government enabled it to contribute significantly to diplomacy at the same time as promoting the interests of science. Chapter 8 details examples of scientific relations being used to mitigate diplomatic tensions, with the Soviet Union, with China, with South Africa and with Argentina, at periods when it was difficult to conduct government-to-government relations. The work could be controversial, especially where there were diverse views within the Fellowship over how most effectively to respond to human rights abuses.

The growing European dimension in scientific affairs, mirroring the growing European dimension in British public life, is the subject of Chapter 9. Challenged by its Australian President Howard Florey in 1965 to be 'good Europeans', the Royal Society launched its European Science Exchange Programme two years later with 16 partner countries; by the end of the century, the Programme had made 10,000 awards. Particular disciplines, notably molecular biology, were also busy establishing European groupings, and the Society was active in the associated debates. It was active, too, in European groupings of research funders and of national academies, and pushed hard to ensure that such groupings were run by scientists rather than civil servants. And it extended its policy advice work to policy-makers at the European level.

The Society's many-faceted role on the global scientific stage, including at the Commonwealth level, is discussed in Chapter 10. In 1900, one seventh of the Society's Fellows and Foreign Members lived outside the UK; by 2010, that figure had grown to one third. That was one response to the globalisation of science during the twentieth century. Other responses are seen in the Society's capacity-building activities, its involvement with global scientific bodies, its publishing operations and its scientific expeditions.

The final chapter offers some reflections on the identity of the Royal Society and how it has developed since the War, in response to its own internal dynamics and to changing external circumstances. The Society has a remarkable history and a global reputation among scientists. These were not enough in themselves to ensure continued useful existence in the complex and competitive conditions of the postwar world. The Society is a private organisation, but it could not live its life privately. It had to engage imaginatively with key groups beyond its Fellowship, to extend its established practices and to find new ways of promoting science. Its 350th anniversary in 2010 in effect celebrated its success in becoming more outward looking over the previous 50 years.

The Society is both a corporate entity and a collection of individuals. For the Society to express a corporate view is a matter of the elected Council (or honorary Officers acting on behalf of Council) agreeing what line to take; it does not imply that every Fellow shares the corporate view, still less that every Fellow has been explicitly consulted. The Annex provides information, in an historical context, about how the Society operates formally: aspects of its governance, including the role of the Council and honorary Officers; finance; and the process of electing Fellows. It includes a detailed list of all Officers and Executive Secretaries in post between 1945 and 2015. Those unfamiliar with the Society may find it convenient to look at this Annex before embarking on the rest of the text.

The sources of evidence for this book are described in 'Sources'. They are of four types. First, the Society's own archives, as valuable for the twentieth century as for the seventeenth and including modern administrative records en route to permanent archival status. I have enjoyed the privilege of unrestricted access to the complete collection. I have also made extensive use of archives from other institutions, including the National Archives at Kew, and of various holdings of personal papers. Second is published material, including that arising from the 2010 conference on the Royal Society in the twentieth century. One of the features of contemporary history is the possibility of consulting surviving participants. I therefore carried out interviews with nearly 60 individuals able to comment from personal experience on the Society as an institution. The transcripts from these interviews constitute a third source of evidence, valuable not only for this project but also for future studies, and some provide unique evidence of the personal impact that the Society can have. They are available for consultation at the Society's Centre for History of Science. A fourth source, informal but vital, is colleagues at the Society who have responded untiringly to requests for information.

To make both the writing and the reading easier, I have omitted formal titles when using people's names, if only because, in quite a few cases, the titles changed as a person's career developed. I have similarly omitted the conventional post-nominal 'FRS'; individuals mentioned in the text who are or were Fellows or Foreign Members are shown as such, with their dates of election, in their index entries.

This is not an 'official' history, in the sense of having been commissioned or vetted by the Royal Society. Nor is it an anniversary history, expected by convention to celebrate its subject. It has, nonetheless, been written with the warm support and encouragement of the Society, which I most gratefully acknowledge. This has made my task far easier and more genial.

I did not appreciate at the outset just how many people would be involved in the solitary occupation of researching and writing this book. It is a pleasure here to record my indebtedness to those who facilitated what turned out, of course, to be a far from solitary experience. Archivists and librarians at the Royal Society, the National Archives and numerous other institutions (see 'Sources') eased my path with their knowledge and good cheer. Royal Society staff generously fielded my requests for detailed information. Joanna McManus at the Society's Centre for History of Science did valuable work on the illustrations; Bill Johncocks in the wilds of Skye produced a very thorough index; and Michael Watson and others at CUP steered the book through the publication process.

Many individuals provided encouragement and specific help through discussion and through commenting on segments of the draft text. With apologies for inadvertent omissions, I should like here to thank in particular Jon Agar, David Boak, Walter Bodmer, Robert Bud, Bob Campbell, Simon Campbell, Lorna Casselton, Peter Cooper, Peter Cotgreave, Ruth Schwartz Cowan, Keith Davis, Laura Dawson, David Edgerton, Brian Follett, Robert Fox, Terry Garrett, Phil Gummett, Hans Hagen, Brian Heap, Julia Higgins, Jeff Hughes, Dan Kevles, John Krige, Peter Lachmann, Tony McBride, Paul Nurse, John Pethica, Rachel Quinn, Keith Root, Simon Schaffer, John Skehel, George Stirling, David Walker, Wang Zuoyue, Peter Westwick, Rapela Zaman, three anonymous and thoughtful referees, and the participants in a conference I organised in April 2010 on the Royal Society in the twentieth century. I should also like to thank most warmly those who consented to be interviewed on the record about their experiences of the Society.

Three individuals – Pat Bateson, Julie Maxton and Stuart Taylor – demonstrated the meaning of friendship by reading the entire draft text. The final product owes much to their insightful comments, challenges and generous encouragement over a sustained period. I am greatly in their debt.

My wife, Geralyn, not only read and commented on the entire draft text but also coped with me during the writing of it and, indeed, during the three decades of work at the Royal Society that preceded the writing. I gladly take this opportunity to value her role in this as in much else.

PETER COLLINS
March 2015

Abbreviations

ABRC	Advisory Board for the Research Councils
ACARD	Advisory Council for Applied Research and Development
ACME	Advisory Committee on Mathematics Education
ACSP	Advisory Council on Scientific Policy
ALLEA	All European Academies
ARC	Agricultural Research Council
ASE	Association for Science Education
BA	British Association for the Advancement of Science
BBC	British Broadcasting Corporation
BSE	bovine spongiform encephalopathy
CAS	Chinese Academy of Sciences
CAST	Chinese Association for Science and Technology
CEGB	Central Electricity Generating Board
CEI	Council of Engineering Institutions
CERN	Organisation Européenne pour la Recherche Nucléaire
CISC	Committee on International Scientific Cooperation
CJD	Creutzfeldt-Jakob disease
CONICET	(Argentina) Consejo Nacional de Investigaciones Científicas y Técnicas
COPUS	Committee on Public Understanding of Science
COSR	Committee on Overseas Scientific Relations
CPRS	Central Policy Review Staff
CSIR	(South Africa) Council for Scientific and Industrial Research
CSA	Chief Scientific Adviser
CSP	Council for Scientific Policy
DES	Department of Education and Science
DfID	Department for International Development
DoE	Department of the Environment
DPRC	Defence Policy Research Committee
DSIR	Department of Scientific and Industrial Research

DTC	Department of Technical Cooperation
EASAC	European Academies Science Advisory Council
EEC	European Economic Community
EFSF	European Fundamental Science Foundation
EIJC	Engineering Institutions Joint Council
EMBC	European Molecular Biology Conference
EMBL	European Molecular Biology Laboratory
EMBO	European Molecular Biology Organisation
EOC	Equal Opportunities Commission
EPS	European Physical Society
ERC	European Research Council
ESEP	European Science Exchange Programme
ESF	European Science Foundation
ESFP	European Science Fellowship Programme
ESRC	Economic and Social Research Council
ETI	(Royal Society) Engineering, Technology and Industries Committee
EUCHEM	European Association for Chemical and Molecular Sciences
EuroHoRCs	European Heads of Research Councils
FA	Football Association
FCO	Foreign and Commonwealth Office
FRD	(South Africa) Foundation for Research and Development
FRS	Fellow of the Royal Society
GM	genetically modified
IAC	(Royal Society) Industrial Activities Committee
IAC	InterAcademy Council
IAP	InterAcademy Panel
IAU	International Astronomical Union
IBP	International Biological Programme
ICSU	International Council of Scientific Unions (from 1998, International Council for Science)
ICT	information and communications technology
IGY	International Geophysical Year
IIASA	International Institute for Applied Systems Analysis
IoP	Institute of Physics
IPCC	Intergovernmental Panel on Climate Change
IQSY	International Quiet Sun Year
IRC	Interdisciplinary Research Centre
IRDA	Industrial Research and Development Authority

IUPPS	International Union for Prehistoric and Protohistoric Sciences
IVF	in vitro fertilisation
MEP	Member of the European Parliament
MRC	Medical Research Council
NAPAG	National Academies Policy Advisory Group
NAS	(USA) National Academy of Sciences
NATO	North Atlantic Treaty Organization
NERC	Natural Environment Research Council
NGO	non-government organisation
NIRNS	National Institute for Research in Nuclear Science
NPL	National Physical Laboratory
NRRC	Natural Resources Research Council
OECD	Organisation for Economic Co-operation and Development
ORC	Overseas Research Council
PGA	Parliamentary Grant-in-Aid
PRS	President of the Royal Society
RDS	Research Defence Society
RI	Royal Institution
RSSAf	Royal Society of South Africa
R&D	research and development
SAC	Scientific Advisory Committee (to the War Cabinet)
SAIS	(Royal Society Committee on) Scientific Aspects of International Security
SCORE	Science Community Partnership for Supporting Education
SEPSU	(Royal Society/Fellowship of Engineering) Science and Engineering Policy Studies Unit
SERC	Science and Engineering Research Council
SGM	special general meeting
SRC	Science Research Council
SRGC	Scientific Research Grants Committee
SZR	(Royal Society) Southern Zone Research Committee
S&T	science and technology
TWAS	Originally the Third World Academy of Sciences, now the World Academy of Sciences for the advancement of science in developing countries
UCL	University College London
UFC	Universities Funding Council
UGC	University Grants Committee
URF	(Royal Society) University Research Fellow

1 Presidential politics and postwar priorities

None of us ... has known what the Fellows really think about him.[1]

The election of a new President of the Royal Society in 1945 turned out to be good deal more interesting than these things usually are. It was one of those occasions when a significant number of Fellows decided not to leave it all to the trusted process of private soundings leading to a recommendation by the Officers[2] duly ratified by Council, but instead set out to make their voices heard directly. It was also one of those occasions when the Society seemed to be at something of a turning point. So the presidential debate was not just about personalities: it was also about the Society's identity and future direction. Indeed, it rehearsed issues that would loom large in the Society's affairs later in the century, and it exposed difficulties that the Society would find itself repeatedly having to face. And, as it turned out, it illustrated how high-level strategic considerations can sometimes be overwhelmed by capricious happenstance.

The Royal Society in 1945

Three of the Royal Society's core leadership team of five Officers were due to stand down at the end of 1945: the President (Henry Dale), the Biological Secretary (A. V. Hill) and the Foreign Secretary (Henry Tizard). A year ahead of the event, that prospect was giving rise to much discussion.

---

[1] Henry Dale to Henry Tizard, 3 June 1945, reflecting on the business of enabling individuals to make professional judgements about each other without rupturing personal relationships: Tizard papers, Imperial War Museum, #427. Some of the material in this chapter has already appeared in Peter Collins, 'Presidential politics: the controversial election of 1945', *Notes and records of the Royal Society of London*, 65 (2011), 325–42, and is reproduced here with permission.

[2] The Officers are five Fellows who fill particular functions in the Society's governance on a volunteer basis; for details, see Annex. In 1945, the President and Foreign Secretary could serve up to five years, and the Treasurer and Biological and Physical Secretaries could serve up to ten years, all subject to annual re-election.

It was not only a matter of administrative disruption. The Society's role in public life also seemed to be at stake at that time. Its leadership had aspired to a corporate role in shaping science policy during the Second World War, but this had proved a frustrating experience, and the Society had found itself working mostly outside the established structures, its ties with government 'informal, discrete, ubiquitous'.[3] Many Fellows recognised that the postwar Society would be operating in a radically new context for science and would need to develop new skills. The position of science had been transformed during the War; its peacetime status had yet to be negotiated. The growth of public spending on science brought with it the spectre of increased government control. There had been vigorous debates before and during the War about centralised planning in science and about how to maximise the social benefits of science.[4] Individual Fellows featured prominently on various sides of these debates, but the Royal Society corporately had kept a low profile. The growing prominence of these issues towards the end of the War challenged the Society's incoming leadership team to engage more openly with public controversy.

One specific trigger for internal debate about the Society's postwar role was a letter to the Society's Officers from the physicists Ralph Fowler and Patrick Blackett in October 1943. They were worried about how fundamental physics would fare in a postwar world that they expected would prioritise the applications of science. This could not, in their view, be left to chance: it needed organised advocacy if the necessary resources were to be secured for fundamental physics, and it needed organised oversight if the extra resources were to be used well.

The defence of fundamental research was an archetypal cause for the Royal Society. In response to the Fowler/Blackett letter, the Society set up a series of committees to examine not only physics but also seven other broad areas of science. The ensuing report, colloquially known as the *Postwar*

---

[3] Philip Gummett, *Scientists in Whitehall* (Manchester University Press, 1980), 30–1, 93–5; Philip J. Gummett and Geoffrey L. Price, 'An approach to the central planning of British science: the formation of the Advisory Council on Scientific Policy', *Minerva*, 15 (1977), 121; Ronald W. Clark, *Tizard* (Methuen, 1965), 273–5; William McGucken, 'The Royal Society and the genesis of the Scientific Advisory Committee to Britain's War Cabinet, 1939–1940', *Notes and records of the Royal Society of London*, 33 (1978), 87–115; Stuart S. Blume, *Toward a political sociology of science* (Collier Macmillan Publishers, 1974), 191; John Peyton, *Solly Zuckerman: a scientist out of the ordinary* (John Murray, 2001), 109–10; Tizard to Dale, 1 June 1945: HD/6/2/4/6; Dale to Tizard, 5 June 1945: HD/6/2/4/7.

[4] See, for example, Peter Collins, 'The British Association as public apologist for science, 1919–1946', in Roy MacLeod and Peter Collins, eds., *The Parliament of science* (Science Reviews Ltd, 1981); William McGucken, *Scientists, society and the state: the social relations of science movement in Great Britain, 1931–1947* (Ohio State University Press, 1984); Gary Werskey, *The visible college* (Allen Lane, 1978), 244.

needs report and circulated to the Fellowship in January 1945, concluded that academic scientific research would have to expand so much in future that it could no longer rely on private benefactions or the block grants allocated by the University Grants Committee but, in the national interest, would require major direct inputs of public money.[5] Blackett wanted the Society to be the body controlling these financial provisions for research.[6] The Society's Council would not go that far, but the *Postwar needs* report did argue – more strongly than it would have done a few years earlier[7] – for a central role for the Society in advising the Treasury on how such money should be spent. It also argued that the Society should receive substantial increases in funding for its own programmes of research grants, travel grants and publication grants. The Society at that stage wanted to be actively involved in the national organisation of science, but through exercising influence rather than serious power and through securing enough resources to pursue its own niche initiatives.

The Society's balancing act was highlighted at a small meeting of Fellows in May 1945 to discuss the *Postwar needs* report. Here A.V. Hill argued, in agreement with Blackett, that the Society was the most appropriate body for 'guiding and stimulating the healthy and balanced development of scientific enquiry taken as a whole'. The Fellows agreed, but thought that healthy and balanced development would be achieved naturally if each university always chose the most distinguished research leaders for its posts. Central planning was unnecessary provided there was sufficient spontaneous support for the less fashionable areas of research.[8] This was management with a very light touch.

The debate about the Society's postwar aspirations was not a purely private matter. The radical science journalist J.G. Crowther speculated in the *New Statesman* in December 1944 how the Society might respond to wartime developments and peacetime opportunities.

---

[5] Royal Society, *The needs of research in fundamental science after the war* (printed January 1945; also at Appendix A to CM 14 December 1944). Also CM 4 November 1943, minute 16; CM 30 November 1943, minute 7; CM 13 July 1944, minute 11 (b); and CM 12 October 1944, minute 6. Fowler died in July 1944, so did not see the outcome of his initiative. The report became known as the 'pink paper', because of the colour of its cover: see interview with Bernard Lovell.

[6] Bernard Lovell, 'PMS Blackett', *Biographical memoirs of Fellows of the Royal Society*, 21 (1975), 102. Also an important talk by Bernard Lovell to the Association of British Science Writers in October 1984, 'Authority in science': D.C. Phillips papers, MS Eng. c.5510, O.121.

[7] Gary Werskey, *Visible college*, 273.

[8] A.V. Hill, 'The needs of special subjects in the balanced development of science in the United Kingdom', *Notes and records of the Royal Society of London*, 4 (1946), 133–9.

What relation is the Society to have to these new and immense scientific activities, many of them conducted and financed by Government? Is it to have a directive function? ... Hasn't the policy of the last hundred years unfitted the Society for the role of statesmanship? If so, shouldn't the Society reform itself again on the original Baconian lines, rather like the Soviet Academy of Sciences, with definite official status, resources and powers?

Crowther was worried that the Fellowship had become so specialised since competitive elections were introduced in 1847 (see Annex) that it lacked the broader skills needed for a major executive role in public life of the type that Blackett sought. He feared that, if the Society were not centrally involved in running science, there might be a separation of scientific authority (resting with the Society) and administrative responsibility (resting with government): 'these huge administrative machines will grow without ideas and possibilities of their own, repulsive to men of intelligence, and finally without brain or soul.' However, he resignedly concluded that the Society, as the 'custodian of scientific quality', would probably stay clear of planning and seek the more modest path of 'fostering and encouraging, with the sustainment of quality'.[9]

The Society's leadership expected, by long custom, to keep the debate about who should succeed Henry Dale under reasonably tight control. The formal process was that, towards the end of the Society's year, the existing Council would determine a slate of eleven current members and ten new members,[10] and, among those twenty-one, the individuals recommended for appointment (or reappointment) to the five Officer posts; this slate would be put to such Fellows as were able personally to attend a formal meeting at the Society on Anniversary Day (30 November); and the newly elected or re-elected Councillors and Officers would take up their posts at the end of that day. Quiet discussions among the most influential individuals about potential nominees would of course start rather earlier, not least to ensure that those identified were in practice willing to take up their intended roles.

This controlling organisational culture rankled with quite a few Fellows. In 1935, ninety-two Fellows (20 per cent of the total Fellowship), animated by Frederick Soddy, had petitioned Council to shorten the terms served by each Officer, to allow for Council members

---

[9] J.G. Crowther, 'The Royal Society', *The New Statesman and Nation* (2 December 1944), 375, a review of Henry Lyons' history of the Royal Society; advance copy in the Blackett papers: PB/8/12. Andrade wrote a vigorous response accusing Crowther of completely misrepresenting the Society – see *The New Statesman and Nation* (16 December 1944), 405–6 – and sent the typescript to Henry Dale: HD/6/8/6/6.

[10] The Society's Charter then stipulated a Council of twenty-one members of whom ten must retire each year.

to be elected by postal ballot, and from a list of names greater than the number of vacancies, and for the Officers then to be chosen by the incoming rather than the outgoing Council. These proposals had been rejected by the then Council, and by the Fellowship as a whole.[11] The demand for greater democracy in the Society's affairs and, associated with that, for greater public engagement by the Society had borne some modest fruit in the following years, but not enough to still the pressure for reform. It just needed a Fellow determined to stir things up.

The Andrade 'memorial'

The physicist Edward Neville da Costa Andrade, known to his friends as Percy (Figure 1.1), was elected a Fellow of the Royal Society a month after Soddy's 1935 petition was submitted to Council. He served two years on Council, finishing in November 1944, and during that time contributed to shaping Council's response to the Fowler/Blackett letter. He was deeply opposed to Crowther's suggestion that the Royal Society might emulate the Soviet Academy's central role in the national planning of science, but short of that he was keen to see the Society contribute strongly to the development of science policy.

Andrade was a man of strong personal likes and dislikes. From his vantage point on Council he caught the early chatter about possible candidates to succeed Henry Dale as President. He did not like what he heard. Undeterred by Soddy's experience, he decided to draft a 'memorial, or what you will' and to collect signatures in support of an alternative candidate. While on Council he had heard 'frequent regrets that Fellows did not more frequently let Council know what they are thinking';[12] he decided to take Council at its word.

He discussed the idea quietly with a few trusted colleagues, including his fellow physicists Patrick Blackett and Henry Tizard. Tizard was then the Society's Foreign Secretary, and Blackett was about to go back onto Council. Once he had completed his term on Council at the end of November 1944, Andrade started approaching his friends and contacts to collect signatures to his memorial. By mid January 1945, ten Fellows had signed; by 19 February, fifty Fellows had signed; and by the time

[11] Jeff Hughes, ' "Divine right" or democracy? The Royal Society "revolt" of 1935', *Notes and records of the Royal Society of London*, 64 (2010), S101–17.

[12] Andrade to A.M. Tyndall, 10 January 1945. Uncatalogued Andrade correspondence, Royal Society archives. Unless otherwise indicated, all cited Andrade correspondence below is from this collection.

Figure 1.1 Percy Andrade. © Godfrey Argent Studio

Andrade finally submitted the memorial on 19 March, eighty-four Fellows had signed it.[13]

This was all done by personal contact, with a strong emphasis on confidentiality. The aim, however, was not to catch the Society's core leadership off its guard: the byword was discretion rather than secrecy. Andrade kept the Officers informed, taking care to do so in such a way that they would not have to respond officially before the memorial was formally submitted. So, for example, in addition to consulting Tizard, he discussed the memorial at length with the Biological Secretary A.V. Hill 'as a wise friend and not as an official of the Royal Society', and he reported to his close friend Charles Sherrington that Hill 'considers our action a perfectly proper and constitutional one, welcomes it, and thinks we are doing the Society a service ... If A.V. Hill thinks that all is well I do not think there can be much wrong.'[14] He also informed John Griffith Davies, the head of the Society's staff. Sherrington had been President during 1920–5, and his early decision to sign was a major fillip for Andrade since it unequivocally legitimised the initiative. He could tell potential signatories that his initiative was, apparently, compatible with organisational culture: 'There is, of course, nothing irregular or Bolshevistic in the Fellows memorialising Council.'[15]

Andrade wanted the Society to be more outward facing, engaging more effectively in public life and injecting science into the highest levels of policy-making: 'The Royal Society completely missed the boat at the beginning of the war and I am afraid that unless we have an energetic and courageous President, who has experience of how to get things done, we shall do the same at the end of the war.'[16] It was essential to have the right leadership team: 'The whole future of the Society is at stake, and if we appoint an ornamental or quarrelsome President the Society will lapse into being a purely honorific body.'[17] Sydney Chapman agreed: 'There is much need for the Royal Society to awaken to the social and national relations of science, and to bestir itself in these matters, just as the British Association [for the Advancement of Science] has in recent years, to

[13] Andrade to C.S. Sherrington, 16 January 1945; Andrade to W.E. Curtis, 19 February 1945. Of the eighty-four signatories, half had been elected to the Society in the previous ten years, compared with 40 per cent of the Fellowship as a whole; three quarters were from the physical sciences, compared with just over half for the Fellowship as a whole; and 70 per cent were based in the Oxford/Cambridge/London triangle, compared with, again, a little over half for the Fellowship as a whole. Only two of Andrade's signatories (E.F. Armstrong and Gordon Dobson) had also signed Soddy's petition in 1935.
[14] Andrade to C.S. Sherrington, 14 February 1945.
[15] Andrade to A.M. Tyndall, 10 January 1945.
[16] Andrade to W.E. Curtis, 31 January 1945.
[17] Andrade to C.S. Sherrington, 8 January 1945.

much good effect.'[18] Dudley Newitt commented darkly: 'I have had the impression during recent years that there have been influences at work in political circles which have tended to deprive the Society of its rightful place in national affairs; and there could be no better time than the present to deliver a counter-attack.'[19] To those concerned about politicisation, Andrade stressed 'None of us wants to see the Royal Society a political body although we do want to see it speak for science when the politicians want advice.'[20]

So the memorial argued that the Society should put its elitism to work in public life. It should 'assume its just place as the voice of British science and exercise that guiding influence on the scientific aspect of our national wellbeing which was contemplated by our founders'. Such status seemed then to be slipping from its grasp. The Society had earlier played a key role in the nation's scientific machinery, but now 'its real influence in national matters would seem to be decreasing rather than increasing'. In international relations, too, 'the prestige of the Society has not increased in recent years'. These trends had to be reversed, or else it would be left, precariously, to 'the various government departments and government-controlled corporations, and to other scientific and professional scientific bodies, to advise our rulers; to see that science, in particular academic science, is justly treated'. The memorial took heart, though, from the *Postwar needs* report, which showed that the Society had 'clearly realised its national responsibility in the matter of scientific research'.[21]

When discussing the matter with potential signatories, Andrade typically quoted the Society's need for better headquarters as an example of a policy issue needing vigorous attention from a new, politically sophisticated and politically engaged President.[22] This was certainly a big talking point within the Society at that time, and for another twenty years. But it seems a touch parochial in view of the high-flown rhetoric of the memorial.[23]

[18] Sydney Chapman to Andrade, 15 February 1945. Chapman had been involved with the British Association in initiatives on the social relations of science.

[19] D.M. Newitt to Andrade, 16 February 1945.

[20] Andrade to S.R. Milner, 19 February 1945.

[21] There are copies of various drafts of the memorial in the Royal Society archives, for example at MDA/B/3.4 and 3.5, HF/1/17/1/30 and PB/9/1/101.

[22] For an account of the accommodation issue and its bearing on the Society's sense of its own identity, see Jeff Hughes, Presidential Address to the British Society for the History of Science, July 2009; Trevor I. Williams, *Howard Florey: penicillin and after* (Oxford University Press, 1984), 327–39; and Trevor I. Williams, *Robert Robinson, chemist extraordinary* (Clarendon Press, 1990), 136–9.

[23] William Wilson, refusing to sign the memorial, told Andrade that the Society should be able to sort out the housing issue 'even if its President were in the final stages of senility'. Wilson to Andrade, 10 February 1945.

It is unlikely that Andrade's initiative was motivated primarily by a wish to secure a better home for the Royal Society. He was determined that the Society should share his aspirations for impact at national level, exercising a guiding influence on science policy. He was also keen that the Society should not allow itself to be outmanoeuvred on the international stage by the British Council, which he saw as trying to usurp the Society's natural position as 'the voice of British science abroad'.[24] But Andrade was motivated, too, by personal reactions to the individuals being mooted as potential Presidents.

The memorial spelt out in considerable detail the attributes required of the President if the Society was to be the voice of British science and to exercise a guiding influence on the nation's scientific affairs. He[25] should understand the machinery of government and not be unduly in awe of the leading figures in government and administrative circles; he should be accustomed to presenting the case for science to politicians; and he should have considerable international experience. He should also, of course, be energetic, sufficiently young in spirit to handle both opposition and apathy, a good speaker and of high academic status.

But could such a paragon be found? And if not, was there any scope for negotiation over just how brilliant a scientist the President himself had to be? Would it be appropriate, in the exceptional circumstances prevailing in 1945 and without necessarily setting a precedent, to settle for an individual whose scientific achievements were of just below Copley Medal[26] status in order to secure the other attributes? The memorial suggested that this would indeed be appropriate, but such a break with the prevailing culture was not lightly to be entertained. Though Andrade was careful in selecting those he approached, one third of them refused to sign the memorial, nearly all because they feared the Society's prestige would suffer if the President was not demonstrably in the very top rank of acknowledged scientific achievement. The argument about exceptional circumstances proved double-edged: if a President thus elected proved successful, it would be all the easier for the practice to become embedded.

[24] Andrade to Cecil Tilley and Owen Jones, 26 February 1945. J.G. Crowther had been appointed the first Director of the Science Department of the British Council at the outbreak of war and used the position to promote his radical politics.

[25] When the memorial was drafted, the election of the first female Fellows was still some months off. The election of the first female Officer lay forty-six years in the future, and the first female President further off still.

[26] The memorial originally cited the Nobel Prize, but several Fellows argued that the Society's own Copley Medal was the preferable benchmark of exceptional quality. See, for example, Henry Plummer to Andrade, 12 February 1945: 'I cannot see why our President should be elected by a Scandinavian body.'

Besides, 'The times truly are critical, but I cannot remember any time when they were not.'[27]

The two specific names being floated most prominently at the outset as potential Presidents were the organic chemist Robert Robinson and the physicist G.I. Taylor – both men of great personal scientific distinction, devoted to their work and unlikely to be sympathetic to Andrade's activist agenda. Andrade dismissed them together: 'Neither of them is a good speaker, neither of them is particularly a man of affairs, and neither of them would, I feel, give sufficient attention to the affairs of the Society' – adding, unconvincingly, 'I have not a grain of personal feeling against Taylor or Robinson ... I am thinking solely of the Society.'[28] For all his protestations to the contrary, Andrade's campaign had shades of being directed personally against Robinson, whom he repeatedly described to potential signatories as 'temperamentally unfitted for this particular post'. Given Robinson's scientific eminence, and given that he had just become a member of the Royal Society Council and that Dale had made him one of the Vice-Presidents, this needed some care.

Andrade's preferred choice for President was Henry Tizard (Figure 1.2). Tizard was a man with serious Whitehall experience, first as Secretary in the Department of Scientific and Industrial Research and later as Chairman of the Air Defence Committee before and during the War. He had also been Rector of Imperial College. At the time of the memorial, he was starting his final year as the Society's Foreign Secretary. However, he had not been awarded any of the Society's medals, let alone the Copley. What Andrade did not know was that Tizard was then hoping to reduce his involvement in Society affairs, not increase it. In response to the problem of three Officers retiring together, he had even offered to resign immediately, in January 1945, so that his successor could start ten months early.[29]

The first draft of the memorial, which Andrade sent to Tizard on 28 October 1944, was too subtly phrased for Tizard to realise its full import. The memorial was intended to signal to Council that Tizard commanded considerable support among the Fellowship as a potential President and should therefore be considered seriously – lack of Copley Medal notwithstanding. So in early February Andrade amended the text to say just that, explicitly and controversially mentioning Tizard by name. Tizard was horrified, and said so to Andrade. Andrade tried to soothe him, extolling

---

[27] W. V. D. Hodge to Blackett, 26 March 1945, PB/9/1/101; Gilbert Cook to Andrade, 9 February 1945.

[28] Andrade to Gilbert Cook, 16 February 1945.

[29] This would give Tizard early release from what had become a burden: 'I don't fill the office well, and am already doing too many things.' Tizard to Dale, 8 January 1945. HD/6/2/1/56. In the event, however, he served out his full term.

Figure 1.2 Henry Tizard. © Godfrey Argent Studio

the virtues of those who had so far signed: 'There is a big job to be done if the Royal Society is to take its rightful place ... The memorial is designed to convince you that you are called on to do the job, to strengthen the hands of your supporters on Council, and to answer any on Council who may say "It will find no support among the Fellows".'[30] Tizard agreed to remain neutral for the moment but demanded that Andrade not involve him in any future discussions on the matter.[31]

Early on in the campaign, Sherrington warned Andrade that the memorial must not be seen as an attack on the current Officers, the Council or, above all, the President Henry Dale.[32] The final version of the memorial was therefore strikingly deferential, leaning over backwards to conform to the prevailing cultural norms:

We hope that it will be clear to Council that our action is dictated solely by a wish to aid, and not to embarrass, Council in the extremely difficult task which it has before it, and that we shall loyally accept whatever decision Council may make in this critical matter.

Privately, however, Andrade was a great deal less complimentary about the Royal Society hierarchy. As rumours swept to and fro at the end of February 1945, he tried to set up a meeting with Albert Chibnall, one of his supporters, to discuss 'some very queer developments which make our action all the more necessary'. Chibnall, for his part, had heard of developments that he thought would make a present of the presidency to G.I. Taylor. A few days later, unable to fix a time for an immediate meeting, Andrade confided his own, different, rumour in a letter:

I think that the prefects are going to try to put in AVH. By a manoeuvre that I will describe to you when we meet they have practically arranged that he shall succeed Dale at the RI [Royal Institution], where Dale is to have two more years ... I do not think that the prefects will stop at anything. There is a lot more that I could say if we could only meet. Please destroy this letter.[33]

The outcome

Percy Andrade's memorial was formally circulated to Council on 4 April 1945. The 'prefects' – the Officers – had of course not sat idly waiting for

[30] Andrade to Tizard, 10 February 1945. Tizard papers, Imperial War Museum, #427.
[31] Andrade to Sidgwick, 15 February 1945.
[32] Sherrington to Andrade, 10 February 1945.
[33] Andrade to Chibnall, 5 March 1945. In the event, Eric Rideal succeeded Henry Dale as Director of the Davy-Faraday Research Laboratory at the RI in 1946, and Andrade succeeded Rideal in 1950. It is not clear whether Andrade had wanted the post for himself in 1945–6.

the memorial to arrive. They were as divided as any other group of Fellows about the relative merits of the various potential candidates and about whether there should be any flexibility about stellar scientific achievement as the overwhelming criterion in selection of the next President. There was no preferred 'establishment' candidate at that stage, despite Andrade's broodings on the matter. But the Officers were united on organisational culture: the need to avoid personal embarrassment to individual members of Council and any repeat of the 1935 experience of a publicly contested – and publicly reported – presidential election.

The Physical Secretary Jack Egerton therefore had a private meeting with Andrade on 16 February, probably at Tizard's request, and persuaded him to drop explicit mention of Tizard from the memorial – or so he thought.[34] However, when on 19 March Andrade sent in the memorial with its eighty-four signatures to the Officers prior to submitting it formally to Council, Tizard's name was still there. Tizard objected strongly and told his fellow Officers that, unless it was removed, he would resign immediately from Council and withdraw his name from all further discussion of the presidency. Andrade and the Biological Secretary A.V. Hill therefore drafted a letter to all the signatories to explain that persisting with Tizard's name in the memorial would be self-defeating. The letter was sent out by the two Secretaries, the offending sentence was removed before the memorial reached Council and overt embarrassment was avoided for the moment.[35]

It was an instructive illustration of how the Society handled awkward internal disputes at that time. The initiative for dealing with the presidential succession rested with the five Officers in the first instance. Three of them were effectively *parti pris*: Henry Dale as the outgoing President and A.V. Hill and Henry Tizard as potential candidates to succeed him. That left Jack Egerton and the Treasurer Thomas Merton. Merton was based in Herefordshire and Egerton in London, so it largely fell to Egerton to manage the quiet diplomacy. Having had his meeting with Andrade, Egerton arranged for six members of Council to have an informal discussion at his Knightsbridge flat on 1 March.[36] Before then, he and Hill checked out two cultural assertions: virtually all Presidents for many years had indeed had the Copley Medal, but the practice of selecting Presidents alternately from the physical and biological sciences (the 'A side' and 'B side', respectively, in Royal Society parlance) dated only from 1915 and thus had 'no

[34] A.C.G. Egerton diary (AE/2), p. 92: 16 February 1945.
[35] A.V. Hill to Andrade, 23 March 1945. Royal Society Archives, MDA/B/3.4. Copy also at Egerton papers, AE 1/11/11. Egerton and Hill were so irritated that they nearly charged Andrade for the postage.
[36] Egerton to Blackett, 21 February 1945: PB/9/1/101; A.C.G. Egerton diary (AE/2.6), pp. 119–23: 1 March 1945.

real basis in the Society's (even recent) history'. This allowed Hill to advocate the merits of his fellow physiologist Edgar Adrian, whom he thought a stronger potential President than, for example, Henry Tizard.[37]

At the 1 March lunch, Egerton alerted the Council members to how, with the impending simultaneous retirement of three Officers, 'the whole future of the administration of the RS is as it were in the melting pot'. He argued the case for not automatically maintaining the A/B alternation in the presidency. He briefed them in some detail on the imminent Andrade memorial, and particularly on the question of whether 'the highest scientific eminence in discovery and research' should be the dominant criterion in selecting the new President or whether 'experience, knowledge of affairs, drive and judgement are qualities needed at the helm at the present time' – suggesting that Hill, Adrian and a third physiologist Joseph Barcroft (then aged 72) would meet both criteria.[38] To provoke discussion, he produced a list of ten presidential possibles: Edgar Adrian, Joseph Barcroft, G.H. Hardy, A. V. Hill, Lord Rayleigh, Robert Robinson, N.V. Sidgwick, G.I. Taylor, Henry Tizard and E.T. Whittaker. By the end of the meeting the feeling was that, on the criterion of scientific eminence, Taylor and Robinson were the leading candidates on the A side, with Adrian and Hill equally strong from the B side.[39] Council members also agreed to explore the possibility of creating a post of 'Chairman of Committees', separate from the presidency. This would allow the selection of a President (e.g. G.I. Taylor) who was scientifically outstanding but was regarded as lacking 'administrative acumen': the Chairman of Committees would then handle those aspects of the post that did not naturally suit the President's talents.

Egerton met with Henry Dale the following Sunday to brief him in detail on how the meeting had gone. Dale took the mainstream view that it would be preferable to maintain the A/B alternation with as scientifically strong a candidate as possible, and was wary of the idea of a separate Chairman of Committees.[40] Egerton convened a further meeting of Council members at his flat on 23 March. Here it was noted that

[37] Of Henry Dale's ten immediate predecessors as President, eight had been awarded the Copley Medal before taking office and the remaining two were awarded it after their terms finished. Dale's successors showed a similar pattern. Dale himself won the Copley Medal in 1937. A.V. Hill to Jack Egerton, 2 February 1945: Egerton papers AE 1/11/11.

[38] Background note for the meeting: Egerton papers AE 1/11/11.

[39] Another of those present, Felix Fritsch, reported the meeting as focusing on three possible candidates: Tizard, Robinson and Taylor. F.E. Fritsch to C.D. Darlington, 19 March 1945: Darlington papers, ms. Darlington c.95.

[40] A.C.G. Egerton diary (AE/2.6), 124: 4 March 1945. See also entries for 8, 20, 23 and 26 March 1945.

there were many reasons for keeping Tizard, with all his experience in dealing with government, on the slate of possible candidates even though that would be seen as a break with the traditional focus on scientific excellence. The meeting ended with five names still in play: Adrian, Hill, Robinson, Taylor and Tizard. Tizard, who absented himself from these meetings, told Egerton that he favoured Taylor or, even better, Hill.[41]

Other conversations were going on between Council members. The mathematician Bill Hodge (a future Physical Secretary), for example, told Blackett (a future President) that he was worried about any departure from the criterion of scientific excellence, which had been a 'tremendous asset' to the Society: 'The President's main job, I consider, is to symbolise the Society's devotion to fundamental science before every other consideration.' Hodge also strongly favoured rearranging the Society's work so that more was delegated and less depended on the President's own political and administrative skills. Blackett agreed that it could not all depend on finding a President who would be outstanding on all criteria. Much more responsibility had to be devolved to the Vice-Presidents, to a Chairman of Council or a General Purpose Committee, to the Sectional and other committees, and to an enlarged and higher-calibre staff. Both Egerton and Hill endorsed these comments, though Egerton thought that the President should at least take the chair at Council meetings.[42]

Hodge told Blackett that, given a choice between Robinson and Taylor, he preferred Taylor: Robinson 'would want to take part in all the business of the Society, whether he knew it or not. Taylor ... would want to do as little as possible, but would take the trouble to get up just those things which he had to do; and I think that would be an admirable kind of President to have.' Blackett agreed about preferring Taylor to Robinson: 'The more I see of Robinson, the more I get the idea that he has a considerable degree of cantankerousness in him.'[43]

Cantankerous, interfering and temperamentally unsuited to the post: as a potential President, Robert Robinson was not perfect. And, for some, that was not all. The fifty-one-year-old Albert Chibnall explained to A.V. Hill that, 'like most of the younger Fellows that signed', his main motive for supporting the Andrade memorial was to stop Robinson: 'He is a very great chemist but I think he would be a fatal choice – to the younger

[41] Tizard had successfully nominated Robert Robinson for the Copley Medal in 1942 and, with Egerton, did the same for G.I. Taylor in 1944. CM 16 July 1942, minute 4 and CM 13 July 1944, minute 4.

[42] Egerton to Blackett, 13 April 1945. MDA/A/3.1.

[43] W.V.D. Hodge to Blackett, 26 March 1945; Blackett to Hodge, 11 April 1945; Blackett notes on the memorial, 11 April 1945. PB/9/1/101.

people, especially *outside* the RS, he represents ICI and big business.'[44] Chibnall himself had worked closely with ICI (Imperial Chemical Industries) on the development of a new fibre and had acquired three patents in the process,[45] but even so he felt uncomfortable with the idea of the Society being led by someone with Robinson's strong industrial connections.

Council formally discussed the question of Henry Dale's successor, and Andrade's memorial, for the first time at its meeting on 19 April 1945 – though naturally not a word about this appeared in the minutes. A.V. Hill, Robert Robinson and Henry Tizard were all present. Hill later wrote a private account of the meeting, recalling that many Fellows thought Tizard would make an admirable President on account of the enormous services he had rendered before and during the War and his wide-ranging acquaintance with leading personalities both in science and in public life. But Tizard, along with 'a good many other Fellows', was still deeply irritated by the memorial and refused to have his name discussed at the meeting. Robert Robinson then proposed Hill. Hill did not want to get mixed up in 'this unpleasant business': 'there was nothing I longed for so much as to get back to my scientific work.' So he, too, said that he did not want to be considered. Then, as Hill tartly put it in his private memoir, 'Robinson's name was mentioned and he tactfully expressed his doubts.'[46] No further progress was made.

That evening, Robinson phoned Dale to reinforce the comments he had made at Council. Born in September 1886 (the same month as A.V. Hill), he did not want to spend the next five years on official duties, having already lost so much research time because of the War. He thought that Hill, being based in London rather than Oxford, could take on the presidency at lower opportunity cost.[47] Hill, complaining that he had done no research for six years, did not see it quite the same way. The presidency apparently seemed to both of them more threat than opportunity.

Tizard's supporters on Council were not prepared to let his cause rest. Blackett began to favour him over Taylor (and either of them over Robinson), and said so to both Egerton and Hodge. He thought that Tizard's scientific achievements were at least the equal of Taylor and Robinson but simply less well known because of wartime secrecy, and

---

[44] Chibnall to Hill, 28 March 1945: AE 1/11/11.

[45] R.L.M. Synge and E.F. Williams, 'Albert Charles Chibnall', *Biographical memoirs of Fellows of the Royal Society*, 35 (1990), 57–96.

[46] A.V. Hill note for the record headed 'P.R.S.': AVHL II 4/68. This was explicitly excluded from his unpublished 1974 three-volume collection of largely autobiographical anecdotes, 'Memories and reflections'.

[47] Robinson to A.V. Hill, 2 May 1945. AVHL II 4/70.

that his impact in Whitehall had been achieved through scientific judgement and originality rather than mere administrative ability. Hodge continued to rate Taylor above Tizard scientifically.[48] Another Council member, David Pye, told Tizard that there would be 'real and widespread regret' were he finally to withdraw his name, and assured him that there would be ways of lightening the burdens of office.[49]

When Tizard had first grasped that Andrade was wanting to push him forward for the presidency, he dismissed the idea as 'ridiculous'. By early May, however, he was willing to concede privately to Dale: 'I realise I must take seriously the mention of my name, and must not dismiss it as ridiculous … these are strange times, which call for unusual decisions … although I realise my own shortcomings acutely it would be rather cowardly to decline to act if it were really the wish of the Society that I should do so.' But he was willing to do so only if he had the demonstrable backing of a large majority of Council. He emphatically was not prepared to take part in a contested election. Dale sympathised with Tizard's position, as he sympathised with Hill's and Robinson's declared wish to get on with their research. But he was faced with the situation that the three apparently strongest candidates – each a member of the 1945 Council – had all stated their wish to withdraw from consideration. Vexed, he pleaded with Tizard not to absent himself from the next Officers and Council meetings on 17 May. He guessed that Council would not accept Tizard's withdrawal.[50]

Both Hill and Robinson formally withdrew from further consideration for the presidency at the 17 May Council meeting, each of them sending in letters that Dale was obliged to read out to Council. Robinson's letter explicitly, and against Dale's pleading, highlighted the memorial 'because he wouldn't be persuaded that the object of the memorialists was not to deprecate the choice of himself in particular' – in which he was, of course, in large measure correct. To compound his irritation, Dale also had to read out a letter from the physicist Frank Smith, on behalf of a group of Fellows objecting to the memorial as 'attempted interference by a self-selected caucus', in order to stop Smith's group writing to the press about the issue.

The Council meeting then turned to Tizard and Taylor. Tizard's supporters made such a strong case for him that Dale was convinced a

---

[48] Blackett to Hodge, 11 April 1945 and 7 May 1945. Hodge to Blackett, 27 May 1945. PB/9/1/101. Robinson's biographer includes Tizard among those he speculates might have been in the running for the presidency in 1945, and suggests that his pre-war work on radar may have counted against him: 'he had strayed too far from the groves of academe to be acceptable to the Society, which collectively had little sympathy with applied science.' Trevor I Williams, *Robert Robinson*, 134.

[49] David Pye to Tizard, 8 May 1945. Tizard papers, Imperial War Museum, #427.

[50] Andrade to Tizard, 10 February 1945; Tizard to Dale, undated draft; Dale to Tizard, 15 May 1945. Tizard papers, Imperial War Museum, #427.

vote then would have handed him the nomination by a 'very large majority' – and not necessarily 'on account of such qualifications as the memorialists emphasised'. But no vote was taken because the case for Taylor
had not yet been made, and the meeting had run out of time.[51] The issue
was formally left 'in suspense'.

The challenge for Dale then was to stop Tizard bolting before the next
Council meeting, on 14 June. Sympathising with Tizard for the 'embarrassment of having your claims advocated by a bunch of interfering busybodies',
he begged him not to 'let these tiresome incidental circumstances wreck the
chance of the Council doing what they regard as the best for the Society'. He
saw Andrade's memorial as a large part of the problem, telling Tizard:

It is almost incredible that this signature of a memorial by a few Fellows, most of
them well-meaning and few, if any, deliberately mischievous, could have done so
much harm ... If those idiots had not meddled, we should long ago have settled
our nomination, and ... I believe you would have been chosen.[52]

It fell on deaf ears. Tizard, even with the prospect of now securing the
large Council majority that he set as a prerequisite, nevertheless
responded to Dale's letter with an 'adamant' refusal to be considered
further. Dale told Hill dolefully: 'We shall have to accept the situation.
Never, in the long history of our presidential elections, has so much
mischief been caused for so many by so few.'[53]

Andrade's memorial had thus caused three of the four strongest candidates to drop out, leaving only G.I. Taylor still in the running. Hill was
furious with Andrade and told him so.[54] In contrast, he wrote sympathe-
tically to Tizard:

I am very sorry indeed, because I had greatly hoped you would be President and
said so to Council. But I ... should certainly have done as you have done in the
circumstances: indeed you may remark that I did it with much less provocation! ...
It is very sad. The more I see of politics the more I like science.[55]

Hill also told Dale of a personal worry about the looming scenario where
he would become Foreign Secretary and Taylor would become President,
but based in Cambridge and 'anxious to avoid all the responsibilities he
could': then based in London, Hill feared that he would end up doing

[51] Largely because the Officers had to attend a ceremony for admission of new Fellows –
including, on this occasion, the admission of Kathleen Lonsdale and Marjory
Stephenson, the first women to be elected in the Society's history. A.C.G Egerton diary
(AE/2.7), 69: 17 May 1945.

[52] Dale to Tizard, 3 June 1945. Tizard papers, Imperial War Museum, #427.

[53] Dale to Hill, 6 June 1945. HD/6/8/7/1.

[54] Hill to Andrade, 10 June 1945. AVHL II 4/3.

[55] Hill to Tizard, 7 June 1945. Tizard papers, Imperial War Museum, #427. Hill was then
in his last month as MP for Cambridge University.

most of the President's work. 'That isn't why I'm giving up everything else to get back to my lab!'[56]

Dale briefed Council members on these developments, and when they met on 14 June they decided to let the matter rest a bit longer rather than going immediately for the one name still available. The critical Council meeting, then, would be 12 July – between the casting and the counting of votes in the general election, which would significantly shape the context within which the new President would have to operate. Egerton, a physical chemist, confided to his diary that he regarded Taylor as the leading general physicist in the country: 'his election to presidency would be perfectly normal and therefore lead to no faction whatever; in fact, in my opinion, to general satisfaction. Charming personality.' However, he also thought that Council should have a wider choice, or at least a fallback position, and on 11 July he suggested the chemist N.V. Sidgwick – who had signed the memorial but had featured on a list of possible presidential candidates – as an additional candidate. At seventy-two, he was thirteen years older than Taylor but 'experienced, much looked up to for mental power and knowledge, much respected abroad for his work and his intellect'. So that made it two names on the slate.

The Officers normally held a formal meeting on the morning of Council day and then had a private lunch (i.e. without any staff or other Fellows present) ahead of the Council meeting in the afternoon. During their lunch on 12 July, they realised that having a chemist other than Robinson as President might create problems, especially as Robinson was already Vice-President. So they raised the matter with Robinson, and, at the last possible moment, he retracted his withdrawal. So Dale was able to present Council with a slate of three candidates for the presidency.

That afternoon Council dealt initially with the Foreign Secretaryship, which A.V. Hill agreed to take on for a year in succession to Tizard, and with the Biological Secretaryship, for which the Director of Kew Gardens, E.J. Salisbury, was selected. Dale then showed Robinson to another room; in Robinson's recollection, 'Sir Henry said "I don't think they will appoint you" but offered no further explanation.'[57] After discussion, Council voted first between the two chemists, Robinson winning by ten votes to Sidgwick's eight.[58] They then voted between Robinson and

---

[56] Hill to Dale, 7 June 1945. HD/6/8/6/5.
[57] Robert Robinson: chapter 7, p. 2 of the unpublished second volume of his autobiography *Memoirs of a minor prophet*, written in November 1974 three months before he died. Royal Society archives, ROR A46.
[58] There were nineteen of the twenty-one members of Council present on 12 July, including Robert Robinson, Tizard and Hill; the two missing were Edward Bailey and Douglas Hartree. With Robinson out of the room, that gives a total of eighteen votes. There was

Taylor, with Robinson coming out ahead by a single vote, nine to eight. A third vote with just Robinson's name produced a unanimous result in favour, which was what was conveyed to the waiting Robinson. A formal letter of invitation followed, which Robinson, seemingly no longer worried by the impact on his research time, accepted by return.

Egerton was disappointed, but he observed that there had been no chemist as President for thirty years and that Robinson was a 'greater chemist than NVS, though not such a man of culture'. So he thought it a reasonable outcome, and took comfort from the fact that he would be able to continue working with his friend A.V. Hill.[59]

In later life, Robinson described his election in the following possibly disingenuous terms:[60]

The question of [Dale's] successor was carefully studied at a number of meetings and in the last of these I found I had been nominated myself . . . I have wondered from time to time how this election came to pass. There were several scientific cliques supporting leaders in war service [Tizard] but unfortunately their loyalty was unidirectional [Tizard did not reciprocate]. A process of cancellations [Hill and Tizard withdrawing] allowed me to slip in on grounds of achievements in scientific researches. I had already been awarded the Davy and Royal Medals and in 1942 the Copley Medal. The latter . . . has come to be regarded as a qualification for nomination to the presidency. [Taylor had the Copley, but Sidgwick did not.]

Soon after the decisive Council meeting, Robinson sent an appreciative letter to Tizard: 'I was genuinely surprised and am overwhelmed of the outcome because I was not well informed about the way events were shaping themselves. I have accepted the nomination . . . If elected [i.e. at the 30 November Anniversary Day meeting of Fellows] I shall owe this success chiefly to your support and I shall never forget your large-hearted and generous conduct.' David Pye, who had tried to dissuade Tizard from withdrawing, later acknowledged to Tizard that he had in fact taken the only course open to him:

I suppose you were right about the PRS affair. Indeed, the line you took was probably the only possible one as things had worked out, but it made me very sad that misdirected zeal should have prevented what I'm certain a large majority would have liked to happen – and which I believe would have occurred if things could have been left to take their normal course.[61]

---

one abstention – possibly by the President (who had favoured Tizard but had mostly tried to stay out of the discussions) – on the second vote.

[59] A.C.G. Egerton diary (AE/2.7), 182 and 182 bis: 11 and 12 July 1945.

[60] See Note 57.

[61] Robert Robinson to Tizard, 15 July 1945; David Pye to Tizard, 20 August 1945. Tizard papers, Imperial War Museum, #427.

Reactions

So, from Andrade's point of view, how did it all go wrong? By the time he launched his memorial, Andrade had served two years on Council and should have known how things worked. He did his homework, he kept the 'prefects' informed of what was going on, he quickly got some big names on side. Many Fellows, including many on Council, were sympathetic to his wish to see the Society actively engaged in public life, and there was considerable appetite among the Fellowship for the Society to 'assume its just place as the voice of British science and exercise that guiding influence on the scientific aspect of our national wellbeing' that the memorial sought. What Andrade also did, though, was first to confuse Tizard about the purpose of the memorial and then to alarm him with its directness. Whether or not he actually wanted the presidency, Tizard knew that open campaigning, even when done by someone else, was not the way to secure high office in the Royal Society, and he would have been horrified at the thought that Fellows might imagine he had been inept enough to instigate the campaign.

The selection process was designed to bolster the authority of the incoming individual by contriving to secure, in the end, the unanimous support of Council for his nomination. Andrade's memorial threatened such courtesies by disrupting the discreet deliberations of Officers and Council. Hence the embarrassment that it occasioned, made all the more acute because Tizard, Hill and Robinson were all on Council at the time.

The downside of such an approach to elections, of course, was that the general Fellowship could easily feel excluded from the decision-making process. Behind both Andrade's and Soddy's attempts to petition Council lay the additional motive of chipping away at the oligarchic nature of the Society's governance. As one of the signatories complained to Andrade:

The Council in no sense represents the Fellows, and it is made so difficult for any Fellow to influence the decisions which are taken in the name of the Society as a whole that most Fellows seem to lose the interest which they might be expected to take in the Society's affairs and activities. Unless the Council can take the first steps in modifying the constitution in such a way as to make itself a Council appointed by the Fellows rather than one acting as though it were divinely ordained, the outlook for the Society is not very bright – even if we have a President who can deal with politicians.[62]

Ten years later, another group of Fellows would meet to discuss, again, the problems of the concentration of power in the Officers' hands, the need for

---

[62] E.C. Stoner to Andrade, 27 February 1945.

a more effective Council and the difficulty that most Fellows encountered in seeking to influence what the Society did.[63] It was a persistent theme.

Henry Tizard stepped down from the Foreign Secretaryship at the end of 1945. In 1948 he re-entered the public spotlight as the inaugural chairman of two major government bodies, the Advisory Council on Scientific Policy and the Defence Policy Research Committee.[64] A.V. Hill, who had forsworn the presidency in order to get back to research and because he hoped Tizard would take it on, did one year as Foreign Secretary before handing that post to Edgar Adrian. His return to full-time academic work was then delayed a further six months by having to stand in as Physical Secretary while Jack Egerton recovered from a nasty skiing accident. Adrian was Foreign Secretary for four years and, in 1950, became President. Hill later speculated that, had he stayed with the Foreign Secretaryship for a full five-year term, he might well have become President in 1950 instead of Adrian. He reflected, a little wistfully, that he was able to do some useful scientific work instead, that Adrian did a good job as President, and that the OM would not compensate for the loss of research time that the presidency would have entailed.[65]

Being a signatory to the memorial was not the prelude to automatic banishment to the outer circle of Society affairs. One third of the Council membership for 1946–7 had signed the memorial. Both N.V. Sidgwick and E.J. Salisbury were signatories – the one touted as a presidential possible just months after signing, the other selected to succeed A.V. Hill as Biological Secretary. Other signatories included Harrie Massey, who led the Society's pioneering work on space science during the 1950s; Harry Melville, who was to become Secretary of the DSIR in 1956 and subsequently the first Chairman of the Science Research Council; and Solly Zuckerman, soon to be Deputy Chairman of the Advisory Council

---

[63] See memo by C.D. Darlington dated 1 January 1955 (probably in error for 1956), blaming the Society's ineffectiveness in public debate on 'the principle of nominated succession in its governing body concealed by the formality of a free election'. This led to discussion with other Fellows in Oxford: Howard Florey and H.H. Plaskett (who both signed Andrade's 1945 memorial), Ewart Jones and Hans Krebs (neither of whom had been Fellows in 1945) and Francis Simon. Darlington himself had been on Council in 1945. Darlington papers, ms Darlington c.95.

[64] Robert Robinson complained to American colleagues that these two new bodies were 'encroaching on the prerogatives formerly enjoyed almost exclusively by the Royal Society and its Officers'. Memorandum of conversation between E.A. Evans, Jr, J.B. Koepfli and Robert Robinson, 27 January 1948: Koepfli papers (Caltech), box 1, folder 1. I am grateful to Peter Westwick for this reference.

[65] A.V. Hill note for the record headed 'P.R.S.': AVHL II 4/68; chapter 12 'Jack Egerton as Secretary of the Royal Society' in A.V. Hill, Memories and reflections. Since the Order of Merit was instituted in 1902, every President of the Royal Society has been made a member. A.V. Hill became a Companion of Honour in 1946 and was awarded the Copley Medal in 1948 to go with his 1922 Nobel Prize.

on Scientific Policy and subsequently Chief Scientific Adviser to the government.

As for the well-intentioned but rancorous Percy Andrade, he remained active in Royal Society affairs as Chairman of the Library Committee from 1944 until his death in 1971, Robinson appointing him Honorary Librarian in 1948. He resigned his post as Quain Professor of Physics at UCL to become Director of the Royal Institution Davy-Faraday Laboratory in 1950, the post he had earlier claimed was being earmarked for A.V. Hill. Here he was again unsuccessful in initiating institutional reform, and he was forced out within two and a half years. Thereafter he worked as a research consultant. He later remarked ruefully to Harold Hartley, 'I had a very high opinion of Tizard ... I tried to get him elected PRS in 1945 ... the times were exceptional ... as it turned out, I merely did myself harm and the Society no good.'[66]

The sequel

The 1945 round of elections left the Royal Society with a set of Officers (especially Egerton, Hill and Salisbury) keen to pursue a relatively activist agenda and to shoulder more of the leadership function within the Society, and a President of the highest personal scientific distinction with strong industrial connections who was determined to keep the Society out of politics as much as possible, especially at the national level. Robert Robinson's term of office was marked by devotion to fundamental research, by a range of initiatives to restore international scientific relations in the aftermath of war,[67] and by a series of Anniversary Addresses that were relentlessly dominated by the latest advances in organic chemistry. Much as Andrade and his co-signatories had feared, the Society was not conspicuously active in public affairs during Robinson's presidency, nor, for that matter, during the presidency of his successor Edgar Adrian.

Ironically, one policy issue on which Robinson did fight vigorously as President was the one that Andrade had highlighted, the question of finding new accommodation for the Society; and here his ambitions for

[66] Andrade to Hartley, 20 June 1964: HART Box 105; Frank James and Viviane Quirke, 'L'affaire Andrade, or how not to modernise a traditional institution', in Frank A.J.L. James, *The common purposes of life: science and society at the Royal Institution of Great Britain*, (Ashgate, 2002), 273–304; Alan Cottrell, 'Edward Neville da Costa Andrade, 1887–1971', *Biographical memoirs of Fellows of the Royal Society*, 18 (1972), 1–20.

[67] John S. Rowlinson and Norman H. Robinson, *The record of the Royal Society of London: supplement to the fourth edition for the years 1940–1989* (The Royal Society, 1992), 5–8; Alexander Todd and J.W. Cornforth, 'Robert Robinson', *Biographical memoirs of Fellows of the Royal Society*, 22 (1976), 425.

a major science centre on the South Bank were thwarted by Edgar Adrian. Robinson was still grumpy about it thirty years later. It was Howard Florey, a signatory of Andrade's memorial, who finally cracked the problem of securing new premises for the Society, in Carlton House Terrace, that matched its aspirations to exercise a guiding influence as the voice of British science.

All Presidents covered in this book were selected as individuals of outstanding personal distinction in fundamental scientific research, signalled for example by the award of the Nobel Prize or equivalents such as the Fields Medal or Crafoord Prize, and in most cases by the Copley Medal. The Society's commitment to the highest standards in science, and its associated wish that such commitment should be epitomised by the President, remained undiminished. This was seen as essential to maintaining the Society's scientific authority. But, as the century progressed, the wish for the President also to epitomise a corporate commitment to engagement in public life gradually became a more prominent (though not predominant) element in the mix, just as the presidency itself gradually morphed from being essentially an honour to being also something of a job – albeit unpaid and part-time. This in turn was seen as essential to the Society's credibility and effectiveness in the arena of public policy.

It required a certain delicacy. Some Presidents, such as Howard Florey, Andrew Huxley, Michael Atiyah and Aaron Klug, were not instinctively public figures but were quite prepared to argue with ministers and to engage in public debate when sufficiently provoked. Some, such as George Porter, Bob May, Martin Rees and Paul Nurse, were natural communicators who made full use of the scope that the presidency provided. Some, more controversially, were associated with particular political parties. When seeking support for his memorial, Andrade had had to assuage fears that greater involvement in public affairs might lead to undue politicisation of the Society. The Society was alert to the need for a long-term perspective: it would have to work with the next government as well as the present one, and key to that was remaining staunchly independent of all governments and all political parties.

Presidents occasionally seemed to sail a bit close to the wind on this. Both Howard Florey and Patrick Blackett were active members of the Brumwell group, which helped the Labour Party develop its science and technology policy during the 1950s. Blackett turned down an invitation in 1964 to be science minister in Harold Wilson's Labour Government, but did then agree to become scientific adviser to the new Ministry of Technology. In the event, he was more than ready to switch the main focus of his attention from the Ministry to the Society when he became President in November 1965, though he remained close to Wilson throughout Wilson's premiership and his own presidency.[68] Alex Todd, strongly Conservative, judged Blackett to have been a 'disaster' for the Society, with a 'wholly false idea of his political power', but thought that Blackett's successor Alan Hodgkin 'did not meddle in the political field' and was therefore an improvement.[69] Todd himself had also declined the offer of a ministerial post when the Department of Education and Science was established in 1964, but used his political connections energetically, especially after the general election returned a Tory Government in May 1979. A later President, Bob May, served two Prime Ministers from different political parties as Chief Scientific Adviser before becoming President, but he was not publicly associated with either party and the Society's political independence was not impugned.

So the 1945 presidential election highlighted some enduring issues. Andrade's campaign proved inept in the short term. However, it evoked very considerable sympathy within the Fellowship both for trying to open up the Society's controlled decision-making processes and for making explicit the case for the Society to engage more vigorously with public life. One important if unintended outcome was affirmation of the Society's commitment to fostering science of the highest quality. In the long term, this proved to be not the opposite of engagement with public life but rather, from the Society's particular perspective, the precondition for it.

---

[68] Bernard Lovell, 'PMS Blackett', *Biographical memoirs of Fellows of the Royal Society*, 21 (1975), 75–85, 101; interview with Bernard Lovell.

[69] Alexander Todd, 'Address at the Anniversary Meeting, 30 November 1979', *Proceedings of the Royal Society of London. Series B, Biological sciences*, 206 (1980), 373–4; private notes made by the Executive Secretary Ronald Keay at an informal meeting of Officers, 3–4 November 1979; also OM/123(79). Todd's practised disparaging of his predecessors as President was to resurface, more guardedly, in his valedictory Anniversary Address and in his autobiography. See also interviews with Brian Flowers and Peter Warren.

2   Running UK science?

> While not becoming an arm of government, we can perform a useful
> national service by maintaining a close collaboration with those who
> control our destinies.[1]

To Howard Florey is attributed the cliché that 'he hoped to get something
done even if he had to carry the Royal Society kicking and screaming into
the twentieth century'.[2] His chance to do so came when he succeeded
Cyril Hinshelwood as President of the Society on 30 November 1960.
David Martin, head of the Society's staff,[3] dug out a copy of Percy
Andrade's memorial to remind Florey of what he had signed up to as a
young Fellow in the closing months of the War.[4] Florey was still keen to
see the Society take an active role in the running of UK science, and his
presidency was marked by spirited attempts to prevent the Society being
ever more marginalised by the growing apparatus of government control.
It was a robust exposure to the art of the possible. By the end of his
presidential term, Florey had crystallised many of the key directions along
which the Society would develop over the rest of the century.

The Royal Society in 1960: opportunities and threats

How had the Society changed since 1945? Most obviously, it was bigger,
in the sense that it had more Fellows. The 1945 decision to increase from
20 to 25 the number of Fellows elected annually had, by the beginning of

---

[1] Howard Florey, 'Address at the Anniversary Meeting, 30 November 1963', *Proc R Soc
Lond B*, 159 (1964), 451. Some of the material in this chapter has already appeared in
Peter Collins, 'A role in running UK science?', *Notes and records of the Royal Society of
London*, 64 (2010), S119-30, and is reproduced here with permission.

[2] Alan Hodgkin, 'Edgar Douglas Adrian, Baron Adrian of Cambridge', *Biographical memoirs
of Fellows of the Royal Society*, 25 (1979), 53.

[3] Assistant Secretary (i.e. head of the staff) from 1947, post re-titled 'Executive Secretary'
(very nearly 'Comptroller') in 1962, died in office 1976. Uniquely for a non-Fellow,
Martin was given a biographical memoir: Harrie Massey and Harold Thompson, 'David
Christie Martin', *Biographical memoirs of Fellows of the Royal Society*, 24 (1978), 391–407.

[4] HF/1/17/1/30.

1960, led to a 27 per cent growth in the total number of Fellows, to almost 600. There were also more staff – the salary bill had increased more than tenfold since the end of the War. The headquarters at Burlington House were bursting at the seams, with some staff housed off-site and serious efforts being made to find alternative premises. The Society was also more active. It was managing Britain's engagement with the International Geophysical Year (IGY), the most ambitious experiment up to that point in international scientific cooperation. It represented British science in numerous international fora. It was spending more public money – £205,000 in 1959–60 (including £64,000 for the IGY), as against £23,000 in 1945–6. It was also spending more of its own money, and seized the occasion of its 1960 tercentenary to launch a major fundraising campaign.

The Society was a small part of a growing whole: less than 0.8 per cent of public spend on research in universities (excluding its special grant for the IGY). In 1955–6, total spend on R&D in Great Britain was about £300 M, or 1.6 per cent of GNP. Of the £300 M, 75 per cent was publicly funded and 59 per cent was for defence purposes. Total spend by the Research Councils was £12.8 M, of which less than 10 per cent was in the form of responsive grants to universities. About £7.5 M of the University Grants Committee's recurrent grant to universities and a further £2 M capital grant were estimated to be spent on research. The national purse strings were about to be loosened by, among other things, the 1957 launch of *Sputnik* and hopes for the contribution that nuclear science and technology could make in both civil and defence spheres. By 1961–2, total national R&D spend had reached £634 M.[5]

As it celebrated its tercentenary, the Society seemed poised for a bigger role in public life. But it was not obvious what that role might be. In J.G. Crowther's phrase, the Society had scientific authority but little directive function. Scientific authority did not translate straightforwardly into political clout, in either advisory or executive capacity. The view increasingly entrenched in Whitehall was that, because of 'the fundamental importance of ministerial responsibility', independent sources of authority had to be kept at arm's length.[6] This made it difficult for a body like the Royal Society, as an institution, to exercise direct influence on policy. Independence implied to some extent being an outsider.

Edgar Adrian, President during 1950–5, thought that the Society should try at least a little to use its inside knowledge of the scientific

[5] *University development 1952–1957* (Cmnd 534. HMSO, 1958); *Annual report of the ACSP, 1956/57* (Cmnd 278), 22; *Annual report of the ACSP, 1961/62* (Cmnd 1920), 38. Note also Treasury minute dated February 1958: TNA T 218/191.
[6] Philip J. Gummett and Geoffrey L. Price, 'Advisory Council on Scientific Policy', 120.

world to engage with government: 'We cannot blame Government departments and Ministers for mismanaging scientific affairs if we are unwilling to assume our responsibility for aiding them.' But his response to the long-term erosion of the Society's 'directive function' in the management of British science was to deny that it had any such aspirations, since that would jeopardise independence:

It seems far better for the Royal Society to keep itself outside the State organisation. The larger this becomes the more important will it be for us to maintain our status as an independent body of scientists whose chief aim is the advancement of knowledge ... The Society ... should not be willing to accept the kind of detailed administrative work which it is sometimes asked to undertake.[7]

There was competition for influence in science policy. The heads of the Department of Scientific and Industrial Research (DSIR), the Agricultural Research Council (ARC) and the Medical Research Council (MRC), appointed by government, officially provided all the scientific advice that was needed in their respective fields and were hostile to any suggestion that government might also seek advice from the Royal Society, let alone give it responsibility for serious amounts of money. Insofar as the generic role of government Chief Scientific Adviser existed in 1960, it was filled by the Secretary of DSIR or the Chairman of the Advisory Council on Scientific Policy (ACSP).[8] The fact that the leaders of all these bodies were usually Fellows of the Royal Society might facilitate informal communication but did not mean that the Society itself could shape policy. Nor did shared Fellowship necessarily mean that they saw things in the same light in their various professional roles.

It was quite a turbulent time in terms of the organisation of publicly funded civil science. The Royal Society was not the only body jockeying for position. Within the structure of government, the four Research Councils – DSIR, ARC, MRC and Nature Conservancy – came under the Minister *for* Science, a post created in October 1959. The minister, Lord Hailsham,[9] saw his role as providing general guidance, helping the

---

[7] E.D. Adrian, 'Address at the Anniversary Meeting, 30 November 1955', *Proceedings of the Royal Society of London. Series A, Mathematical and physical sciences*, 234 (1956), 157–9.

[8] See Hailsham's speech to science correspondents on becoming Minister for Science, 21 October 1959: TNA CO 927/697.

[9] Styled 'Viscount Hailsham' following the death of his father in 1950, he disclaimed his title and reverted to plain 'Quintin Hogg' after Macmillan's resignation in 1963, in order to return to the House of Commons with an eye to the leadership contest. For simplicity, I have referred to him as 'Hailsham' throughout. In 1959 he effectively had political, but not financial, responsibility for those areas of civil science that did not fall squarely under major departments other than DSIR.

scientific establishment deal with Whitehall, and acting as political champion for science.[10] The Research Councils were willing to accept Hailsham's political support so long as he did not exercise even his limited powers to intervene in their affairs, but they were worried about his Office becoming too powerful.[11] The lawyer Hailsham, for his part, objected to being talked down to by the scientists.[12]

There was no centralised Science Budget as such: individual Research Councils (and the Royal Society) negotiated their annual budgets direct with the Treasury. This created problems for the Treasury, which lacked any coherent mechanism for making judgements about scientific priorities. The Prime Minister, Harold Macmillan, and the Treasury wanted a full-scale Ministry *of* Science with a proper budget, which Hailsham resisted.[13] The main source of funding for academic research, meanwhile, was the University Grants Committee (UGC) rather than any of these other bodies. The dual-support system, with the UGC providing core support for university research activities and the Research Councils providing supplementary project support, was then in its early days, and there was much vexed discussion about where UGC responsibilities ended and Research Council responsibilities began.

Below the surface, and sometimes above it, relations between the key players could be tetchy. Academics were suspicious of the Research Councils, which they saw as more interested in their own units and institutes than in funding individual university researchers. DSIR, where academic scientists held least sway, was least esteemed, and the ACSP Chairman Alex Todd, for example, was known to have a low opinion of it.[14] The Royal Society had a long struggle with DSIR over control of space research, which it eventually lost.[15] Alex Todd vied with

---

[10] Hailsham briefing to science correspondents, 21 October 1959: TNA CO 927/697.

[11] For example, William Penney (ASCP member, Deputy Chairman of the UK Atomic Energy Authority and Royal Society Treasurer 1956–60) to Roger Makins (Chairman, UKAEA, and previously Joint Permanent Secretary at the Treasury), 8 March 1961: TNA AB 16/4247.

[12] See, for example, Hailsham to Frank Turnbull, 2 October 1961, complaining about the ARC Secretary Gordon Cox baulking at showing him ARC agendas in advance on the grounds that they were too technical: 'If I have any more lip from Cox I will use my powers of direction.' TNA CAB 124/1392.

[13] Macmillan memo to the Cabinet Secretary Norman Brook, 4 May 1959, and subsequent exchanges with Brook and Hailsham, the latter anxious to ensure that any new post was not to the detriment of his existing portfolio: TNA PREM 11/2723.

[14] Internal Treasury minute, 12 April 1960, reporting a comment from Frank Turnbull: TNA T 218/666.

[15] The Society strongly resented the DSIR takeover of its work on space research. The Treasury (e.g. J.A. Annand, Arnold France, Richard Griffiths) and Hailsham's top official (Frank Turnbull) had a sneaking sympathy for the Society in this context, recognising that it had legitimate complaints about DSIR's administrative incompetence;

the Secretary of the DSIR Research Council, Harry Melville, for Hailsham's confidence. Hailsham disliked the DSIR Council with its confused reporting lines, and he disliked the UGC, which was outwith his control. The Treasury was not all that impressed with Hailsham, but then it was cheerfully disparaging, at least in private, about most parts of the system.[16] It tended to see the scientific landscape as a series of turf battles between competing factions, with its special scorn being reserved for independent and quasi-independent groups like the Royal Society and ACSP.

Alex Todd (Figure 2.1) in 1980 alleged that his predecessors as President of the Royal Society had been caught off guard by the government's tightening control of science policy, especially when ACSP was abolished in 1964, new ministries were established and the research council system was reformed.[17] The Society was not caught off guard. This chapter documents how the Society tried strenuously to avert the dénouement that the Trend Committee ushered in, and how during the preceding years it fought to secure as strong a role as politically possible. But it was up against public accountability. As the government machinery for dealing with science became more complex and more comprehensive, the Society's role appeared to diminish to one of filling any remaining advisory or directive gaps. Thus it did important work in securing funding for, and shaping the early development of, areas like computing, space research and the IGY.[18] But it was remorselessly being edged out. In their more candid moments its leaders were prepared to admit as much. For example, Cyril Hinshelwood confided in 1959 to the influential Solly Zuckerman, then ACSP Deputy Chairman:

but they were even more worried about an independent body gaining too much control over a major government-funded programme, and hence backed DSIR claims. See various exchanges at TNA T 218/520, TNA CAB 124/1389, TNA CAB 124/1710. Also Neil Whyte and Philip Gummett, 'Far beyond the bounds of science: the making of the United Kingdom's first space policy', *Minerva*, 35 (1997), 168.

[16] See, for example, Richard Griffiths to Arnold France, 1 February 1961: 'The real conflict even today is between the various Research Councils, ACSP and the Steering Group [on space research] in relation to the scientific advice they tender to the Minister. The Royal Society are the niggers in the wood pile and will always set out to destroy any Research Council (or other body of importance to them) they cannot control.' TNA T 218/520.

[17] Alex Todd, 'Address at the Anniversary Meeting, 1 December 1980', *Proceedings of the Royal Society of London. Series B, Biological sciences*, 211 (1980), 1–13.

[18] Matthew Godwin, *The Skylark rocket: British space science and the European Space Research Organisation 1957–1972* (Beauchesne, 2007); Harrie Massey and M.O. Robins, *History of British space science* (CUP, 1986); Ken Pounds, 'The Royal Society's formative role in UK space research', *Notes and records of the Royal Society of London*, 64 (2010), S65–S76; interview with Ken Pounds.

Figure 2.1 Alex Todd. © The Royal Society

We are very much perturbed at the serious shift of the centre of gravity to the continuing detriment of the Royal Society's position. The Government Grant to the Society for research now amounts to a trivial fraction of the steadily mounting grant to the DSIR ... The actual positive argument which is put forward for the relative weakening of the traditional function of the Royal Society is that of public accountability. This is a kind of fashionable catchword which sounds impressive but means extremely little.[19]

The edging out had to be done carefully, since the Society had enough social clout to make trouble at the highest levels. Ministers on the whole wanted to keep on good terms with the Society, and senior civil servants on the whole wanted to avoid making that too difficult for them. Both DSIR and Treasury recognised that they ought to maintain smooth relations between the Society and the Lord President (who had minister-ial responsibility for DSIR): 'Ministers are unlikely to want to do anything that upsets the Royal Society.'[20]

This played to the Society's advantage when, in 1956, Treasury offi-cials wanted to kill off the Society's Scientific Investigations Grant scheme altogether in favour of a new scheme to be run by the DSIR Research Council[21] – a proper Whitehall body, staffed by proper civil servants and with full ministerial accountability. The Physical Secretary David Brunt, negotiating for the Society, was able to insist that the price of the Society's acquiescence in an increase for the DSIR was a parallel increase in its own scheme. The upshot was an increase in the DSIR scheme from £78,000 in 1956–7 to £175,000 in 1957–8, while the Royal Society's £30,000 Scientific Investigations scheme was increased to £50,000 (and to £75,000 in 1960–1). The Society's scheme of grants for international travel – the Treasury thought 'it might be possible to consider a concession here without embarrassing DSIR' – was also increased, from £16,500 to £31,000. These were still pretty modest sums, of course, but important for the Society because it thus remained a player in the grant business.

So 1960 was a point of opportunity for science, overlain with institu-tional struggle. The government and the Civil Service were coming to terms with the need for an unprecedentedly large peacetime effort in civil science, continuing the immediate postwar preoccupation with produ-cing more skilled scientists and technologists, and experimenting with different forms of organisation to handle increasingly expensive forms of research. In this situation, the Royal Society had to think hard about just how to make the most of its particular strengths.

The Royal Society's claim to serious attention was founded, in the first instance, on its core skill of recognising excellent science and excellent

---

[19] Hinshelwood to Zuckerman, 22 December 1959: OM/1(60).

[20] Treasury minute, 25 July 1956, and related exchanges: TNA T 218/191 and T 218/198.

[21] Harry Melville, *The Department of Scientific and Industrial Research* (George Allen and Unwin, 1962), 69–71; also Chapter 3.

scientists. It came naturally both to Robert Robinson and to Edgar Adrian to focus the Society during their presidencies on epitomising excellent science. The real key was the quality of the Fellowship. Adrian famously said of the Society in 1955: 'I am convinced that the most important thing it does now is to exist and to perpetuate its existence by electing new Fellows ... able to justify the claim to represent all branches of science at the highest available level.'[22] Cyril Hinshelwood, in his 1960 Tercentenary Address, echoed Adrian in stressing that 'the right choice of its members is a matter demanding the most anxious care ... and is perhaps the most important thing it has to do.'[23] A Fellowship of exceptional quality was one condition for the Society's effectiveness in any venture it undertook. Like Adrian, Hinshelwood also stressed a second condition, its independence. The Society was very willing to 'lay an account of its stewardship before the world', but strictly on its own terms.[24]

If it was going to put its elite experience to work and contribute in new ways to running UK science, one place to start could be grant giving. At first sight, that might seem an odd choice for a small independent body, but the successful defence of the Scientific Investigations Grant in 1956 suggested that this might be a long-term niche for the Society. It was also a role where the Society's independence and its unique access to outstanding scientists would be significant assets.

This was reinforced, in a roundabout way, when the Prime Minister Harold Macmillan had a discussion with the nuclear physicist John Cockcroft on how to strengthen Britain's basic research effort.[25] It was December 1958, 14 months after the *Sputnik* launch, and Cockcroft had just returned from a visit to the Soviet Union. He suggested that a good deal could be accomplished by giving an extra £3–4 M to university research, on top of the budget increases already in train for the Research Councils. Macmillan asked his officials for ideas about how best to use such a sum. This started a number of hares running, in the Treasury, DSIR and elsewhere, with both Todd and Hailsham promptly preparing detailed responses for Macmillan and with senior Treasury officials trying to forestall any ideas of a free-for-all. Todd, supported by Hailsham, trained his ACSP spotlight on better provision for equipment (for which he wanted

[22] E.D. Adrian, 'Anniversary Address, 1955', 157.
[23] David Martin took the same line in 'The Royal Society today', *Discovery*, 21 (7 July 1960), 293: 'The regeneration of the Fellowship is quite the most important annual activity of the Fellows.'
[24] Cyril Hinshelwood, 'The Tercentenary Address at the formal opening ceremony, 19 July 1960', *Notes and records of the Royal Society of London*, 16 (1961), 22–3.
[25] See numerous exchanges at TNA T 218/170 and TNA PREM 11/2794. Todd's paper was dated 17 January 1959 and Hailsham's, transmitting Todd's to Macmillan, 27 January 1959.

£2 M p.a.), skilled technicians (£1.5 M p.a.), senior research posts with no undergraduate teaching responsibilities (£0.5 M p.a.) and improved buildings (£1 M in total). That in turn led to a debate about how to ensure that any new money would actually find its way to scientific research in the universities. The UGC was committed to the philosophy of the block grant and would not earmark funds for specific purposes; ARC and MRC were willing to earmark but were thought not to have the interests of the universities at heart; DSIR was not highly regarded, and grant giving, though growing rapidly,[26] was not its primary business. There might be an opening here for the Royal Society to exercise a modest directive function, turning its independence to advantage.

The Scientific Research Grants Committee, 1959–61

One of Hailsham's first actions on being appointed Minister for Science in October 1959 was to commission a study from the ACSP on the balance of scientific effort, in response to growing concern that research was lopsidedly concentrating on a few, expensive areas of physical science to the detriment of other fields.[27] In the same context, he approached the Royal Society with a request for a survey of the needs of scientific research. The ensuing report, *The encouragement of scientific research in the UK*, was submitted to him on 21 June 1960.[28] The Society made full use of the opportunity to exceed its brief and press its case for an explicit role in the running of British science.

In addition to their frustration at being increasingly squeezed out of the action, the Society's Officers were convinced that the main funding agencies were not really committed to pure academic research. DSIR, ARC and MRC were all seen in their various ways as primarily concerned with their own institutes, giving lower priority to their responsibilities for academic research.[29] There was also an anxious debate about the virtues

---

[26] By 1960–1, the four Research Councils together were spending £2.225 M on grants for university research, up from £0.948 M in 1957–8.

[27] TNA DSIR 17/714.

[28] OM 17 December 1959, minute 2(d); CM 21 January 1960, minute 20; CM 16 June 1960, minute 22; OM 7 July 1960, minute 2(b). The report was eventually published, almost as an afterthought, a year later: *The encouragement of scientific research in the UK* (Royal Society, 1961). Hailsham's commission, consciously or not, had echoes of the Society's January 1945 report on the needs of research in fundamental science after the war.

[29] In 1960 the DSIR was responsible for fifteen research organisations, covering areas from buildings to water pollution and including the National Physical Laboratory. The great bulk of its £12 million budget went on these; less than £2 million went on 'grants for special researches' and 'grants to students, etc.', that is spend in universities. Harry Melville, Department of Scientific and Industrial Research, 69–71, 195. At the same

(from the research perspective) of having a plurality of funding sources and the dangers (Treasury perspective) of wasteful duplication. Nor was there any external mechanism for controlling how much each university spent on research from the quinquennial block grant that it received from the UGC. This mattered because of the rapidly escalating costs of research as more and more sophisticated equipment and materials, including computers,[30] came to dominate work in more and more fields of research. And the spiralling costs of high-energy physics were beginning to intrude into all discussions of the management of science.

Seeking to encapsulate the mood within the Society's circles, and alert to the potential for enhancing its influence, the Assistant Secretary David Martin wrote to the Physical Secretary Bill Hodge on 4 April 1960 with a proposal. This was that the government should establish a new 'Scientific Research Grants Committee' (SRGC), both to coordinate national spending on grants for scientific research and to offer its own grants. The Research Councils could continue their support for applied science as before, but 'in pure research the RS advice should be integrated into national spending more so than at present'. This would be achieved by having the SRGC chaired by the President of the Royal Society and including two other Officers and a number of independent Fellows in its membership, as well as the heads of the Research Councils, the UGC and the ACSP. It should have the same constitutional position as the UGC, reporting direct to the Chancellor of the Exchequer, and have 'a considerable measure of independence of Whitehall accounting procedures'. An SRGC along these lines would be an effective mechanism for the Society to exercise that 'guiding influence' on UK science to which Andrade's 1945 memorial had aspired.

Hodge thought this a splendid idea, but also identified potential problems. The most obvious was the likely attitude of DSIR, ARC and MRC. Hodge predicted, correctly, that they would be hostile to external interference, especially in respect of research that they supported outside the university system. On the other hand, ACSP would probably be supportive: 'indeed, Alex [Todd] might find it a useful channel through which to carry out his wishes.' The key was how SRGC used its own budget. Martin and Hodge saw this as being used for pump-priming new fields (like seismology) and supporting work that fell into gaps in the existing funding structure. For major items like buildings or telescopes, SRGC should be empowered to put a case direct to the Treasury, as the Society was accustomed to doing. The

time, ARC ran 29 research institutes and MRC supported a major institute and 83 research units and groups. *Annual report of the ACSP 1959/60* (Cmnd 1167), 32.

[30] Jon Agar, 'What difference did computers make?', *Social studies of science*, 36 (2006), 869.

SRGC secretariat should be closely linked to the Royal Society so that the President 'could go in frequently and keep his eye on things'.[31]

Drafted at a time of growing rhetoric within the Labour Party about the need to 'redefine our socialism in terms of the scientific age',[32] the Royal Society's *Encouragement* report began by bluntly dismissing the idea of a central, government-controlled master plan for research as 'largely illusory': 'We are far more likely as a nation to maintain our honourable position, not only in pure science, but in the long run, in applied science, if we pay due attention to the nursing and strengthening of spontaneous new growths.' Indeed, 'in many ways it is distasteful to think that fundamental research needs organisation at all'. The report explicitly repudiated the idea of the Society setting up and running its own experimental establishments 'in the manner of an Academy of Sciences'. Instead, it detailed the Society's aspirations for a stronger role in running UK research by identifying a series of shortcomings in existing arrangements that could be rectified by giving the Society the necessary funding and responsibility:

- For support of key individuals: 'When the work cannot be supported from other existing sources the Royal Society itself should be in a position to obtain special funds to use, as it deems fit in the best interest of advancing science.'
- For international exchanges: 'It sometimes happens that a major advance takes place abroad ... the part of the parliamentary grant-in-aid to the Royal Society relating to travelling expenses should be increased to at least £50,000 [from £32,000].'
- For large collaborative projects in areas like space research and the IGY: 'The organisation required for such projects can often best be achieved under a non-government body such as the Royal Society, and financed by Parliamentary Grants-in-Aid.'
- For research generally, the Society's annual Scientific Investigations Grant should be increased from its 1960–1 level of £75,000, now 'grossly inadequate for its purpose', to £250,000, and the associated Treasury strings should be loosened.
- Finally, and additional to the sums already mentioned: 'Since major scientific advances are essentially unpredictable, one of the needs of research ... is a substantial sum of money, of the order of £1 million per annum, in the hands of bodies with the broadest possible outlook and

[31] Letter from David Martin to Bill Hodge dated 4 April 1960, and Hodge response dated 25 April 1960: HF/1/17/2/24.
[32] Philip Gummett, Scientists in Whitehall, 4; David Edgerton, 'The White Heat revisited: the British Government and technology in the 1960s', *Twentieth century British history*, 7 (1996), 53–82.

sympathy, and having the duty to detect and encourage scientific originality wherever it may appear.'

The *Encouragement* report did not bid for the Society to take over the grant-giving role of DSIR, ARC and MRC – though of course the Society would do so if asked – but, rather, suggested that it 'might with advantage take a greater share in the support of research in universities by means of grants', and more generally that it should 'play a larger part in the determination of the policy of grant-giving'. Something along the lines of David Martin's Scientific Research Grants Committee 'would serve a very useful purpose' in this context.

Hailsham had been expecting something much more specific (and politically harmless) – for example astrophysics needs a large computer, oceanography needs to attract more researchers, 'borderline' (i.e. cross-disciplinary) subjects such as biophysics and molecular biology need special support. The *Encouragement* report provided that as well, but inevitably it was its other recommendations, both financial and organisational, that caught the attention.

The Society was canny enough to realise that it needed to prepare the ground before submitting its report to Hailsham. It discussed the report fully with ACSP Deputy Chairman Solly Zuckerman and privately briefed Hailsham's office on its likely contents even before it had been signed off by the Society's Council.[33] The report went to Hailsham both as the response to his commission and as the Society's contribution to the ACSP study on the balance of scientific effort. That enabled Hailsham to say that he needed ACSP's input before he could respond fully, though he promised a meeting with the Officers within six months. Todd was due to spend August to November 1960 in Australia and would not be able to chair a meeting of ACSP on this topic until December, so that stretched the timetable further. Meanwhile, all parties got to work analysing the Society's proposals and defending their territorial interests.

The Treasury recognised that the status quo was not perfect. There was thus a degree of sympathy for the Society's proposals, and recognition, too, of the Society's competent track record in managing science. But this was coupled with suspicion of its apparent ambitions and resolute insistence on the accountability issue that had so irritated Hinshelwood. The head of the Treasury's Arts and Science Division, Richard Griffiths, minuted his boss Otto Clarke:

This paper by the Royal Society is an important one. It represents a take-over bid by the Royal Society for much of the patronage of scientific research, particularly in universities, which now rests in the hands of DSIR and to a lesser extent MRC

[33] Roger Quirk to Frank Turnbull, 13 June 1960: TNA CAB 124/1389.

and ARC. It directly attacks DSIR in its suggestion that a government department is not the best place for the detection, encouragement and coordination of scientific research ... I find myself in considerable sympathy with what the Royal Society have to say on the major questions of organisation.

I cannot see, however, how the Treasury and Parliament can be expected to accept the implied suggestion that an increasing element of the scientific research in this country should be financed, not by Parliamentary Votes to the Research Councils, but by increased grants-in-aid to the Royal Society.

In general I think it is all to the good that the Royal Society have come out with this paper. Many of the points they make are extremely sensible.[34]

Frank Turnbull, newly moved from the Treasury to run the Office of the Minister for Science, briefed Hailsham along comparable lines, mixing praise for the Society (at least in comparison with DSIR) with persistent iteration of the accountability issue, and adding the suggestion that a new research council for biological sciences might solve many of the problems.[35] Hailsham then informally consulted the Secretaries of the Research Councils, and asked for considered comments from their full Councils. Hailsham had initially wanted to secure a modicum of extra funding for the Society, but by the time he met the Society's Officers on 26 October 1960, he was on the defensive: 'the constitutional position demanded that there should be ministerial responsibility for the spending of public money', all funding issues were matters for the Treasury rather than him, extensive discussions were already under way about the organisation of research, and the UGC was still strongly opposed to earmarking in any form. He did, though, raise the possibility of increased finance for more Royal Society professorships and readerships, as a means of strengthening university science – subject of course to Treasury approval.[36]

The Research Councils, with most to lose, were distinctly hostile, taking the *Encouragement* report both as personal criticism and as a threat to their empires. They concurred with the anti-planning philosophy espoused at the beginning of the report, but otherwise had little positive to say. They bluntly rejected the implied criticism that they were neglecting important opportunities for new research. They opposed the Society's claim that its long-established Grant Boards constituted the best screening mechanism for pure research proposals. And they lambasted the proposed SRGC as 'unnecessary, impractical and politically unacceptable'. They did, though, allow that if the Society could secure additional

[34] Internal Treasury minute, R.C. Griffiths to R.W.B. (Otto) Clarke, 28 June 1960. Note also draft by D.R. Collinson, 24 June 1960. Both at TNA T 218/520.
[35] Turnbull to Hailsham, 7 July 1960. See also briefing for meetings on 25 July (with the Research Council Secretaries) and 27 July (with Todd), and notes on their outcomes: TNA CAB 124/1389.
[36] OM/90(60), C/146(60).

funding – though not at Research Council expense – it might enlarge 'those activities which it is specially well fitted to carry out, e.g. the creation of posts of professorial or reader status in particular fields, the encouragement of international scientific cooperation and the exchange of scientists between this country and countries overseas'. That neatly defined one, non-threatening, view of the proper role for the Society: it should deal with specialised support for individuals, not with mainstream funding structures.

The biochemist Hans Krebs alerted Howard Florey to this hostility a week before Florey succeeded Hinshelwood as President of the Royal Society at the end of November 1960. In addition to enclosing a copy of the Research Councils' response, Krebs reported that he and two other Fellows had mounted a vigorous defence of the *Encouragement* report at a meeting of the ARC Council on 15 November and had secured agreement that there were indeed real problems that had to be addressed. Conveniently, most of them were the fault of the UGC, which was seen as too focused on expansion of student numbers rather than on promoting research and which, Krebs suspected, did not effectively fight its corner with the Treasury. One of the most pressing problems was the stream of first-rate scientists emigrating to the United States in pursuit of better research facilities and lower teaching loads. Krebs, tipped off by Alan Hodgkin about parallel debates within MRC, also wrote to Harold Himsworth, the long-serving MRC Secretary, protesting at his negative reaction to the *Encouragement* report and informing him that ARC had eventually agreed to take a more constructive approach; Himsworth subsequently toned down his criticisms of the report. As Krebs pointed out, if Hailsham seemed to be offering to help scientists secure more resources, it was not the time to be claiming that everything was already perfect.[37]

As Florey (Figure 2.2) took over the Royal Society presidency, then, top of his in-tray were resolving the controversies sparked by the *Encouragement* report and building good relations with, in particular, the Research Councils and the Treasury. His approach was to refocus the Society's ambitions for a directive function:

I do not like the idea of the Royal Society undertaking to have an overall control, as it were, of grant giving bodies. I think this would only introduce more bureaucracy and result in great delay in getting the resources to the people who actually do the

---

[37] The Research Councils' collective response, drafted by Gordon Cox of the ARC, and Krebs' correspondence with Florey (22 November 1960) and with Himsworth (17 November 1960) are at HF/1/17/2/6. See also MRC Council meeting, 21 October 1960, minute 252: TNA FD 13/102.

Figure 2.2 Howard Florey. © The Royal Society

research and I should think that one of our main functions is to put ourselves in the place of those who do the research and to see that they get the tools with the minimum of fuss and bother. In this way we might try to use all the machinery that at present exists without trying to put ourselves in an overriding position.[38]

When the ACSP finally had a substantive discussion on the *Encouragement* report, on 7 December 1960, there was sympathy for the Society's key message about the lack of machinery for an overview of grant giving. Todd himself was very much in favour of having an over-arching mechanism for reviewing grants. He also favoured giving the Society a number of research professorships and readerships. But he warned Florey that he was 'having to thread his way very carefully so as not to tread too hard on the ... Research Councils'.[39] The upshot of the meeting was a decision to look further at giving the Society 'a greater opportunity to create senior research posts in science, and to foster international liaison, travel, and conferences in the interests of scientific advance'. Improved grant giving would be addressed separately.[40]

Arguments about how to establish an appropriate framework for grant giving bubbled on through the early months of 1961. Todd, as Chairman of the ACSP, briefed Hailsham on 31 January 1961:

It is an open secret that there has been recurrent friction between the Royal Society and DSIR – the most recent trouble being over space research. In my view the essential basis of this friction – and it is a real one – is the fear that the control of scientific research may be passing into the hands of a Government department which is not regarded as properly constituted to exercise such a function. The same comments apply if in lesser degree to the activities of MRC and ARC.

Given that DSIR's main function was 'to bring science into industry', Todd wanted it to focus on that and reassign its grant-giving work to a new body, to which he would also assign ARC and MRC grant giving 'outside their immediate fields of action'. He went on to outline something not too different from the Royal Society's SRGC.[41] Harrie Massey, then a member of the DSIR Research Grants Committee and recently a member of the Royal Society Council, similarly told Florey that physicists

[38] Private and unsigned aide-memoire by Florey dated 5 December 1960: HF/1/17/2/24.

[39] Note by Florey of a meeting in the Athenaeum with Solly Zuckerman on 7 December 1960, following the ACSP meeting. They were later joined by Todd. HF/1/17/2/24. The Treasury brief for the ACSP meeting stressed: 'We must continue to resist strongly the implication that an increasing amount of scientific research should be financed directly by increased grants-in-aid to the Royal Society.' TNA T 218/520.

[40] ACSP meeting, 7 December 1960, minute 4.

[41] A copy of Todd's paper was circulated to the Trend Committee as ESO (62)11 on 19 July 1962.

had no confidence that DSIR was constituted to serve the best interests of university science. The department's primary commitment was seen to be to its own numerous applied research establishments; university interests were not properly represented on its main council, and it was left to second-rate people with little understanding of science to conduct negotiations with the Treasury.[42]

The Treasury, meanwhile, interpreted a separate paper by Todd, proposing a series of national institutes to manage expensive equipment for academic researchers, as part of a conspiracy by ACSP and the Office of the Minister of Science 'to wrest away from DSIR control of the significant parts of the Government's assistance to university research'.

Naturally the scientists led by the Royal Society want to get all power in this field including financial power into their own hands ... We are in the gravest danger of being manoeuvred by the present proposals into a position where effective financial control of the Government's scientific expenditure, even with a Science Budget, will remain an impossibility for years ahead.[43]

The issue, again, was accountability – as Richard Griffiths put it, power without financial responsibility.

Back at the Royal Society, David Martin was busy drafting and redrafting his original SRGC proposal into a memorandum, *Grant-giving to universities for scientific research*, in the process trying again to define a set of problems to which the Royal Society could provide the solution.[44] The primary problem now crystallised as insufficient openings for early-career researchers of proven ability, especially in borderline areas (e.g. between physics and biology). Associated with this claim was the asserted existence of a brain drain. As well as calling for the creation of up to 20 Royal Society professorships and readerships, adequately supported with funds for research and carrying responsibilities for training postgraduate students, the Society therefore launched a groundbreaking attempt to quantify and characterise the brain drain.[45]

Martin's second problem was the structure of DSIR, which had a very wide range of responsibilities, especially in relation to industrial research, and consequently 'cannot devote as much attention to the grant-aiding of scientific research extra-murally as a body solely devoted to such a

---

[42] Note by Howard Florey of a discussion with Harrie Massey, 14 February 1961: HF/1/17/2/24.

[43] Richard Griffiths to Arnold France, 1 March 1961: TNA T 218/520.

[44] For successive drafts, see memo from Martin to Florey, 16 January 1961, HF/1/7/2/24; OM/12(61), 19 January 1961; OM/36(61), 28 February 1961.

[45] OM 18 January 1961, minute 3. Brian Balmer, Matthew Godwin and Jane Gregory, 'The Royal Society and the "brain drain": natural scientists meet social science', *Notes and records of the Royal Society of London*, 63 (2009), 339–53.

purpose'. He therefore proposed that its grant-giving role be transferred to an SRGC, though 'awards of a technological character' should stay with DSIR. MRC and ARC should remain unchanged, but 'there are certain fields not catered for in any of the existing grant-giving bodies which the proposed new body could take on'. These included short-term, large-scale activities such as scientific expeditions that fell outside the province of existing grant-giving bodies, and the need for a source of small amounts of money rapidly administered to provide special apparatus or to assist small scientific expeditions.

Florey agreed with Martin,[46] and on 24 February 1961 convened a meeting of Officers together with Patrick Blackett, John Cockcroft, Peter Medawar, Alex Todd and Solly Zuckerman to push the matter forward. They rehearsed the now familiar arguments, adding that the UGC had insufficient money for buildings or technicians or tenured posts. They accepted the UGC unwillingness directly to earmark grants for research. DSIR was, rightly, preoccupied with the 'application of science to industrial practice'; funding for research grants and postgraduate training should therefore be moved from DSIR and given to a new body. So the Society should propose a new body to handle DSIR's grant-giving and postgraduate training roles, the financing of national institutes set up to assist university research,[47] and financial support for programmes such as space research involving a number of different universities. However, 'at this stage no proposals should be forwarded suggesting the precise role of the Royal Society in any new organisation'. They did, though, agree that the Society was 'particularly well equipped' to handle appointments to research posts and the conduct of international scientific relations.[48]

Martin fed this discussion into his *Grant-giving* draft. He proposed a new grant-giving body to deal with all the shortcomings identified, including the need for additional senior research posts in universities.[49] He was careful to highlight the positive elements of how the UGC, DSIR, ARC and MRC operated, and made no mention at all of a possible role for the Society.[50] Florey discussed Martin's draft confidentially with

---

[46] Memo by Florey dated 15 February 1961 [OM/29(61)], especially the copy at HF/1/17/2/6.

[47] Such as the National Institute for Research in Nuclear Science (NIRNS).

[48] OM/37(61), marked 'personal and strictly confidential'. See also OM 16 February 1961, minute 2(b).

[49] OM/36(61), 28 February 1961; OM 2 March 1961, minute 2(d); CM 23 March 1961, minute 3. *Grant-giving to universities for scientific research: a statement communicated to the Minister for Science by the Council of the Royal Society* is annexed to the 23 March 1961 Council minutes.

[50] Interestingly, though, the next day the DSIR Research Grants Committee was having an inconclusive discussion about handing over its senior research appointments scheme to the Royal Society, and was worrying about the Society's ability to handle all branches of

Arnold France (who represented the Treasury on ACSP and had over-sight of research expenditure) before it went to the Royal Society Council. France stressed that the Treasury wanted reform, but that a new body funded by grant-in-aid rather than by a Vote to an accountable minister was beyond the art of the possible. He did, however, greatly appreciate the private exchange of views, noting that Florey had put the relations between the Royal Society and the Treasury 'on a new and much better footing; certainly there is a possibility of greater frankness between us and a very friendly feeling at present. I should like to try to preserve and even develop this still further.'[51]

But this new-found mutual frankness did not mean mutual agreement. The Society's *Grant-giving* memorandum was discussed at the 3 May 1961 ACSP meeting. The Treasury brief for that occasion pulled no punches:

We look askance at ... the Royal Society's new grant-aided body ... yet one more Research Council, with executive powers and a grant-in-aid, outwith direct Ministerial control, seems to us a non-starter ... Our present tactic is to wait first and see reactions to the proposals by Sir Alexander Todd and the Royal Society.

To rub it in, a follow-up brief observed:

Recent months have shown a considerable intensification of the Royal Society's efforts to win itself a bigger and better place in the sun. Quite apart from the Society's increasing strength in the space research field, its recent proposals [for major investment in oceanography and in international collaboration] reflect a considered attempt to make the Royal Society one of the major pillars in UK scientific policy. Any steps towards this end are to be deplored at the moment, when what is really needed is a reassessment of the place and function of the existing Exchequer agencies for science.[52]

The 3 May ACSP meeting duly proved an uncomfortable occasion for the Society. Arnold France declared its *Grant-giving* proposals unacceptable, and Research Council representatives defended their corner vigorously.[53] Todd as Chairman therefore subsumed the Society's concerns into a larger context – 'the whole question of financing research by Her

technology. DSIR Research Grants Committee meeting, 24 March 1961: extract from minutes at HF/1/17/2/24.

[51] Arnold France, Note for the record, 17 March 1961: TNA T 218/538. Also HF/1/17/2/24; OM 23 March 1961, minute 2(d). One outcome of the Society's improved relationship with the Treasury was its successful negotiation a year later, against the odds, of a grant of £125,000 towards the capital cost of the acquiring of new premises in Carlton House Terrace, and agreement that the Treasury would cover the annual rent: OM/70 (62) and OM/70a(62).

[52] Briefs dated 18 April, 28 April, 2 May 1961: TNA T 218/520.

[53] Florey had earlier said of the Research Council Secretaries: 'They would be wise to pay more attention to what is being said by knowledgeable people.' Letter to David Martin, 25 April 1961: HF/1/17/1/10.

Majesty's Government would need careful consideration and possibly some reorganisation' – and suggested that in the meantime it was incumbent on all parties to make existing arrangements work as well as possible. So they agreed to a series of bilateral meetings to thrash out what exactly the current shortcomings were and how they could be addressed.[54]

Florey and his colleagues prepared assiduously for these bilaterals. But his meeting with the MRC Secretary Harold Himsworth on 20 June proved an ill-tempered affair.[55] Himsworth in effect accused Florey of sabotaging the Research Councils and siding with those who wanted an all-powerful Minister of Science, which he held to be against Research Councils' interests. He was also hostile to the creation of more Society research appointments, and to the idea of regular meetings between the Research Councils and the Society. He wanted to send the Society's *Grant-giving* memorandum to his own Council, but objected to it being circulated to the Royal Society Fellowship.[56] In short, they found little on which they could agree.

Florey reported drily to the Royal Society Council on 13 July 1961 that, while there had not yet been any formal response from the government to the *Grant-giving* memorandum,[57] the Secretaries of the Research Councils 'were not enthusiastic about it'.[58] But the SRGC proposal did not die a quick death. When Florey convened an informal meeting of Fellows at his college in Oxford three months later, the view was still that the Society should continue to press the case both for the SRGC itself and for the Society having an important role in it.[59] And the *Grant-giving* memorandum featured strongly in the Society's dealings with the Trend Enquiry in 1962–4.

The Trend Enquiry, 1962–4

Lord Hailsham announced in Parliament on 7 March 1962 the appointment of an Enquiry, led by Burke Trend, Second Permanent Secretary at the Treasury, into the administrative organisation of civil science. Howard Florey immediately sought a meeting with him to discuss how

[54] OM 18 May 1961, minute 2(d); also OM/63(61), OM/71(51) and OM/72(61).
[55] Note by Florey on meeting with Himsworth, 20 June 1961: HF/1/17/2/24.
[56] Because the memorandum was submitted to the minister and marked confidential, its circulation was restricted initially to a few individuals. Even in October, a decision about circulating it to all Fellows was deferred because it was still being considered in official circles. CM 12 October 1961, minute 20.
[57] There never was a response, other than implicitly through the Trend Committee. Hailsham refused to discuss the matter when Florey pressed him on it at a private meeting on 30 March 1962. OM/37(62).
[58] CM 13 July 1961, minute 9.   [59] OM/120(61).

the Royal Society might get involved, arguing that it was composed of real, active scientists and was 'most anxious that their point of view should not escape the attention of this committee'. At the meeting, on 30 March, Hailsham told Florey somewhat disingenuously[60] that the members had not yet been appointed, but he 'rejected absolutely and several times' Florey's request that the Royal Society should be allowed to nominate one person to sit on the committee.

Florey was then in the middle of a spat with Hailsham on another matter. The ACSP had asked the Royal Society for a report on biology, with a focus on disciplines such as microbiology and molecular biology that were thought to be falling between the gaps of the existing Research Council structure.[61] The Society's report was submitted to the ACSP in December 1961. It leant over backwards to applaud the achievements of the Research Councils, but nevertheless concluded that 'large fields of biology are not covered by MRC and ARC, and the Biology Subcommittee of the DSIR Research Grants Committee has not the authority nor is it ideally constituted to deal with the wide range of problems presented in the proper development of biological research'. This was taken as further stoking the controversy arising from the Society's *Grant-giving* memorandum and the SRGC proposal, even though it was consistent with Frank Turnbull's suggestion the previous year for a research council for the biological sciences. The Research Councils complained that they had not been properly consulted in its preparation.[62] C.H. Waddington, a member of ACSP, briefed Florey that he had had to defend the biology report at the 7 February 1962 ACSP meeting[63] against the Secretaries of the Research Councils, 'who seemed to feel it was an attack on them'. Florey wondered whether it was time to reform the ACSP.[64]

The Society's biology report had as troubled a ride with Hailsham as it did with the ACSP. When they met on 30 March to discuss Trend, Hailsham told Florey that he regarded the Society's biology report as an

---

[60] The more so as the Trend Committee itself stated that it was appointed by the Prime Minister in March 1962. Cmnd 2171, *Committee of Enquiry into the organisation of civil science* 1963 [the Trend Report], 9.

[61] ACSP 1 February 1961, minute 3(ii); Royal Society, *Report of the ad hoc Biological Research Committee* (November 1961). See also Note 35.

[62] Meeting between Hailsham and Harry Melville, 8 May 1962: Hailsham's brief for 12 July 1962 DSIR Research Council meeting. Both at TNA CAB 124/1766.

[63] Another Fellow, Rogers Brambell, found himself having to defend the report at the UGC meeting on 15 February: TNA UGC 1/10.

[64] Letter from Waddington to Florey, 12 February 1962; reply from Florey, 14 February 1962: both at HF/1/17/2/2. Florey later commented to Waddington that 'Todd has little idea about biology although I imagine he would be the last to admit such a thing': letter dated 2 April 1962: HF/1/17/13/3. See also Waddington to Todd, 30 March 1962: TNA AB 16/4247; and Penney to Todd, 30 March 1962: TNA AB 16/4247.

attack on the UGC.[65] 'I was so staggered by this that I didn't take the point up,' recorded Florey. He commented to Waddington that Hailsham was 'still quite ill-informed about the Society' and had 'an unshakeable belief that the RS is an inefficient body which should have nothing to do with giving scientific advice ... The Minister is being fed information about our report which is far from the truth.'

Hailsham sent Florey a follow-up letter on 3 April, which started in conciliatory mood but soon returned to the attack. He accused the Society of being 'amateurish' in comparison with the increasing sophistication and professionalism of modern government, and expressed his disappointment that it had not made more use of the opportunities arising from the creation of his post. Then, reacting to Florey's complaint that Royal Society reports written at his request had received no official response, he stated that it was from the ACSP's discussions of the *Encouragement* and *Grant-giving* papers and similar material that the decision to set up the Trend Committee had emerged. He promised that the *Grant-giving* memorandum would be circulated to Trend, adding 'it is quite obvious that the Royal Society occupies a position which makes it essential that their views should be brought to the notice of the Committee'.[66]

The Trend Committee held its first meeting on 9 May 1962. Already then, with two very senior Treasury officials (Burke Trend and Thomas Padmore), and Frank Turnbull, on board, the Committee took the view that the Royal Society might provide scientific advice but that a directive function would be inappropriate: 'it should not, as an institution, be asked to carry any of the onus of the decision in regard to the allocation of scientific resources.' Alex Todd nevertheless circulated his 31 January 1961 paper, broadly sympathetic to the Society's analysis of current shortcomings in the system and to its SRGC proposal, to the Trend Committee. But the Committee stuck to its initial position on the Society's role. At its 20 December 1962 meeting, the Committee confirmed that there was 'no real case for any major reform in regard to the support for university research, e.g. on the lines proposed by the Royal Society involving the creation of a new research council'.[67]

---

[65] The Chairman of the UGC, Keith Murray, reassured Florey that he did not feel attacked: OM 5 April 1962, minute 1.

[66] OM/40(62); original letter at HF/1/17/13/3. Also OM 8 March 1962, minute 3(a). The Trend Enquiry was prompted by several factors, including (according to Todd's autobiography) the widespread feeling that Britain was not keeping pace with technological innovation in industry. In a minute to Norman Brook on 12 January 1962, Burke Trend specifically identifies the confidential epilogue to the 1961 Zuckerman report on management and control of R&D as the decisive stimulus for the establishment of his committee: TNA T 218/540. See also Matthew Godwin, *Skylark rocket*.

[67] ESO(62)20th meeting, 20 December 1962.

The Society put a great deal of effort into its written submission to the Trend Committee, with Patrick Blackett, Harrie Massey and Frederic Williams preparing an initial draft on the back of an extensive consultation of the Fellowship.[68] The biggest issue in the Society's view was the absence of a policy-making body of sufficient strength, with the corollary that major policy decisions about civil science were in effect being made by the Treasury. Without explicitly calling for the abolition of the ACSP, the Society argued the case for a new Civil Science Board to advise both on general policy issues and on the distribution of funds to the Research Councils. In the expectation that the allocation function would move from the Treasury,[69] the Civil Science Board would be tasked with advising the relevant minister, who would have charge of an overall budget for civil science. The Society pulled back from calling explicitly for an entirely new ministry, suggesting without obvious enthusiasm that the minister in question 'might be the Lord President of the Council' (i.e. Hailsham, whose antipathy to the proposal was known). Because so much of the funding for research in universities came via the UGC, the Society argued that the UGC and Civil Science Board should come under the same minister, in effect creating what it at one stage called a Ministry for Universities and Science.

The second major focus of the Society's submission was the proposal to establish a new Science Research Council (SRC). This would take over the DSIR's grant-giving and postgraduate training functions, oversee research institutes like NIRNS set up to work with universities, monitor the provision of senior research posts in universities and deal with the funding of buildings and other major capital expenditures related to academic research. SRC was to give special attention to engineering and to areas of biology outside agriculture and medicine.

As for its own future role in civil science, the Society was conspicuously reticent. It did remind the Trend Committee that, being 'the senior body of scientists both in this country and in the world', it had 'exercised an important influence on the whole development of science'. But it refrained from bidding for a specific piece of the action, preferring to wait until the new shape of civil science had become clearer before making detailed suggestions. At this stage it was content simply to mention, in a

---

[68] See papers Trend/14(62) and Trend/15(62), copies at HF/1/17/19/2. The final text was published as Royal Society, *Evidence to the Committee of Enquiry into the organisation of civil science (Trend Committee)* (1963).

[69] Edward Bridges advised the Society that the Treasury was keen to divest itself of responsibility for detailed decisions about research spend. He had recently retired as Permanent Secretary to the Treasury and was then serving as Chairman of NIRNS. He had been elected to the Society under Statute 12 in 1952 and held it in high esteem.

carefully worded sentence, its wish to continue its 'association with international scientific affairs and with other scientific activities such as giving grants for investigations and publications, electing research professors and fostering research generally' – in other words, activities that by general consensus were within its established remit.

The Society gave oral evidence to the Trend Enquiry on 27 March 1963. It strongly argued the case for a single science budget controlled by a minister advised by a suitably independent board, and agreed that its own public funds should come through this route. As for membership of that board, appointments should be personal rather than representative: it 'should certainly not include the Secretaries of the Research Councils'. The Society, naturally, 'would probably be able to assist by suggesting names'. The proposal for a new Science Research Council was primarily intended to improve the arrangements for grant giving – in effect, to deliver the objectives of the previously mooted SRGC. And, finally, Florey made a strong play for recognition and continuation of the Society's international functions in any new structure.[70]

The Trend report was submitted to the government in September and was published on 30 October 1963, shortly after the Robbins report on higher education. Hailsham promptly briefed a meeting of Royal Society Fellows about both reports. Much of Trend was broadly in line with the Society's thinking, though not with Hailsham's.[71] Blackett, for example, suggested that 11 out of the 15 main recommendations could be accepted by the Society without further ado.[72] However, an analysis by David Martin of Trend's direct comments about the Royal Society highlighted where the Society stood in the new dispensation.[73] The President would need to be consulted when various senior scientific appointments were being made.[74] The research professorships scheme (see Chapter 3) should be expanded. And there was recognition of the Society's continuing role in promoting non-governmental international scientific cooperation.[75] But that was it. There would be a new advisory body in place of the ACSP, but no special role for the Society in it. There would be a Science Research Council but, again, the Society would not

[70] ESC(63)10th meeting, 27 March 1963.
[71] At the 20 December 1962 meeting of the Trend Committee, Turnbull reported Hailsham as being opposed to the idea both of an overarching science budget and of a top-level committee to advise on the allocation of funds between the Research Councils.
[72] Trend/4(63), paper dated 29 November 1963.
[73] Memo dated 30 November 1963: HF/1/17/2/57.
[74] At least in some instances, Todd as Chairman of ACSP was ahead of Florey as PRS in the pecking order of consultation. See letter from Frank Turnbull to Todd, 31 January 1964, about the chairmanships of SRC, NRRC and IRDA: Todd Box 5/21.
[75] Cmnd 2171, para 116 – not mentioned in Martin's analysis.

have the starring role in its management that David Martin had envisaged in his original blueprint for the SGRC. Moreover, special projects such as the proposed Anglo-Australian telescope at Siding Spring, on which the Society had invested protracted negotiations, were to be abruptly transferred to the SRC – to the dismay of those involved.[76] Space research, and the associated Royal Society staff, had already been appropriated by the Office of the Minister for Science.[77] The Royal Greenwich Observatory, with which the Society had been very closely involved since 1710, was, similarly, taken out of its hands.[78] There was no mention of the Society's existing role in the National Physical Laboratory, despite its long historical connections with it.[79] The Society, it seemed, was being further marginalised: there were fewer gaps left for it to fill.

Florey sent some preliminary comments to Hailsham on 7 December 1963, and formal comments followed two weeks later after the Society's Trend team had met.[80] He supported the proposal that the Society, like the Research Councils, should deal with the minister rather than directly with the Treasury in securing its public funding – provided that the relevant minister was responsible only for higher education and research and not for the whole of education.[81] He agreed, naturally, that he should be consulted on senior scientific appointments. There was extensive comment on how the successor body to the ACSP should work, and particularly how to ensure that its advice would reach the minister without interference from civil servants. Florey told Hailsham that he was already in discussion with the Treasury over increasing the numbers of Royal Society Research Professors. And on international relations, he

[76] Interview with Bernard Lovell. See also Bernard Lovell, 'Authority in science', talk to the Association of British Science Writers, October 1984: copy at David Phillips papers, MS Eng. c.5510, O.121.

[77] Paper dated 4 May 1962 for the Trend Committee, ESO(62)2: TODD Accession 811, box 4.

[78] Bernard Lovell, 'The Royal Society, the Royal Greenwich Observatory and the Astronomer Royal', *Notes and records of the Royal Society of London*, 48 (1994), 283–97.

[79] On the problems arising from the Society's role, see notes of 6 July 1962 and 13 March 1963 meetings between Hailsham and Harry Melville: TNA CAB 124/1766. The Society was reduced to appointing three members of an NPL Advisory Board, and even this petered out in 1978. John S. Rowlinson and Norman H. Robinson, *The record*, 12, and interview with Michael Atiyah.

[80] Florey to Hailsham, 7 December 1963: Trend/7(63). Copies of this and subsequent exchanges also at TNA CAB 124/1875. The Society's formal response of 20 December 1963 is at appendix B to the Council Minutes of 16 January 1964.

[81] This was a key point for Florey: see, for example, his letter in *The Times*, 19 December 1963. On hearing that the decision was likely to go the other way, he wrote to the Society's Foreign Secretary Patrick Linstead, Rector of Imperial College, on 23 December: 'I suppose the struggle now will be to see if the research in universities is not allowed to be submerged by schoolteachers.' HF/1/17/2/57.

highlighted the importance for science of initiatives at non-governmental level and the Society's role in facilitating them.

In the early months of 1964, the Royal Society convened substantial groups of Fellows to debate the detailed plans for a Natural Resources Research Council,[82] for the new Science Research Council and for the future of the National Physical Laboratory.[83] The recommendations emerging from these meetings – in some instances – had significant impact on the detailed implementation of the Trend proposals at a time when the existing research agencies were vigorously defending their territories.

On more general policy issues, Florey and Martin met with Maurice Dean (Second Secretary at the Treasury) and Frank Turnbull on 13 March 1964 to discuss the Society's views of the body to replace the ACSP, and a week later went over the same ground with Hailsham, Dean and Turnbull.[84] They agreed that the new body should be called the Council for Scientific Policy (CSP) and agreed, too, how it should be constituted. Briefed by Edward Bridges, the Society argued persistently that CSP should have both a full-time chairman and a well-resourced secretariat.[85] This was prerequisite for a strong and effective body able to take on the established bureaucracy and able, also, to tackle a wide range of major policy issues from an independent perspective rather than focus mostly on allocating budgets to Research Councils. The Society explicitly sought a change from 'the type of organisation which tends to give dominance to permanent officials in formulating general policy'. Dean told Florey privately that he, personally, was coming round to Florey's view but that there were those who thought that in pushing this line, the Society was making a covert takeover bid. Florey assured him that the Society had no wish to become an agency of government, but simply

[82] After much vexed debate, the NRRC emerged as the Natural Environment Research Council (NERC), the name being suggested by Graham Sutton during a meeting with Maurice Dean. Sutton subsequently became the first chairman of NERC. Letter from Sutton to David Martin, 16 July 1964: Trend/4(64).

[83] Florey reported to Hodge in May 1964 that Hailsham and Dean had visited NPL and 'the impression was spread that the Royal Society was about to be brushed off as an historical accident in its connection with the NPL . . . I think we are now within measurable distance of being in the hands of the permanent civil servants without scientific training.' In the end, the restructured NPL went to MinTech, and the Royal Society connection was enshrined in a series of ex officio appointments and related arrangements. HF/1/17/15/6–8; also CM 30 November 1965, minute 6 and appendix.

[84] OM/29(64) and OM/27(64). Briefing notes for the 20 March meeting are at PB/6/2/4/5. OM/31(64), drafted by David Martin and dated 6 April 1964, gives the list of initial tasks for the CSP. OM/31a(64), dated 15 April, added 'the increasing need for computers in universities and governmental institutions' to the list. The Officers discussed the outcomes of these meetings on 9 April 1964: OM/35(64).

[85] Richard Griffiths in the Treasury applauded this. Memos to Ronald Harris, 26 and 27 February 1963: TNA T 218/603.

wanted to ensure that the views of the 'working scientist' were taken fully into account in the evolution of the new arrangements. Research Council heads, on the other hand, wanted a much weaker CSP chaired by a minister; Dean opposed this.[86] Blackett fanned the flames, inveighing against the 'important opposition in Whitehall from those who do not want the proposed new Council to have any teeth, and who in the past have successfully prevented the ACSP from growing any'.[87]

The Government's broad acceptance of the Trend proposals had been announced by the Prime Minister, Alec Douglas-Home, on 6 February 1964. The upshot of all the subsequent detailed negotiations was announced by Hailsham on 28 July, as was the intention to introduce the necessary legislation in the next parliamentary session.[88] That session, of course, was overtaken by the general election[89] and the start of Harold Wilson's Government in October 1964. In the event, most of the organisational proposals went ahead as planned; the major change was the creation of the Ministry of Technology and the absorption into it of Trend's mooted Industrial Research and Development Authority, which had been intended to deal with the industrial work of the defunct DSIR. The chairmanship of the CSP was confirmed as a part-time appointment. Florey was duly consulted about appointments to the new Research Councils, protesting vigorously when this appeared to be a rubber-stamping exercise.[90] He also found himself protesting vigorously at a proposal to move Research Council headquarters from London 'to some place like Manchester or Edinburgh'.[91] Despite the promise of a new and overtly science-friendly Labour Government, the Society remained vigilant in defence of what it saw as the interests of UK science.

---

[86] Notes on a meeting, 21 May 1964: STRUUK/1(64). Also OM 14 May 1964, minute 2 (b), for a meeting on 24 April with Research Council heads at which they reiterated their opposition to a full-time chairman of CSP, and OM 18 June 1964, minute 2(f).

[87] P.M.S. Blackett, 'Wanted: a wand over Whitehall', *New Statesman*, 11 September 1964, 346–50. His comment was directed mainly at the Civil Service rather than the Research Councils.

[88] Hansard, cols 285–9, 28 July 1964. Hailsham later wrote of his time as Minister for Science: 'I believe I was a success ... because 10 years after I demitted office I was honoured by being made a Fellow of the Royal Society, which ... almost overcame me with pleasure.' Lord Hailsham, *The door wherein I went* (Collins, 1975), 181.

[89] Alex Todd, anticipating a Labour victory, resigned his chairmanship of the ACSP on 30 September. Alexander Todd, *A time to remember: the autobiography of a chemist* (Cambridge University Press, 1983), 172–3.

[90] See correspondence between Florey, Lord Bowden (new Minister for Education and Science) and Harry Melville (new Chairman of the SRC), 23 December 1964–11 January 1965: HF/1/17/2/26. Also OM 14 January 1965, minute 2(a).

[91] OM/21(65), dated 3 March 1965; OM 4 March 1965, minute 2(a). Turnbull colluded with Florey to marshal the case for staying in London: Turnbull to Florey, 23 February 1965, RMA914.

The Royal Society in 1965: options

Where did the Royal Society stand after all that? Twenty years on from Robert Robinson's contested election to the presidency, and with the major functions in the running of UK science now freshly allocated to agencies more closely under government control and conforming with Treasury notions of accountability, what were to be the Society's priorities?

The weekend before the October 1964 general election, the Society's Officers met in the seclusion of Florey's Oxford college (Queen's) to reflect on the Society in the post-Trend world.[92] They specified the Society's key attributes as commitment to the fostering of fundamental research, commitment to the highest standards of scientific excellence and staunch defence of its own independence. On these its identity rested. If there were no 'directive functions' left for the Society once Trend had been implemented, maybe that was no bad thing after all. The 'definite official status, resources and powers' that Crowther in 1944 had associated with directive function were, perhaps, too close to the Soviet model of a national academy for the Society's taste.

The Society did, though, have formidable scientific status. This underpinned everything from the constant reviewing of the process for annual election of Fellows, to Edgar Adrian's search for 'Olympian dignity and exclusiveness' in the Society's accommodation,[93] and to the grandeur of the tercentenary celebrations (see Chapter 11). It allowed the Society to speak with authority for the practising scientist, as it had sought to do during the Trend Enquiry. It also meant that the Society's leadership knew, and had direct personal access to, everyone involved in running the country's scientific affairs, and they in turn nursed their relations with the Society. Even the most bullish ministers and permanent secretaries recognised that it was wise to invest at least some effort in keeping the Society sweet.

The Officers accepted that the directive function was not going to happen on any significant scale. The new structures then imminent would, relentlessly, entrench the move to virtually all public expenditure on scientific research being handled by government agencies.[94] Florey acknowledged in his 1964 Anniversary Address, devoted entirely to such identity issues, that 'the direct relation of the Royal Society to government

---

[92] OM/73(64).   [93] Alan Hodgkin, 'Edgar Douglas Adrian', 53, 56.
[94] The mid 1960s have been described as 'the high point for the nationalisation of science' in the UK. David Edgerton, 'Science in the United Kingdom', in John Krige and Dominique Pestre, eds., *Science in the twentieth century* (Harwood Academic Publishers, 1997), 759–76.

has been diminishing during the development of the scientific activity of the twentieth century'. Bereft of a significant formal locus in the new dispensation, the Officers highlighted a series of other missions for the Society. They all built on existing strands of activity and exploited the Society's defined attributes.

The most prominent strand concerned relations with government, which would now focus more on giving policy advice than on running parts of the system. Florey recognised that the Society could not compete with what he described as an 'efflorescence of government-sponsored Councils, Committees, Boards and Authorities'. But Fellows had a duty 'to do all they possibly can to assist government in the present somewhat critical stage of the country's development'. The Society would ensure that government was aware of what practising scientists were thinking, and would continue to influence the selection of scientists for important appointments. It would also engage as fully as possible in advising on matters where its scientific expertise was relevant. 'While not becoming an arm of government', Florey had suggested the previous year, 'we can perform a useful national service by maintaining a close collaboration with those who control our destinies.'

Second, the Society would continue to take the lead on non-governmental aspects of international scientific relations. Trend's recognition of the legitimacy of this function was helpful, and the Foreign Office's recent creation of an Assistant Secretary post to specialise in scientific affairs was a good augury. So, too, was the appointment of Maurice Dean as Permanent Secretary of the new Department of Education and Science, where he quickly signalled his willingness to provide funding for the Society's international activities.

Third, and again helped by Trend, was expansion of the Society's role in selecting and funding a small number of research professorships and readerships.[95] This was to have major consequences for the Society's role in UK science in later years.

The fourth strand of the Society's identity, highlighted both at the Oxford Officers meeting and in Florey's 1964 Anniversary Address, was its wish to increase its role in respect of applied science, technology and industry. How to do this effectively was to be a long-running challenge for the Society.

The final strand was the Society's relation to other scientific societies. It channelled financial support from its Parliamentary Grant-in-Aid (PGA) to a number of societies for various purposes, and hoped to increase this work if the funding could be found. But this fizzled out in the mid 1980s.

[95] Letter from David Martin to Florey, 30 October 1964: HF/1/1/17/2/40; note of an additional Officers meeting on 13 November 1964: OM/83(64).

Money was, of course, key to delivering all these aspirations. By the end of 1964, the Society had market investments valued at about £3.2 M. This and related sources provided the Society with private annual income (excluding publication sales) of nearly £150,000. It also received £250,000 in various grants from the government, including PGA of £220,000. Its total disposable income of £400,000, while useful, was indeed a trivial fraction of the £52 M that ARC, MRC, NERC and SRC were able to spend in 1965–6, the first full year of the new Research Council system.[96] Moreover, though the Society had raised £416,000 towards the estimated £500,000 needed to move into its new premises at Carlton House Terrace, it could not easily launch a further fundraising initiative in support of the ambitions just outlined until the house move was paid for.[97] So fostering constructive relations with the new Department of Education and Science, which now had charge of the Science Budget, was a top priority, and the Society managed to secure increases in its PGA to £400,000 in 1965 and £560,000 in 1966.

At the end of 1965, Howard Florey handed to his successor, Patrick Blackett, a Society that he described as having greater confidence than before. It had a clearer sense of its identity and purpose in the rapidly developing world of science and science policy. How that identity played out in the following decades is the subject of the remainder of this book.

[96] Select Committee on Science and Technology, *HC 522: Research councils* (HMSO, 1971), 9. See also Philip Gummett and Roger Williams, 'Assessing the Council for Scientific Policy', *Nature*, 240 (8 December 1972), 329–32.

[97] The eventual cost was £850,000. John S. Rowlinson and Norman H. Robinson, *The record*, 14.

3   Supporting individual researchers

> In these days of rampant egalitarianism our concern for an elite in
> science may be regarded by some as outmoded. But it is not. In science
> the best is infinitely more important than the second best.[1]

Beyond allusions to the 'authority of experiments', the founding Charters
said little about exactly how the Royal Society of London for Improving
Natural Knowledge, as it was formally known, was to set about the task
implied in its name. The options were, anyway, always shaped by its
resources: extensive in terms of networks and access, much more limited
in terms of finance. The options were also shaped by the existence of other
players. In the twentieth century, and particularly in the second half of the
century, the Society's finances grew quite considerably – but not as much
as the number and scale of other organisations. In the increasingly competitive business of funding research – the most obvious route to improving natural knowledge – the Society could flourish only as a specialist
organisation providing niche services to elite segments of the market. Alex
Todd understood that very clearly. The post-Trend dispensation recognised that such a role fitted the Society's identity well.

One of the Society's most enduring characteristics is commitment to the
idea that progress in scientific research is most likely to be secured by giving
the most promising researchers freedom and scope to follow their own
instincts. Its approach to supporting research therefore majored on supporting the best individual researchers it could find, largely without regard
to their areas of research. In this it differed from Research Councils, which
were required to work in particular areas and had to some extent to
negotiate their disciplinary priorities with their political paymasters. The
Society's constitutional independence and its mix of public and private
funding allowed it to set its own course, and the fact that its public funding
amounted to less than 2 per cent of the Science Budget meant that it
attracted less political scrutiny than the much larger Research Councils.

[1] Alexander Todd, 'Address at the Anniversary Meeting, 30 November 1977', *Proceedings of
the Royal Society of London. Series B, Biological sciences*, 200 (1978), xiv.

With no institutes to maintain, it had flexibility to experiment as new needs and opportunities arose. Its relatively small scale meant also that the Society could give a degree of personal attention to those it supported, especially holders of its various research fellowships, and it could use its networks to help create opportunities for career development.

Research professorships

Money to pay the salaries and other costs of individual research scientists on a sustained basis started to become available to the Royal Society, from private sources, at the end of the nineteenth century. The Joule Memorial Fund, launched in 1890, was the first of a number of donations in this period to be assigned to the creation of research posts.[2] The initial pattern was to support studentships, and then more senior (and more expensive) research fellowships. The first research professorship was launched in 1922 following a legacy of £65,000 from Lucy Foulerton. She had already directed a legacy of £20,000 from John Foulerton to the Society for support of studentships, and took care to stipulate that both awards were explicitly open to male and female candidates equally.[3] The then President of the Society, Charles Sherrington, extolled this increase in the Society's capacity to support individual researchers, which he described as the Society's 'fundamental object'.

Looking enviously at the Rockefeller Foundation and the American culture of private philanthropy in support of scientific research, Sherrington speculated that the Foulerton bequests might herald 'the beginning of a trend towards wider public interest in and sympathy with research' in Britain. The following year, 1923, he was therefore particularly delighted to receive an unrestricted gift of £100,000 from the shipbuilder Alfred Yarrow. He claimed this as a vote of confidence in the Society and persuaded his colleagues to use it to create further research professorships.[4] In their first two decades, the Foulerton and Yarrow professorships supported E. H. Starling, A. V. Hill, Edgar Adrian, Alfred Fowler, G. I. Taylor and Owen Richardson – all destined to win the Copley Medal and/or the Nobel Prize.

[2] Henry Lyons, *The record of the Royal Society of London* (Royal Society, 1940), 125, 131–3.
[3] It was at this time that the Society took Counsel's Opinion on whether the Charter and Statutes allowed the election of women to the Fellowship, following an enquiry from the Women's Engineering Society. The Opinion was affirmative.
[4] Charles Sherrington, 'Address at the Anniversary Meeting, 30 November 1922', *Proceedings of the Royal Society of London. Series A, Mathematical and physical sciences*, 102 (1923), 373–88; Charles Sherrington, 'Address at the Anniversary Meeting, 30 November 1923', *Proceedings of the Royal Society of London. Series A, Mathematical and physical sciences*, 105 (1924), 1–16; CM 15 February 1923, minute 5.

Sherrington's hopes were fulfilled to a certain extent. By the outbreak of the Second World War the Society had at its disposal enough private funding to support twenty named research posts at various levels of seniority, and by 1960 this had reached about thirty, precise numbers fluctuating with the value of the associated investments. Total annual expenditure on these appointments in 1960 was £51,000. At that stage, the Parliamentary Grant-in-Aid to the Society was over £150,000, but none of it was for research appointments. Research appointments, destined to be the Society's dominant draw on public funds by the end of the twentieth century, were initiated and for a long time sustained by private funding.

This was a time of serious expansion in the university system, even before the 1963 Robbins report triggered a new wave of universities. Numbers of full-time university students in 'pure science' and 'technology' doubled to 27,000 between 1947 and 1952. By 1961 they had reached 46,000, with expectations of a further doubling by the late 1960s. The number of academic staff was also on the rise, though more slowly, from the pre-war figure of 4,000 (all disciplines) to nearly 9,000 in 1951 and 13,000 in 1961.

One of the unintended consequences of this growth was a squeeze on time for research. The University Grants Committee was seen as concerned primarily with the teaching function of universities, unable or unwilling to protect the research time of academic staff. It was, though, perturbed that the ratio of professors to other academic staff was falling rapidly, and feared that that this would have an inhibiting effect on the recruitment or retention of the best staff as opportunities for promotion appeared to diminish.[5]

The Royal Society was alert to the opportunity. When, at the end of 1958, the Prime Minister Harold Macmillan asked for ideas on how to strengthen the national research effort (see Chapter 2), Alex Todd highlighted the leadership that could be offered by holders of senior academic posts with minimal teaching duties, and called for an annual spend of £0.5 M to increase the availability of such posts. Such thinking was very much in the air as the Society was preparing its controversial 1960 report *The encouragement of scientific research in the UK*. Indeed, the proposal that it be enabled to build up its repertoire of such posts was the only part of that document that secured broad assent among scientific policy-makers. As Charles Sherrington had argued nearly forty years previously, research professorships constituted a niche that the Society was particularly suited

---

[5] By 1961, the ratio of professors to other grades had dropped from the pre-war level of 1:4 to 1:7.4. UGC, Cmnd 534, 84; UGC, *University development 1957–1962* (HMSO, Cmnd 2267, 1964), 35, 145, 154–5.

to fill, and they did not threaten the territory of the Research Councils or other bodies.

The challenge was to convince the Treasury. Hailsham thought that this lay within the art of the possible. But one can always find reasons for not doing something, especially if that something requires new money. In this case the wheels of obstruction started turning as soon as Todd's response was submitted to Harold Macmillan in January 1959. Hailsham's senior civil servant Frank Turnbull (previously at the Treasury) suggested privately to his opposite number at the Treasury, Bruce Fraser, that specific funding for research posts would be a 'glaring breach of university autonomy'. Fraser cautioned Norman Brook, the Cabinet Secretary, against having 'two classes of Dons, the less distinguished doing most of the teaching', adding darkly two months later: 'This is a matter entirely for the Universities and there is a good deal of difference of opinion about it.' In a note to the Prime Minister on 15 May 1959, the Chancellor of the Exchequer, Derick Heathcoat-Amory, argued that he had already provided additional funding for science, and rubbished Todd's proposal with the classic comment, 'This is an interesting idea and I have quite an open mind about it.' Prolonged debates about the optimum balance between teaching and research provided further occasion for inaction.[6]

Undeterred, and with Hailsham's support, Howard Florey pushed the case for public funding for research appointments once he assumed the Royal Society presidency on 30 November 1960. Both ACSP and UGC wanted such appointments to be handled by the Society rather than by DSIR: the Society had already established a track record with research appointments, and it could imbue them with prestige inaccessible to DSIR. For the UGC, 'Royal Society sponsorship would make the scheme more distinctive, less potentially embarrassing, and perhaps also more attractive.'[7] The Treasury on the other hand was worried about control, particularly given 'how much the Space Research Programme is controlled by the Royal Society and how little of it can be said to be under direct Government (departmental) control'. In Treasury terms, the growing support for a scheme of publicly funded research professorships run by the Royal Society was therefore seen as 'sinister'.[8]

Florey cut through this by nurturing his alliances, notably with the Whitehall veteran Keith Murray, Chairman of the UGC. By the end of

---

[6] All at TNA T 218/170. Internal Treasury memos make clear that the sense of Heathcoat-Amory's comment was derogatory. See also TNA T 218/666 and TNA PREM 11/2794.
[7] UGC meeting, 20 April 1961: TNA UGC 1/9.
[8] Treasury briefings for the 7 June 1961 meeting of ACSP: J.A. Annand, 5 and 6 June 1961: TNA T 218/520.

his first year as President, he felt secure that, on this issue at least, scientific leaders would present a united front towards any Treasury attempts to play one off against another. So he wrote formally to the Chancellor of the Exchequer to make the case for research professorships.[9] He emphasised the disproportionate importance of outstanding research leaders, and referred to the phenomenon of first-class individuals heading off to the United States because of poor promotion prospects and lack of decent equipment in the UK. He argued that the Society could 'perform a service to the country by lending its prestige to filling this gap in the provision of senior scientific posts', and that no other organisation could do it as well. The bid was for ten publicly funded 'Royal Society Research Professors' – both full salary costs and a significant allowance for research expenses. 'We regard this matter as urgent and presenting one of the most important contributions that HMG could make through the Royal Society to the advancement of British science.'

The immediate reaction in the Treasury was mixed. The Third Secretary Arnold France, generally well disposed towards the Society, thought 'we probably ought to do it', but noted that 'this will require careful handling'. Richard Griffiths, who reported to Arnold France and was freer with his opinions, characteristically discerned a conspiracy: 'This no doubt is another attempt by the scientists to remove themselves from competition, within the universities, from the claims of non-scientists.' But the Treasury's consultations evinced strong support from the UGC, and 'basic, though grudging' agreement from the Research Councils that the Society had made its case.[10] Hailsham, claiming that he had been championing research professorships for well over three years, expressed his support for the Royal Society proposal to Henry Brooke, Chief Secretary to the Treasury. Then in the middle of a spat with the Society over its report on the development of biological research (see Chapter 2), Hailsham added that a positive response would help him to mend fences.[11] The Treasury, too, needed to make a mollifying gesture since it was in the process of imposing major cuts on the UGC's budget bid for the next quinquennium: anything that could offset impressions of

[9] 'The need for research professors': CM 30 November 1961, appendix A; also TNA 224/2970. The letter benefited considerably from Keith Murray's input: see exchanges between Florey and Murray, September–November 1961. TNA UGC 7/167.

[10] Memos by France and Griffiths, December 1961, and J.A. Annand to Richard Griffiths, 20 March 1962. MRC argued that its research units (of which it then had 80) were functionally equivalent to research professorships, so that there was no gap needing to be filled in its own area of research. J.D. Whittaker to E.M. Church, 15 February 1962. TNA T 224/2970.

[11] Hailsham to Brooke, 21 March 1962: TNA T 224/2970. See also follow-up letter from Hailsham to Heathcoat-Amory, 24 March 1962: TNA T 218/170.

meanness or philistinism would be welcome. So, five months after Florey had put in his bid, Henry Brooke accepted it in principle.[12]

It then became a matter of agreeing terms – the size and cost of the scheme. The upshot of numerous exchanges with David Martin was an offer from the Treasury for five fully funded research professorships, including generous allowances for research assistants and an equipment grant of £2,500 per post. The Society accepted the offer as a step in the right direction.[13]

This established a useful precedent for the Society. It was quick to build on it. Even before the first of the new appointments had been made, the Society's written submission to the Trend Committee in early 1963 highlighted its wish to develop such activity. In a report that otherwise left meagre pickings for the Society, Trend endorsed an expansion of the research professorships scheme. Within a month after publication of the Trend report, Florey wrote to the Treasury to ask for three more publicly funded research professorships. With continued support from Hailsham, these were duly approved in February 1964 on terms similar to the first five.

By the mid 1960s, research posts were a central element of the Royal Society's strategy for improving natural knowledge, and were recognised as such when the Officers reviewed their post-Trend options in October 1964. The Society had at its disposal

- eight publicly funded research professorships for 'really outstanding research ability' (two further professorships followed in 1967, fulfilling the original aspiration for ten such posts);
- three privately funded research professorships (a fourth was added in 1973);
- four privately funded senior research fellowships ('Category B') for 'proved merit in research for persons of senior status';
- seven privately funded research fellowships ('Category C') for 'proved merit in research for younger persons'; and
- nine privately funded junior research fellowships ('Category D') for 'young persons who show exceptional promise'.

All these awards were made to individuals, not to the institutions in which they held their awards. That gave post holders considerable clout in negotiations with their institutions, since they were free to move with their funding to another institution. Moreover, the Society expected that they would devote themselves more or less full time to research, and their institutions could not instruct them to take on other duties. With a

[12] Brooke to Florey, 1 May 1962: TNA T 224/2970.
[13] OM/57(62), C/109(62), OM 21 June 1962 and CM 12 July 1962, minute 21 plus appendices. The scheme was announced in Parliament and attracted significant press coverage: Hansard, 28 June 1962, 661, col 163.

significant contribution towards research costs included in the package, they had real freedom to pursue their research ideas. Research professors could also tackle relatively long-term projects, since their five-year appointments were renewable, subject to satisfactory performance, till retirement. All the Society required in return was an annual report. Such conditions epitomised the Society's view on how to foster real progress in science.

The first intake of ten publicly funded Royal Society Research Professors, appointed between 1963 and 1968, covered fields from applied mathematics (James Lighthill) and information-processing systems (Hugh Longuet-Higgins) to marine biology (Eric Denton) and animal behaviour (Robert Hinde).[14] Three of the ten appointees returned to the UK from the United States to take up their posts, and one returned from Israel. Their age at appointment ranged from thirty-eight to fifty (average forty-one). Three were not FRS at the time of appointment. The aim was to support those who had already demonstrated their prowess as researchers and who would be able to exploit the opportunities that the post provided, some by continuing and some by making a radical change in their career paths.

At first, research professorships were not publicly advertised. Instead, Fellows were invited to suggest possible candidates and Council then selected the strongest. So Robert Hinde, for example, was astonished to receive, right out of the blue, a letter offering him a professorship: 'It was just like a jewel being thrown into my lap.' The post gave him scope to do whatever research he wanted and to spend most or all of his time on it. The prestige attached to the post made it easier to raise additional funds from Research Councils and other bodies. The bureaucracy associated with the post was minimal, and the small scale of the scheme meant that it could be administered with a personal touch that was much appreciated – due largely to Ursula Maunsell, a member of Royal Society staff who managed research appointments for thirty years.[15]

The CSP was a strong supporter of the research professorship scheme, and its chairman, Fred Dainton, saw it as something that only the Society, with its independence and broad disciplinary coverage, could run.[16] By 1976 the number of publicly funded research professorships in the Society's gift had risen to fourteen. However, attention was then beginning to move to options at more junior levels, and plans for a further increase to sixteen were dropped. Even at the professorship level, the search was for

[14] OM/88(68).
[15] Interview with Robert Hinde. For similar comments, see interviews with Michael Atiyah, Bob May and Martin Rees. The practice of publicly inviting applications for research professorships did not start until 1982.
[16] OM/62(68); OM/52(70).

'young, really able candidates' who were not already professors: references to candidates being of FRS standing were deleted from the eligibility criteria. The next two research professors to be appointed, Richard Gardner (developmental biology) and Alan Fersht (protein engineering), were both ten years younger than the long-term average of forty-four years old for these positions, and neither was yet FRS.[17]

From then on, research professorships were maintained at an approximately constant level. In the early 1990s the terms of tenure were gradually reduced to promote greater turnover.[18] The scheme as a whole was reviewed in 1994. The original problem of relative lack of openings at professorial level was recognised to be no longer an issue. Instead, the problem to which the scheme was now a response was the increasing pressures of academic life, which were impeding the research activities of senior staff. Research professorships were therefore valued as enabling those 'exceptionally talented at research to be free from the restrictions of administration and teaching', and the Society continued to accord the scheme high priority. But not the highest. The 1994 review recognised that professorships had been supplanted by research fellowships for younger scientists as the Society's flagship research appointments scheme.[19] A major review of overall Royal Society strategy in 2011–2, however, concluded that research professorships were still a niche area where the Society could have real impact.

Early-career fellowships

It was in February 1971 that the far-sighted Shirley Williams, previously a DES minister in the Wilson Government, warned: 'For the scientists the party is over.'[20] What was over, she argued, was the heady experience of sustained escalation in research budgets, government optimism that this would somehow translate into economic growth, and general public support for science. What lay round the corner was budgetary constraint, the controversial Rothschild Report on how to focus scientific research on public benefit, and a public wish to bring science and technology under greater control. The Science Budget had enjoyed an annual growth rate of more than 10 per cent in real terms

---

[17] OM 4 November 1976, minute 6; OM 3 March 1977, minute 2(i); CM 31 March 1977, minute 5; C/66(94).

[18] See interviews with Peter Lachmann and with Martin Rees, a future President, who was one of the first to be affected by the reduced tenure.

[19] C/27(94) and CM 21 April 1994, minute 28.

[20] Shirley Williams, 'The responsibility of science', *Times Saturday Review*, 27 February 1971, 15.

through the 1960s; Williams told the scientists that during the 1970s, by contrast, the Science Budget was going to grow more slowly than the economy as a whole and that they were going to have to adapt to a markedly harsher environment.

Less than eight years earlier, the Robbins report had heralded a further expansion of the university system beyond what had already been achieved since the War. Total student numbers doubled between 1961–2 and 1971–2 and were expected to increase by a further 30 per cent in the following five years, with postgraduates growing significantly faster than undergraduates. Academic posts this time grew almost in proportion.[21] When the brakes came on, however, the newly created posts had mostly been filled and the newly graduating PhDs faced increasingly severe competition in establishing themselves in academic careers. Openings were becoming fewer while candidates to fill them were still rapidly increasing. Coupled with the sharp slowdown in the Science Budget predicted by Shirley Williams, this engendered a sense of unwonted pressure among academic scientists. The issue became increasingly prominent as the 1970s wore on.

Against that background, there was a switch in the Royal Society's approach. In November 1976 it decided to prioritise the more junior Category C and D fellowships. That was where the greatest need lay. The initial idea was to use public funds 'to stimulate the formation of new research groups in potentially interesting new fields around able young scientists'.[22] A group might comprise one 'relatively young' researcher (i.e. thirty to thirty-five years old) and two assistants. This idea formed the centrepiece of the Society's budget bid for 1978–9. The bid acknowledged the political realities spelt out earlier by Shirley Williams, but argued that 'the economic recovery of the country is bound to depend heavily on scientific and technical advance backed by adequate investment'. Mrs Williams, by then back in power as Secretary of State at the DES, was sympathetic, as was the Advisory Board for the Research Councils (ABRC).[23] As a starter, £108,000 was added to the Society's £2.3 million PGA for 1978–9, enough to support two small research groups.[24]

---

[21] UGC, *University development 1962–1967* (HMSO, Cmnd 3820, 1968); *University development 1967–1972* (HMSO, Cmnd 5728, 1974).

[22] CM 31 March 1977, minutes 5 and 20. Fellows in the medical sciences, however, opposed both the idea of funding research groups and the focus on young scientists: CM 21 July 1977, minute 41 (v).

[23] CM 16 December 1976, minute 20; OM 3 March 1977, minute 2(b). The ABRC replaced the CSP – in October 1972. It lasted until March 1994 and the implementation of the *Realising our potential* White Paper.

[24] OM/64(77), OM/78(77). OM 30 November 1977, minute 2(e).

Alex Todd rehearsed the argument in his 1977 Anniversary Address. He was worried that, with stagnation in the academic sector, thwarted talent would seek its livelihood overseas and British industry would have to import its know-how rather than rely on home-grown expertise. He wanted to use the Society's research posts both to support outstanding individuals and to help them build up research teams 'on whose work our technological future will ultimately depend'. He was undeterred by the thought that the Society's resources were small in comparison with the overall scale of the challenge. Unfashionable though it might be 'in these days of rampant egalitarianism', said Todd, the Society's 'concern for an elite in science' was wholly justified: 'the best is infinitely more important than the second best ... and a country which ignores or forgets it does so at its peril.'[25]

By June 1978 doubts about the research groups scheme had set in, reinforced by evidence of the complexity of the factors affecting academic careers. However, the fact that five conventional Category C research fellowships attracted 300 strong applications demonstrated that there was an unmet need for career openings. In the end, just one 'research group' award was made, to the physicist R.F. Willis working at the Cavendish Laboratory.[26] The new PGA money was switched – ironically – into creating additional senior research fellowships ('Category B'), around which research groups could be developed. And alongside bidding for PGA for research groups, the Society went out vigorously for private funding: between 1976 and 1981 it increased its repertoire of privately funded research appointments below professorial level from thirty-one to forty-seven, supported by donations from a variety of major industrial companies and some charitable foundations and individuals.[27] Such independent means helped the Society to maintain its niche as a provider of research appointments.

The cramped academic job market, the skewed age distribution of staff (a result mostly of the rapid post-Robbins expansion) and limited options for developing research careers continued to provoke general concern. The European Science Foundation concluded that, across the university sector in Western Europe as a whole over the following fifteen years, job openings would be 'exceedingly low'. A survey of job

[25] Alexander Todd, 'Anniversary Address 1977', xiv.
[26] OM 10 November 1977, minute 3(a); OM 30 November 1977, minute 2(e); OM 12 January 1978, minute 2(c); OM 9 February 1978, minute 2(a); CM 15 June 1978, minute 12; *Report of Council for the year ended 31 August 1978* [1979 *Yearbook*, p. 291]; OM/143(77), OM/145(77).
[27] OM/105(76); OM/152(76). See also John S. Rowlinson and Norman H. Robinson, *The record*, 59–61, 193–208. The proportion of the Society's staff effort going on research appointments rose from 12 per cent in 1974–5 to 22 per cent in 1981–2: OM/34(81).

advertisements in *Nature* found 'no end in sight to jobs gloom'.[28] Alex Todd told the Royal Society that all this was the fault of the Robbins report and the 'misguided euphoria' of the 1960s.[29] He particularly blamed the notion that the post-Robbins expansion should be modelled on traditional institutions and traditional students with traditional expectations. Tenure of employment now exacerbated the lack of openings for aspiring academics. Todd called for greater selectivity in allocation of research funding, greater concentration of facilities, and greater institutional diversity.

Up to Margaret Thatcher's success in the May 1979 General Election, it was possible to maintain an expectation that the university system would continue to expand, at least modestly. Most universities were therefore ill prepared for what followed the election.[30] In 1979–80, the UGC still accounted for 74 per cent of the total government spend on research carried out in universities. Indeed, it nominally spent more on research than all Research Councils combined.[31] University research was thus very vulnerable to any changes in UGC policy or budget, the more so as inflation was then at 11 per cent and climbing. The immediate post-election budget, in June 1979, highlighted this vulnerability, introducing cuts in public expenditure and signalling that wider-ranging changes were in the offing.

Conveniently but illogically, the element of UGC funding intended to support academic research was distributed according to student numbers, which were made to serve as a non-selective proxy for research activity. With the Royal Society beginning to push the case for core research funds to be distributed selectively to the best researchers, and with growing concern about the continued feasibility of the 'well-found laboratory' that was supposed to be provided through the UGC block grant, the ABRC and UGC launched a joint study under Alec Merrison on support of university scientific research in February 1980. Before Merrison could report, however, the strain on university funding intensified considerably. The March 1981 Budget announced game-changing cuts in university funding, and

[28] *Nature*, 276 (21/28 December 1978), 743; C/23(79).

[29] Alexander Todd, 'Address at the Anniversary Meeting, 30 November 1979', *Proceedings of the Royal Society of London. Series B, Biological sciences*, 206 (1980), 369–80.

[30] John Sizer, 'A critical examination of the events leading up to the UGC's grant letters dated 1 July 1981', *Higher education*, 18 (1989), 639–79.

[31] The DES spend on research in 1979–80 was £279 M through the Science Budget (i.e. the Research Councils, Royal Society and one or two other bodies) and an estimated £359 M through the UGC via the block grant to universities. Of the £279 M, £102 M was allocated by the Research Councils to universities in the form of grants, studentships, etc. Research spend by other government departments in universities totalled £24 M. Cabinet Office, *Annual review of government funded R&D 1983* (HMSO, 1984).

letters from the UGC on 1 July 1981 spelt out the implications for individual institutions. The cuts in the UGC block grants to universities were eventually estimated at 17 per cent between 1980–1 and 1983–4, coupled with a cut in student numbers of only 5 per cent and thus even greater pressure on research time.[32] The UGC Chairman, Ted Parkes, told the Public Accounts Committee in March 1981 that the cuts would mean compulsory redundancies among academic staff, exacerbating the difficulties faced by those trying to start academic careers.

The Royal Society leadership discussed these developments closely with the ABRC and UGC, and with the Secretary of State at the DES, Mark Carlisle. Its line was: to accept the fact of the cuts, at least in public; to press for them to be implemented selectively so as to protect the best research; and to advocate maintenance of the dual-support system. That was also the line the UGC took, for example in its letters of 1 July 1981 telling each institution what its block grant would be. These letters referenced advice from the Royal Society among others, and the President Andrew Huxley wrote to *The Times* to explain that the Society's advice had been in favour of selectivity and to applaud the UGC's commitment to that approach.[33]

Huxley's letter also commented on the 'disastrous' prospect that recruitment of bright young researchers might dry up completely. From the Royal Society perspective, that was the key issue. A group appointed by the Society to monitor the impact of the cuts on scientific research considered the possibility of seeking further funds from industry to create new research fellowships as part of a 'rescue operation' for outstanding researchers.[34] The largest Research Council and the one channelling most of its resource direct to universities, the SERC,[35] was also concerned about the problem of creating job openings, and reviewed possible initiatives. A past Chairman, Sam Edwards, was invited to a meeting of the Royal Society's monitoring group on 1 October 1981. He argued for a scheme to appoint twenty new researchers at lecturer level for each of the next twenty years; the scheme would be run by the Royal Society or a similar body with funding transferred from the UGC.

The problem of 'new blood' was very much in the air at that time. In March 1982 Keith Joseph, stimulated by a day's visit to the Rutherford Appleton Laboratory focused not on politics but on the latest science,

[32] John Sizer, 'A critical examination', 671. Also interview with John Ashworth.
[33] Letter published 23 July 1981.
[34] PUF/1(81) [the Society's Committee on University Funding]; CM 16 July 1981, minute 20.
[35] The Science and Engineering Research Council changed its name from Science Research Council on 30 April 1981.

startled the SERC Chairman, John Kingman, by asking him privately how he would spend some new money. Kingman's instinctive response was to highlight the plight of the coming 'lost generation of young scientists' thwarted in their aspirations to follow careers in academic research. It was not the first time Joseph had heard such a comment. He proved sympathetic, inviting Kingman to work up a proposal with Ted Parkes at UGC.[36] He had been equally encouraging of the Royal Society's aspirations in this area at a lunch with the Officers the previous month.

ABRC, UGC, SERC and the Royal Society worked closely together[37] to make the best of a Secretary of State willing to listen to reasoned and evidenced argument. Ted Parkes kept up the pressure in public, telling the Parliamentary and Scientific Committee on 20 July 1982: 'The problem of the recruitment of young blood is the single greatest difficulty facing universities during the rest of the 1980s.' A survey of Fellows of the Royal Society strongly endorsed this assessment.[38] There were in fact two, overlapping, drivers: one arising from the March 1981 cuts, and the other arising from the post-Robbins recruitment bulge and subsequent distortion of the age profile of the academic staff. The concerns articulated by the Merrison report (published in June 1982) about the health of the dual-support system were a further consideration. A mix of potential solutions emerged, each playing to the strengths of its respective proponent. They were all essentially palliative – short-term measures to keep bright young talent in play until normal service was resumed and mainstream tenured posts became available – and all depended on new money.

Initially, ABRC's preferred approach was to work through the Research Councils. It proposed to allocate earmarked funding for new posts to each Council[39] in proportion to the perceived need in each subject. The Councils would then solicit applications from university departments, and the departments would fill their allotted posts through normal recruitment processes, though possibly with a Research Council nominee on the appointing panel. Appointees would spend much of their time on research, but also take on the normal breadth of other academic activities.[40]

---

[36] Interview with John Kingman.

[37] Peter Swinnerton-Dyer was one of the key figures in these interactions: the Society's assessor on ABRC and very active in both bodies, and shortly to be appointed Chairman of the UGC in succession to Ted Parkes. The Society's Biological Secretary, David Phillips, succeeded Alec Merrison as Chairman of ABRC in January 1983. There were numerous other overlaps.

[38] PUF/42(82). See also Paul Flather, 'The missing generation', *Times higher education supplement* (24 September 1982), 8–9; Jon Turney, 'Science suffering says FRS survey', *Times higher education supplement* (3 December 1982), 1.

[39] Except the Social Science Research Council, then under attack from Keith Joseph.

[40] ABRC, *The Science Budget: a forward look 1982* (October 1982), especially annex 5.

However, ABRC then agreed to fold its proposals in with those being developed by UGC. This had the advantage of covering the full range of disciplines. It would also make it easier to manage the transition between a new-blood post and a permanent tenured lectureship, since both would be funded via the UGC block grant. In the joint scheme, the UGC would handle the funding and invite universities to submit bids for new posts. ABRC and the Research Councils would advise UGC on the disciplines and departments most in need, and the Research Councils would help UGC assess bids. Any additional money for a new-blood scheme had to be used selectively. That implied a degree of dirigisme, and thus a significant cultural shift for the UGC from its long-standing policy of not earmarking funds within the block grant that it gave to each university.[41]

SERC was thus involved in delivering as well as devising the UGC new-blood scheme. In February 1981 it had contemplated an extension of its own Senior Fellowship scheme, under which it had made twenty-six awards during 1975–9. These were intended for outstanding scientists and engineers, without age limit, to devote themselves full time to research. However, it judged that its appointees were generally not quite of the calibre of holders of Royal Society research appointments, and decided to scale back. Two months later, after the major Budget cuts, SERC also decided to reduce its Advanced Fellowship scheme, which was then annually appointing twenty-six researchers aged between twenty-eight and thirty-five to research posts for up to five years.[42] It reasoned that there would be fewer mainstream posts to which Advanced Fellows could expect to be appointed on completion of their terms.

In contrast, the Royal Society found in the cuts an opportunity to seek a greater role in research appointments. Indeed, it attached such importance to this that it sacrificed the 130-year-old Scientific Investigations Grant (£0.8 M in 1982–3) in order to help pay for it. The ABRC – notably John Kingman – was strongly supportive. The Society had in mind a new scheme, with a total of 100 five-year appointments. The post holders in this scheme would have greater teaching responsibilities than usual for Royal Society research appointees, the selection criteria (contrary to normal Royal Society practice) would include the specific needs and merits of the departments in which the appointments would be held, and there would be an explicit assumption against concentrating

[41] UGC papers 11/SW/82, 25 September 1982, and 234/82, 28 October 1982: TNA UGC 1/164 and 1/166. UGC paper 266/82, 2 December 1982, and UGC 2 December 1982, minute 4: TNA UGC 1/167 and 1/168. ABRC 1 December 1982, minutes 12–14: copy at D.C. Phillips papers, MS Eng.c.5526, O.220.

[42] SRC 12–81, 18 February 1981; SRC 39–81, 15 April 1981.

appointments in golden triangle institutions.[43] Like UGC, this implied a culture shift for the Society, diluting its stringent focus on individual excellence by factoring in other criteria as well. The bid was included in ABRC's budget advice to Keith Joseph, submitted in August 1982. But as the much larger though otherwise similar proposal from ABRC/UGC then gathered momentum, the Society adjusted its aim and reverted to established custom. Its scheme of 'University Research Fellowships' (URFs)[44] as finally made public on 17 December 1982 had a clearer emphasis on the research excellence of the individual but retained reference to the age profile of the department and its capacity to provide a strong research environment. Disciplinary priorities were no longer a consideration. Applicants would be the individuals themselves, not their host departments, and would be assessed and selected directly by the Society. They had to be less than thirty-three years old[45] (compared with thirty-five for the UGC new-blood posts), would be appointed for up to ten (5 + 3 + 2) years, and would spend the bulk of their time on research. The golden triangle inhibitions were dropped.[46]

So two schemes emerged in parallel. From the point of view of the post holder, the UGC new-blood lectureships offered a more secure route to tenure while the Royal Society University Research Fellowships offered greater freedom and prestige, more research time and the prospect of personal mentoring from senior Fellows. In the first year, the URF scheme attracted nearly three times as many applicants per post as the new-blood scheme.[47]

The UGC new-blood scheme ran for three years, with 242 posts allocated in 1983–4, 350 posts in 1984–5 and 200 posts in 1985–6, 82 per cent of the 792 posts being for natural science broadly interpreted. In the wake of the 1982 Alvey report *A programme for advanced information technology*, the UGC added a parallel new-blood scheme with an

---

[43] See the Society's Forward Look interview with ABRC, 18 May 1982: OM 52(82), OM/59(82) and OM/66(82). John Mason to Alec Merrison, 25 May 1982: OM/64(82) (copy at TNA UGC 1/162). Further details in OM/78(82), dated 6 July 1982. CM 15 July 1982, minute 21. 'Golden triangle' refers to the academic institutions of Oxford, Cambridge and London.

[44] The full name was 'Royal Society 1983 University Research Fellowships', in recognition of its original context. The '1983' was dropped in 1991.

[45] CM 30 November 1982, minute 15. The scheme gradually became more flexible, as did the UGC new-blood scheme, as it was accepted that such a low age limit discriminated against those taking time out for family responsibilities or similar reasons. Career stage replaced biological age as the criterion.

[46] OM/114(82), C/209a(82).

[47] Five hundred and thirty-eight applicants for 30 places (CM 21 April 1983, minute 12) as against 1,400 applicants for 210 natural science places (UGC paper 66/83, 24 March 1983: TNA UGC 1/170).

additional 147 posts specifically for IT over the three years.[48] Having had to find large sums for restructuring and redundancy costs arising from the 1981 cuts, the government then stuck with its resolve to terminate both UGC new-blood schemes after 1985–6.

The Royal Society URF scheme, similarly, was planned as a limited initiative, with appointments being made over a five-year period, on a decreasing scale as the fifth year (1987–8) approached and with a view to having a total of 100 URFs in post by the end of that time. However, there was soon a rethink. David Phillips at ABRC took every opportunity to impress on Keith Joseph that 'the rarest resources in scientific research and those of the greatest importance are scientists of genuine originality and the novel ideas they generate'. Joseph made sure that the message reached the Prime Minister. By the end of 1983, Peter Swinnerton-Dyer, then running the UGC, was suggesting to the Society that its URF scheme might be extended indefinitely: it was big enough to make a real contribution to the needs of outstanding scientists, but not so big as to distort normal career paths.[49] Here was a challenge to the Society: just how big a role did it want to have in research appointments at national level?

There was at first a degree of ambivalence in the Society's attitude to URFs. The scheme was the sort of activity that it could undertake really well, there was a demonstrable need for it, and prominent association with youth was good for public relations. But there were limits to how far the scheme could grow without losing the elite character that was the Society's hallmark. There were anxieties, too, about a single substantial scheme distorting the overall balance of Royal Society activities.[50] The biggest concern, though, was that it left the Society financially exposed. Almost before the first appointments were made, the Society was worrying about the long-term commitments it was taking on with the URFs and the extent to which the scheme was underfunded during a period of high inflation. Continuation, let alone expansion, depended on costs being fully met by the government. In February 1984, Officers agreed to prioritise URFs over other research appointment schemes in their budget discussions with the ABRC. By late 1985 they accepted that there was no chance of the government extending the UGC new-blood schemes beyond 1985–6.[51] Buoyed by the growing reputation of the URF scheme,

[48] See UGC *Annual Surveys* for 1981–2 to 1984–5.
[49] David Phillips to Robin Nicholson, enclosing a copy of his memo to Keith Joseph, 29 July 1983; Keith Joseph memo to the Prime Minister, 15 September 1983: D.C. Phillips papers, MS Eng.c.5517, O.177 and O.178. RS Committee on University Funding, 10 November 1983, minute 2 [PUF/21(83) and C/231(83)].
[50] OM 13–14 July 1985, minute 14.
[51] Andrew Huxley to Keith Joseph, 28 October 1985, and Keith Joseph's reply, 4 December 1985: OM/150(85).

the Society took the bold step at the beginning of 1986 of repositioning it from a limited response to the post-1981 'emergency situation of extreme scarcity of new appointments' to a more open-ended programme focused on developing the careers of the most talented researchers and, incidentally, partially mitigating the problems of the surge of retirements of academics appointed during the post-Robbins expansion.[52]

With the consistent support of both David Phillips and Keith Joseph in subsequent budget rounds,[53] and with periodic bouts of brinkmanship over costs, the Society started building up its URF scheme beyond initial projections. Growing concern about the brain drain reinforced the momentum behind the scheme. By the beginning of 1987, the Society was committed to making new appointments to the scheme beyond 1987–8 in order to maintain 100 URFs in post for the long term. Prompted privately by David Phillips, the Society then raised its sights further, to a target of 200 URFs in post by 1992,[54] and Brian Follett made it a key aim of his 1987–93 term as Biological Secretary to deliver this growth. Phillips publicly stoked the case for URFs in an ABRC policy document published in May 1987 that, in the context of promoting creativity, extolled URFs as a way of enabling talented young academics to concentrate on research and receive long-term support. The Society, in its public response to that document, stressed the need to think of URFs as a long-term measure rather than a temporary expedient.[55]

This view took some defending. The Society was clear that the prime selection criteria were the quality of the applicant and the value of giving bright young scientists the opportunity to establish themselves in research careers. This played well with a government determined to increase selectivity in research funding. But in 1988, for example, the Society had to fend off pressure from the government's Chief Scientific Adviser, John Fairclough, to focus URFs on Interdisciplinary Research Centres (IRCs), the latest fashion in science policy. It did so robustly: 'It would be

[52] OM 11 December 1985, minute 3(i); CM 16 January 1986, minute 15. Royal Society, *Corporate plan: a strategy for the Royal Society 1986-1996* (February 1986). Forward look interview with ABRC, 14 March 1986: ABRC(FL)(86)28. ABRC meeting, 28 October 1987, minute 3 and annex 3.

[53] In budget allocations and in debates with the Prime Minister, Keith Joseph supported the Society's case for expansion and noted the view of the Chief Scientific Adviser, Robin Nicholson, in 1985 that the new-blood money 'was excellently handled by the Royal Society; less well by the Research Councils'. Note for the record, 8 October 1985: D.C. Phillips papers, MS Eng.c.5508, O.113.

[54] OM/29(87); OM 5 March 1987, minute 3(c). Robert Honeycombe to David Phillips, 15 February 1988: D.C. Phillips papers, MS Eng.c.5501, O.55. The target of 200 URFs was met in 1993.

[55] ABRC, *A strategy for the Science Base* (HMSO, May 1987). The Society's response was published in October 1987.

completely inappropriate, and a misunderstanding of our philosophy, to try to direct these people towards IRCs, which you seem to favour.' In 1990, the Secretary of State, John MacGregor, pushed for URFs to be focused on priority disciplines. The SERC Chairman, Bill Mitchell, wanted them focused on fields with recruitment gaps. The ESRC Chairman, Howard Newby, told ABRC that the scheme could be faded out as academic recruitment patterns returned to normal.[56] During a difficult public expenditure survey in 1991, David Phillips had to raid ABRC's flexibility margin to stop the URF scheme dropping from 185 to 135 posts over the following two years. He appreciated the URF scheme as 'well managed in a personal, non-bureaucratic way' and as benefiting from the Royal Society cachet, and was keen to see it prosper.[57] Numerous URFs testified to the benefits they derived from the scheme's freedoms and flexibilities and from the boost to their self-confidence that came from the Society's backing.

By the end of the 1980s, the URF scheme was the biggest career-oriented research fellowship scheme on offer in the UK. In 1990 the Society tried to make it even bigger, bidding to have 300 URFs in post by 1994. That proved to be too much, too soon, and was turned down by ABRC.[58] Undaunted, the Society's new President, Michael Atiyah, launched a major inquiry, *The future of the Science Base*, in 1991, which concluded, after much consultation and statistical analysis, that there was a case for creating 500–750 fellowships on URF lines, with the awards still being made to individuals rather than departments and primarily on the basis of individual talent. In the most important government statement on science policy since 1971, the May 1993 White Paper *Realising our potential* gave a strong though unquantified endorsement to the Society's vision. After some vacillation over whether it should run the whole of such a scheme by itself (and risk greater entanglement with the processes of government accountability) or share it with another organisation (and risk eventually losing the whole scheme to that organisation), the Officers signalled in an equally unquantified manner their 'wish to run

---

[56] Roger Elliott to John Fairclough, 15 March 1988: David Phillips papers, Bodleian, MS Eng. c.5519, O.193. Correspondence between Brian Follett and John MacGregor, December 1989 – March 1990; OM 11 January 1990, minute 3(c). Also Brian Follett to David Phillips, 16 October 1990 and 7 June 1991: D.C. Phillips papers, MS Eng. c.5501, O.56.

[57] Note on a meeting with Michael Atiyah and Brian Follett, 10 June 1992: David Phillips papers, MS Eng. c.5501, O.56. On the value of URFs, see interviews with Patrick Bateson, Sam Edwards and Brian Follett.

[58] Royal Society, *The Royal Society – the next 10 years* (March 1990); ABRC meeting, 20–22 April 1990, annexes H and K; ABRC(FL)(90)8.

a greatly enlarged URF scheme'.[59] In the event, the scheme reached the 300 mark by the turn of the century and remained at about that level for the following decade. The Chairman of the House of Commons S&T Select Committee, Ian Gibson, suggested in 2002 that the URF scheme could be handed over to the Research Councils, but found no support.[60] By 2012, when the scheme had been running thirty years, a total of 1,066 scientists – over 20 per cent of them female – had held or were still holding URF posts, and twenty-one of the earlier appointees had so far subsequently been elected to Fellowship of the Society.

The URF scheme was financially much the largest of the Society's research fellowship schemes, but it was by no means the only one. By 2010, the Society was running nine distinct schemes with a total of 700 fellows in post and a budget of over £43 M from a combination of public and private sources. The second-largest fellowship scheme, the Wolfson Research Merit Awards, was an interesting example of a mixed economy. The Wolfson Foundation approached the Society in 2000 with a proposal to top up the salaries of researchers of 'outstanding achievement and potential' in order to attract them to, or retain them in, the UK. Recipients had to have their basic salaries already covered by their employing universities.[61] The Foundation offered to fund the scheme at £2 M p.a., provided the government matched that sum – which it duly did. In 2010, the scheme was supporting 170 awardees with sums of £10 K to £30 K p.a. for periods of up to five years.

Research professorships were open equally to men and women, though it was nearly forty years (1960) before the first female research professor (Dorothy Hodgkin) was appointed, and another twenty-three years before the second (Anne Warner) followed. The URF scheme, targeted at an earlier stage of research careers where a higher proportion of the pool of potential candidates was female, did rather better: 28 per cent of fellows appointed in the first three years of the scheme (21 out of 76) were female, considerably higher than the 12 per cent of established

---

[59] OM/30(93); OM 21 April 1993, minute 3(e). An explicit target of 400 was current for a while but then quietly dropped.

[60] House of Commons Select Committee on Science and Technology, *Government funding of the learned scientific societies* (fifth report of session 2001–2, HC 774-II), Ev 7; Editorial, *Research fortnight*, 15 May 2002, 3. At their own initiative, some 370 past and present research fellows published a letter in support of the Society continuing to run the scheme: *Daily Telegraph*, 31 July 2002, 19.

[61] Such a scheme had been proposed by George Porter in 1986, and discussed with the Prime Minister, as a response to the brain drain. The ABRC considered it in detail in October 1987, but it was not taken forward at the time. OM 12 June 1986, minute 3(p); OM 18 December 1986, minute 3(c).

lecturers at that time who were female.[62] An element of flexibility with the upper age limit of 33 was introduced to encourage applicants who had taken time out from research.

The disproportionate difficulties that women faced in developing research careers gained prominence during the 1990s. The Cabinet minister responsible for science, William Waldegrave, appointed a Committee on Women in Science, Engineering and Technology in March 1993, its members including the Society's Foreign Secretary Anne McLaren (Figure 3.1) and Patricia Clarke (who had earlier alerted the Society's Council to the practicalities of making the URF scheme as accessible as possible to women). The Committee's report, *The rising tide*, was published in 1994 and pushed the issue significantly up the policy agenda. The Society's input to that report explicitly rejected the use of quotas and positive discrimination to redress the balance, and sought rather to ameliorate some of the circumstances that hindered female scientists progressing in research careers. It therefore stressed the need for greater flexibility in terms of part-time appointments, home working, childcare arrangements, mentoring, and enabling scientists on a career break to keep in touch with developments. The Society then started to build these desiderata into the running of its existing fellowship schemes.

In 1995, the Society also trialled a new research fellowship scheme targeted specifically at scientists in the first few years after their PhD who for family or similar reasons had a special need for flexible arrangements. The scheme was open to men and women alike, but was named after Dorothy Hodgkin to emphasise the expectation that most applicants would be female. To get it off the ground, the Society used its private funds to support the first four appointees, and then secured public funding for a further twelve. This proved successful, and the scheme was expanded from 1997, still with mixed private/public funding, including some funding specially raised from industrial sources.[63] By 2010, some 200 researchers had been appointed to Dorothy Hodgkin fellowships, of whom about 93 per cent were female, and the scheme was well regarded. But the special flexibilities of the scheme had by then become relatively commonplace, and there was no real case for maintaining it as a distinct element of the Society's repertoire. The scheme was substantially curtailed, but the evocative name was retained.

---

[62] The 12 per cent figure is for lecturers in the category 'Biological, mathematical and physical sciences' for 1985–6, from table 26 of *University statistics 1985/86*, vol I. This is slightly more senior (and therefore less female) than the pool from which URFs were typically drawn, but gives the general idea.

[63] OM/61(93), OM/11(94), C/25(94); SPB 13 July 1995, minute 8; Royal Society, *Corporate update and PES submission* (June 1995), Royal Society, *Into the new millennium: a corporate plan for 1997–2002* (May 1997).

Figure 3.1 Anne McLaren. © The Royal Society

The Royal Society's research fellowship schemes were never just about research: they recognised the breadth of skill and aptitude needed for a successful career in science. An important feature of the schemes was the extent of interaction between the research fellows and the Society. This was a two-way benefit. The Society had direct contact with a much wider segment of the scientific community, while the research fellows, in turn, gained skills and contacts relevant to the wider aspects of their careers. They were also encouraged to take an active part in Royal Society life, for example contributing to strategic discussions about the future of the Society and accessing the private facilities at Carlton House Terrace. Training was provided on such matters as commercial exploitation of research and media and communications skills. There were opportunities to work with schools and to observe policy processes at first hand by working alongside Members of Parliament and civil servants. This wide-ranging approach was applauded by the government.[64] And it was greatly appreciated by the research fellows themselves, who testified to the empowerment and practical support they experienced from their appointments.

Grants

The Royal Society first secured public funding to enable it to offer research grants in 1850, over a century before it secured public funding for research professorships. The funding was £1,000 p.a. and was intended primarily for 'private individual scientific investigation'.[65] From 1888, the Society established an elaborate system of discipline-based boards to assess applications for what was by then £5,000. It used the existence of these boards to bolster bids to run other funding initiatives, for such purposes as scientific publications (from 1895), payment of international subscriptions (from 1919), and support for the libraries of scientific societies (from 1949).

In 1946, the Scientific Investigations Grant totalled £21,000. It was administered with an extremely light touch, the Treasury noting approvingly: 'It is definitely not the policy of the Society to apply a meticulous check upon the expenditure of the grants, it being considered that the standing of the grantees renders this both unnecessary and undesirable, but the Society expects to receive reports from grantees.'[66] By 1960, after some horse-trading with DSIR, the Grant had reached £75,000 (in real

[64] *Realising our potential: a strategy for science, engineering and technology* (Cm 2250 1993), para 7.22. In 2010–1, the Society made a total of 148 new appointments to research fellowships of various kinds, from over 1,500 applications.

[65] Henry Lyons, *The record*, 185–91; Roy M. MacLeod, 'The Royal Society and the government grant: notes on the administration of scientific research, 1849–1914', *Historical Journal*, 14 (1971), 323–58.

[66] 'Administration of grants by the Royal Society', *Notes and records of the Royal Society of London*, 5 (1947), 2–4.

terms, double the 1946 figure).[67] This sum was still very small in relation to what was becoming available through other channels. Efforts to treble the Grant at the beginning of the 1960s proved unsuccessful, but by 1970 it had reached £194,000. A proposal to close the Grant altogether was resisted on the grounds that it 'was much appreciated by many Fellows, especially on the biological side'.[68]

The Scientific Investigations Grant more or less kept pace with inflation through the 1970s. In 1982, it stood at £0.8 M, 18 per cent of the Society's PGA and the biggest single item in it. It was still small fry: by then, SERC alone was spending £68 M on grants.[69] Demand for both Royal Society and SERC grants was increasing, and there was much concern over the funding being insufficient to cover all alpha-rated applications. But the Society took the view that this was a lesser problem than the career prospects of young scientists being stymied by the 1981 cuts in the UGC budget. ABRC accepted its proposal to transfer money from the Scientific Investigations Grant to the new URF scheme, and the former was wound down over the following two years.[70]

It was not long before the Grant began to be missed. With an upper limit of £12,500 per award, it had mostly been used for small items of equipment, and evidence soon accumulated that there was a real need for such funding.[71] The UGC block grant, under intense pressure, was increasingly unable to underpin the 'well-found' laboratory, and other funding agencies were wary of diverting their own resources to make good UGC's shortcomings. In pursuit of a more strategic approach to research funding, ABRC pushed for more programme grants at the expense of project grants; and, in pursuit of administrative efficiency, SERC raised the minimum grant application it would consider to £25,000.[72] These moves, coupled with the introduction of Interdisciplinary Research Centres, were seen by the Royal Society as evidence of increasing dirigisme in science policy. George Porter made it a centrepiece of his 1985–90 presidency to oppose this trend. It got quite heated.

---

[67] See Chapter 2 above. For comparison, the Scientific Publications Grant in 1960 had an annual budget of £10,000, and the Libraries Grant just £2,000.

[68] OM/43(66), OM/56(66), OM/62(66); OM 16 June 1966, minute 3(b); CM 14 July 1966, minute 6. See Nicholas Kurti to Patrick Blackett, 2 June 1966 (RMA914), for the suggestion that the grant be closed on the physical side.

[69] Cabinet Office, *Government funded R&D 1983*, table 7a.

[70] OM/54(84), OM 17 May 1984, minute 3(b); CM 14 June 1984, Appendix D. Overseas Field Research grants, and other specialised schemes, were merged into the resurrected Research Grant Scheme in 1993.

[71] Peter Hirsch to David Phillips, 27 November 1985: D.C. Phillips papers, MS Eng. c.5528, O.238.

[72] ABRC, *A strategy for the Science Base* (HMSO, May 1987), para 1.37.

This was hardly a new theme in science policy. The scientific community (represented by the Royal Society) saw itself as defending unfettered individual creativity as the prime source of improved natural knowledge; ABRC and others charged with extracting resources for science from the Treasury, in contrast, had to handle political demands for greater 'relevance' in academic research and greater accountability for use of public money. Relevance and accountability came to imply both more top-down setting of research priorities to deliver desired practical outcomes and, at the same time, reduced administrative costs.

Of course it was rather more nuanced than that, and in some ways the differences between the two perspectives were more apparent than real. There was a considerable degree of shared outlook and considerable overlap of personnel: half the members of ABRC were FRS, mostly with experience of serving on the Society's Council, and the ABRC Chairman David Phillips had previously served as the Society's Biological Secretary. ABRC was fully committed to the Society's URF scheme with its emphasis on the importance of individual creativity. The Society, for its part, was fully committed, in the words of its 1663 Charter, to 'promoting the sciences ... of useful arts', and George Porter was fond of commenting that there were only two kinds of science: applied science and not-yet-applied science. It was a matter of ensuring that the compromises necessary to defend the Science Budget politically did not cramp the flair of the best researchers.

It was not just a private debate among scientists. At a Sunningdale retreat for Permanent Secretaries in October 1986 to which both David Phillips and George Porter were invited, the head of the Treasury, Peter Middleton, argued that there was a 'world science mountain', that basic science had 'precious little' to do with industrial culture – 'everyone seems to agree that we have quite enough science for industry to use at present' – and that 'in relation to exploitation, basic research is manifestly a bad buy'.[73] A month later, the Science Budget was cut, and ABRC publicly voiced its 'dismay'.[74] No more than any other group could scientists assume that what was self-evident to them was self-evident to anyone else.

---

[73] See acerbic exchanges between David Hancock (Permanent Secretary at DES) and Peter Middleton in the weeks after the retreat: D.C. Phillips papers, MS Eng.c.5526, O.224. John Fairclough had cautioned the Cabinet Secretary, Robert Armstrong, about inviting Porter to participate in the retreat. 'World science mountain' was an allusion to the phenomenon of overproduction in the European Community, leading for example to so-called butter mountains and wine lakes.

[74] ABRC, *Science Budget: allocations 1987–88* (3 December 1986, published February 1987).

Schemes handing out small amounts of money purely on criteria of research excellence were clearly going to be a hard sell in these circumstances. Nevertheless, Bill Stewart, who had just completed a term on the Society's Council and was shortly to become Chief Executive of the Agricultural and Food Research Council, urged George Porter to reintroduce a grant scheme targeted specially at providing small items of equipment – in effect, resurrecting the Scientific Investigations Grant.[75] Porter used a lunch with the new science minister Robert Jackson to press the case. Jackson proved remarkably supportive,[76] and the Society included a proposal for a £2.5 M scheme in its bid to ABRC for 1989–90. This was so worthwhile, Porter told his Anniversary Day audience on 30 November 1987, that the Society's Fellows would willingly give their time and expertise to assessing the applications.

With the Society's Council behind him, Porter wrote formally to David Phillips in February 1988 to build the momentum for responsive-mode funding in general and for a new scheme of small grants in particular.[77] The message was reiterated at the subsequent ABRC meeting, and in correspondence with John Fairclough.[78] David Phillips attended a special meeting of the Royal Society Council on 3 May 1988 where the arguments were thoroughly rehearsed. In theory, the objectives of a small-grants scheme should have been met through UGC funding of the 'well-found laboratory'; in practice, a different solution was needed because the UGC was unable to deliver that level of resource. The idea of a special scheme at this stage was a temporary expedient, not a long-term solution to the erosion of the dual-support system. It turned out to be an unusually benign budget round, and the upshot was that the Society received £2 M p.a. from 1989–90 to launch what became formally known as the Research Grant Scheme. The Society celebrated its role 'in influencing a change of policy toward responsive-mode funding for scientific research'.[79]

A review of the Research Grant Scheme during its third year of operation found that it was being used in much the same way as the old Scientific Investigations Grant. With an upper limit set initially at £10,000, awards went mostly to enable young researchers starting up, or established researchers changing fields, to acquire equipment.

[75] Bill Stewart to George Porter, 18 June 1987: OM/73(87); OM 9 July 1987, minute 3(i).
[76] OM 8 October 1987, minute 3(r); OM/102(88).
[77] George Porter to David Phillips, 12 February 1988: Phillips papers, MS Eng.c.5501, O.55 and CM 11 February 1988, Appendix B. Also OM/10(88) and CM 14 January 1988, minute 13.
[78] Report on ABRC 'Forward Look' meeting, OM/44(88); Roger Elliott to John Fairclough, 15 March 1988: David Phillips papers, MS Eng. c.5519, O.193.
[79] ABRC, *Allocations of the Science Budget 1989–92*; CM 30 November 1988, minute 6.

The Scheme met a real need, filling the gap between what departmental funds could afford and what Research Councils would readily consider. It was, thought David Phillips, 'a useful safety valve in the system', even though not the Society's highest priority. ABRC continued supporting it, despite the fact that SERC in 1991 rescinded its policy of a £25,000 minimum threshold on grants. Plurality of sources was no bad thing.[80] The temporary expedient was set to stay, its minimalist bureaucracy seen as one of its strengths. By 2010–1, it was still running at about £2 M p.a., which, with the upper limit then £15,000, was enough for 180 new grants. It was a pragmatic way to support individual researchers.

[80] SEPSU, *An analysis of the Royal Society Research Grant Scheme* (Royal Society, 1991). Brian Follett to David Phillips, 18 September 1991, and meeting between David Phillips, Michael Atiyah and Brian Follett, 10 June 1992: David Phillips papers, MS Eng.c.5501, O.56. ABRC meeting 23 October 1991, minute 6.

4   The applications of science

> There is no quicker way for the Society to lose its influence than to be regarded as representing only basic science and academic research.[1]

The Royal Society has historically had an ambivalent attitude to its role in the applications of science. The 1663 Charter specified that the Society's studies were 'to be applied to further promoting by the authority of experiments the sciences of natural things *and* of useful arts',[2] and a variety of useful or curious inventions embellished its early history. Yet by the eighteenth century the Royal Society was distinctly cautious in its dealings with the useful arts,[3] and by the mid twentieth century the Society was seen, and mostly saw itself, as concerned very largely with academic science.

Paradoxically, the mid twentieth-century Society (like the eighteenth-century Society) included a significant number of Fellows who had developed careers in industrial research, and many more – the Presidents Robert Robinson and Cyril Hinshelwood prominent among them – who consulted extensively for industry while holding academic posts.[4] But this was little reflected in their interactions with the Society. Robinson's Anniversary Addresses, for example, focused very strongly on the latest developments in basic organic chemistry and ignored the applications of science altogether. Hinshelwood followed this precedent, despite being

---

[1] Bill Penney (see note 22 below) to Harold Hartley (note 19 below), 20 March 1963: HF/1/17/1/17. Some of the material in this chapter has already appeared in Peter Collins, 'A Royal Society for technology', *Notes and records of the Royal Society of London*, 64 (2010), S43–54, and is reproduced here with permission.

[2] Henry Lyons, *The record*, 251. Emphasis added.

[3] David Philip Miller, 'The usefulness of natural philosophy: the Royal Society and the culture of practical utility in the later eighteenth century', *British Journal for the History of Science*, 32 (1999), 185–201.

[4] Sally M. Horrocks, 'The Royal Society, its Fellows and industrial R&D in the mid-century', *Notes and records of the Royal Society of London*, 63 (2010), S31–41. Between 1932 and 1960, 38 scientists working in industry were elected to the Society – 6 per cent of all new Fellows in that period.

personally keen to see the Society engage more with applied science.[5] Basic science dominated Royal Society affairs.

In each postwar decade, the increasing national investment in research was coupled with the increasingly urgent expectation that it should deliver national benefit – defined much more in terms of economic and military competitiveness and, somewhat vaguely, quality of life, than in terms of cultural enrichment and national prestige. The Society's leaders were well aware that corporately they could not remain aloof from that agenda. But they had first to be persuaded, and then to persuade the mass of their colleagues, to push it up the priority list. And they had to work out what they could actually do in practice that would make any difference. On this there were, broadly, three options: the Society could spend money on schemes to advance applied science directly, it could offer advice to government and others on policy in relation to applied science, and it could exercise its uniquely powerful accolade function to promote more positive attitudes towards applied science. It did all of these things, to some extent at least. First, though, it had to overcome its ambivalence.

The social status of technology in the 1960s

There was pressure to do so from the Society's friends in government. One such was Harry Melville, who had signed Andrade's 1945 petition and in 1956 had become Secretary of DSIR – and was thus in charge of a substantial portion of civil research spend. Melville was keen to get more of the best people to pursue careers in technology (meaning mostly but not only engineering), and to that end he urged the Society in November 1960 to use its capacity to confer prestige more deliberately. The less palatable alternative, he warned, was that someone else would establish a new organisation for the purpose. This was already being mooted,[6] prompting the influential Harold Hartley to remark to David Martin: 'Heaven forbid, but we must take this seriously.'[7] So the Physical Secretary Bill Hodge, who in 1945 had been arguing that the President's main function was to 'symbolise the Society's devotion to fundamental science before every other consideration',[8] now told

[5] Harold Hartley commented that Hinshelwood 'was disappointed at finding very little support' for his wish to promote the cause of technology within the Society: Hartley to Gordon Sutherland, 4 January 1964: Sutherland E25. Sutherland papers quoted by kind permission of the Syndics of Cambridge University Library.

[6] See, for example, John Oriel, 'Too many learned societies?', *New scientist* (7 April 1960), 854–6.

[7] Martin to Hinshelwood, 9 April 1960: RS archives, 'Officers and Council – Cyril Hinshelwood 1958–61'.

[8] See Chapter 1.

Melville that the Society was indeed 'the proper place in which to recognise outstanding achievements in technology'.[9]

The ultimate means for the Society to do this was to elect more people to the Fellowship for their achievements in applying science. To do so with minimal opportunity cost for other candidates meant increasing the number of Fellows elected annually. The new President Howard Florey decided to test the waters. He convened an informal meeting of Fellows in Oxford on 7–8 October 1961, and they warmly endorsed a proposal to increase annual elections from twenty-five to thirty.[10] A Special General Meeting (SGM) was then called for 15 March 1962 to take the mood of the Fellowship more formally. However, with over ninety Fellows present (15 per cent of the total Fellowship), the majority in favour at the SGM was so small that Council felt itself unable to proceed with any increase in the number of annual elections.[11] It settled instead for making a few modest procedural changes.[12] It also rejected, not for the last time, a proposal to establish a category of associate membership.[13] This outcome considerably weakened Florey's position when, during a wide-ranging and ill-tempered meeting with Hailsham a fortnight later,[14] the minister berated him for the lack of clinicians and engineers in the Fellowship. At that stage it was typically estimated that 10 per cent of the Fellowship were applied scientists, but the persistent impression was that the Society was essentially for pure scientists.

Two developments then fanned the flames. One was the publication by the Society in February 1963 of the report of a group chaired by Gordon Sutherland, presenting hard data on the emigration of scientists from the UK. This generated great interest (and immediate journalistic coining of

[9] OM/89(60), OM 89a(60); OM 30 November 1960, minute 3(b). In a coordinated move, the science minister Hailsham made similar points to Hinshelwood.

[10] OM/120(61).

[11] In the short term, anyway. As Florey commented to Harold Hartley on 30 January 1963, 'I think within a year or two the whole matter of the numbers of Fellows to be admitted every year can be raised again.' HF/1/17/1/17.

[12] CM 12 April 1962, minute 9. Abdus Salam was one of those opposed to any increase – doubtful of any backlog of top-quality candidates in his own field, at least, and 'categorically against' the very idea of general candidates (the category under which those based in industry tended to be elected), their existence in his view a source of 'deep resentment' among unsuccessful mainstream candidates. Salam to Martin, 12 March 1962: HF/1/17/1/17.

[13] The idea was floated again by Solly Zuckerman the following year (Zuckerman to Sutherland, 21 December 1963: Sutherland E31), and by Peierls and Kurti in 1971 (OM/93(71) and C/3(72); OM 11 November 1971, minute 3(d) and CM 13 January 1972, minute 14). It has been raised periodically since, always with the same result.

[14] OM/37(62): 'Notes by the President on his meeting with Lord Hailsham on 30 March 1962'.

the phrase 'brain drain'),[15] and intensified the sense of crisis about the UK's industrial competitiveness. Sutherland's committee continued through the rest of 1963 to monitor government action on the issue and to keep up the pressure for positive action.[16]

The other development was publication by Oxford University the same month of a report called *Technology and the sixth form boy*. This documented evidence that engineering and technology were poorly esteemed by (male, let alone female) sixth-formers, and consequently attracted less than their fair share of the most talented school-leavers. In the years before sixth form, there was a strong bias towards arts rather than science, and within the latter a bias towards pure science rather than technology.[17] Hailsham weighed in again, this time setting up an interdepartmental committee 'to consider what might be done to create a climate of opinion more favourable to technology generally, and so to improve the quality of people coming forward for technological education'.[18]

There was pressure on the Society from the inside as well as the outside. Harold Hartley[19] discussed with the Treasurer Alex Fleck in March 1963 'the anxiety that I know a number of people have been feeling about the present standing of engineering in the UK and the possibility that the Royal Society might give more support'. He was particularly concerned that 'pure science is getting more than its share of the best boys, while on the Continent the reverse is taking place'. Fleck suggested that Hartley assemble a small group of senior Fellows to talk to Florey. Christopher Hinton (then Chairman of the Central Electricity Generating Board) offered to host a dinner at the Connaught Hotel for the group on 27 March 1963. Discussion at the dinner – attended by Christopher Hinton, Harold Hartley, Arnold Hall, Christopher Kearton, Bob Feilden and Alex Fleck – focused on how to get more engineers and technologists

---

[15] Brian Balmer, Matthew Godwin and Jane Gregory, 'Brain drain'.

[16] CM 7 November 1963, minute 6.

[17] 'Sixth formers attitudes to technology', *New scientist* (31 January 1963), 239–42; G van Praagh, 'Technology and the sixth-form boy', *Nature*, 199 (7 September 1963), 958; PV Danckwerts, 'Science versus technology: the battle for brains', *Nature*, 200 (19 October 1963), 219–20.

[18] *Annual Report of the ACSP 1963–64*, Cmnd 2538 (HMSO December 1964). Also SP(64) 9 *Status of engineering*, dated 26 February 1964: copy at TODD Acc 811, box 2.

[19] A physical chemist who had supervised Cyril Hinshelwood, Hartley was already 85 in 1963 but still active in Royal Society circles and much valued for his wisdom and his ability to bring people together. Alongside his academic work he had extensive experience of government and industrial affairs and was much consulted by Prince Philip, Duke of Edinburgh. A.G. Ogston, 'Harold Brewer Hartley', *Biographical memoirs of Fellows of the Royal Society*, 19 (1973), 348–73; HRH The Prince Philip, 'Research and prediction. The inaugural Hartley Lecture, 21 May 1974', *Notes and records of the Royal Society of London*, 29 (1974), 11–27.

into the Society, though by tweaking the existing system rather than by any radical change.[20] Both Hartley and Fleck made sure to brief Florey and the other Officers about these moves early so that 'there would be no suggestion that we were indulging in any hole-and-corner talks'.[21]

Fleck's predecessor, Bill Penney,[22] sympathised strongly. He pointed out that scientists in the Civil Service and in industry 'already regard the Royal Society as not for them', and warned of the dangers: 'There is no quicker way for the Society to lose its influence than to be regarded as representing only basic science and academic research.'[23] He was abroad at the time of Hinton's dinner, but told him 'I have been explaining similar ideas to some of the senior Fellows of the Royal Society for the last few years and have found increasing support'. Penney also complained that basic science – especially high-energy physics – was commanding large sums of money without any serious questioning of how to prioritise public spending on research so as to improve the competitive position of British industry.

Florey asked Alex Fleck to chair a committee on 'what action the Royal Society might take to heighten the esteem of the technologist as a scientific contributor to the national welfare'.[24] At its first meeting, in June 1963, the Fleck Committee affirmed the 'widespread feeling that the UK was losing its position in the world in technology relative to its position in science'. But the Committee argued that responsibility for improving the public image of technologists lay with the professional institutions, and that reaching the public was the role of bodies like the British Association. The Society's business was with the most outstanding technologists, not with technologists in general. And even that only up to a point: 'science and not technology must properly remain the dominant note of the Royal Society.'[25] Gradual development within established boundaries remained the order of the day, even while the Trend Committee was

[20] Minute of the dinner drafted by Fleck, 1 April 1963; Hinton to Fleck, 2 April 1963: Hinton papers, H21.

[21] Harold Hartley to Florey, 20 March 1963: HF/1/17/1/17. Florey was invited to Hinton's dinner but was unable to participate.

[22] Deputy Chairman of the UK Atomic Energy Authority, Penney had stepped down as Royal Society Treasurer on 30 November 1960, to be succeeded by Alex Fleck, the recent Chairman of ICI. The Society's finances, at least, were in the hands of industrial leaders.

[23] The ACSP's reputation was not any better in this context. Penney's experience of it was that 'matters of fundamental science lead to good discussions but matters of applied research, development and industrial scientific questions are left severely alone'. Penney to Hinton, 18 March 1963: Hinton papers, H21.

[24] OM 4 April 1963, minute 2(d)(iii), OM 9 May 1963, minute 2(d)(i) and Council 9 May 1963, minute 29.

[25] TC/1(63), TC/3(63), TC/4(63), TC/6(63), TC/8(63).

preparing to shake up the structure of civil science in the UK and the Labour Party was contemplating creation of a Ministry of Technology.

This was all too leisurely for some. Gordon Sutherland decided to raise the stakes by going public. As well as being Director of the National Physical Laboratory, Sutherland was a senior member of the Royal Society Council, active in debates about the Society's election processes and about the brain drain. The day he stood down from Council, 30 November 1963, he published a letter in the *Guardian* under the heading 'A Royal Society for technology'. It started with the need for more first-class intellects to take up careers in technology.[26] Young people were 'imbued with the idea that "pure" science and "pure" research leading to Fellowships in the Royal Society' were the ideals towards which they should strive. The Royal Society had had a profound influence in keeping Britain in the forefront of pure science; so, with the aim of repeating the trick for applied science, Sutherland proposed that a completely new body be formed for the elite among engineers and technologists. Sweden had separate academies for science and for engineering, the latter with twice as many members as the former, so it could be done. Almost as a throwaway line, Sutherland suggested that his new body be called 'the Edinburgh Society (if the Duke of Edinburgh would consent) as a partial recognition of what he has done to stimulate the appreciation of applied science in this country'.

Sutherland's personal eminence and his closeness to the leadership of the Royal Society ensured that his proposal was widely discussed, generating an extensive postbag for himself and many letters to the national and technical press. Much of the press comment was supportive. Among Sutherland's engineering friends in the Fellowship, however, the most common response was to agree with his argument but to urge that the Royal Society itself should do the job envisaged for the new organisation. He should at least wait, they argued, till the Fleck Committee had reported before finally concluding that the Society would not rise to the occasion.[27]

One issue was whether any separate technology organisation should be fostered by the Royal Society or, instead, by the engineering institutions then trying to collaborate under the aegis of the Engineering Institutions

---

[26] In this context, the terms 'engineering' and 'technology' were often used interchangeably. Howard Florey, though, never lost an opportunity to emphasise that technology included a good deal of biological science and was not synonymous with physical engineering.

[27] Hall to Sutherland, 9 December 1963: Sutherland E24. See also Holder to Sutherland, 16 December 1963: Sutherland E26; Zuckerman to Sutherland, 21 December 1963: Sutherland E31; and other material in the Sutherland papers.

Joint Council. The EIJC leadership assured Hailsham, when they met with him on 13 December 1963, that Sutherland's proposed elite organisation did not clash with their own desire to have a Royal Charter related to their goal of establishing common academic standards across the engineering profession.[28] That was one potential obstacle sorted. A second was addressed, if not necessarily sorted, when Harold Hartley sent the Duke of Edinburgh copies of Sutherland's *Guardian* letter and related material to keep him in the picture.[29] Hartley was a long-time confidant of the Duke of Edinburgh on matters to do with science and technology, and would be key to any attempt to involve him in the initiative to establish a new organisation. Sutherland let the Royal Society know, through Fleck, that Hartley had taken this step. Florey thought it 'a great pity that Harold Hartley has rushed along to the Palace, this makes it more difficult for the Royal Society as we clearly don't want to get involved in Palace politics'.[30]

By early 1964, Sutherland had been persuaded that the Royal Society should be given a further chance to show that it could itself fulfil the roles that he envisaged for his new organisation. It had first refusal, despite its general ambivalence towards engaging corporately with the applications of science. The same strategy of first refusal was just then also being pursued in the United States, where the National Academy of Sciences was acting as the reluctant progenitor of the National Academy of Engineering.[31]

The Royal Society response

It was obvious that the Royal Society had to do something, and it had to do it quickly. It might have first refusal, but others would move in if it remained half-hearted. Gradual evolution would not suffice. The Society's established functions were already under pressure from the Trend Committee; it could not lightly surrender further territory. In the first instance, it was up to the Fleck Committee to produce a strategy that would be acceptable to the existing Fellowship, satisfy external critics and be sustainable in the long term.

---

[28] Sutherland to Hartley, 17 December 1963: Sutherland E25.

[29] Hartley to David Checketts (Equerry to the Duke of Edinburgh), 17 December 1963: copy at Sutherland E33.

[30] Sutherland to Fleck, 24 December 1963: Sutherland E23; copy to Florey at HF/1/17/6/1. Florey to Sutherland, 2 January 1964, and Sutherland to Florey, 5 January 1964: HF/1/17/6/1.

[31] See correspondence with Fred Seitz, Julius Stratton and others at Sutherland E21–28; OM 17 December 1964, minute 2(f)(ii); *The National Academy of Engineering: the first ten years* (NAE, 1976).

(i)   Elect more technologists

As so often in Royal Society life, the election process was the linchpin. The computer pioneer Frederic Williams, a member of the Fleck Committee, put it simply: 'The only way I can see the Society directly influencing the status of technologists is by electing more of them.'[32] The disadvantage was that this was necessarily a slow process.

The Fellowship was not readily going to elect additional technologists if that meant electing fewer 'pure' scientists. So the question of increasing the number of new Fellows elected annually was back on the agenda, less than two years after the previous, failed, attempt to secure an increase. Florey again took soundings and found a grudging acceptance of the need for an increased technology dimension to the Society's work. Charles Goodeve, Director of the British Iron and Steel Research Association and one of the signatories of Andrade's petition, commented: 'I think that the majority of Fellows are conscious of the fact that the RS is in danger of missing much and even becoming weaker if it remained as specialised as it is in the face of what is bound to happen both on the behavioural and technological sides ... We are primarily a Professors' Club.'[33]

The proposal, as before, was to increase annual elections from twenty-five to thirty. At an informal dinner for members of Council in March 1964, Fleck argued the case for pushing the increase up to thirty-two elections per year, which he felt could be safely done without debasing the Society's currency. This was to cover both the technology issue and the impact of the growth of science. The dinner agreed to thirty-two, but rejected a suggested one-off bonus issue of earmarked places for technologists. The increase to thirty-two was duly agreed at an SGM on 12 June 1964, with those expressing concern about it balanced by those wondering whether the increase was big enough.[34] The additional places were intended primarily for applied scientists on both physical and biological sides. This did indeed result in some extra applied scientists being elected in the following years,[35] though the engineers had to share this benefit with other applied disciplines. Implementing the decision, Council appointed an Applied Science Candidates Committee for an

[32] TC/7(63).
[33] Goodeve to Fleck, 20 January 1964: RMA30A; Martin to Florey, 20 January 1964: HF/1/17/2/56. Florey was as interested in promoting the behavioural sciences as in promoting technology.
[34] See Ronald Keay's manuscript notes on the meeting: RMA30A. In contrast to the previous SGM, only forty Fellows attended: some of the heat had gone out of the debate.
[35] John S. Rowlinson and Norman H. Robinson, *The record*, 24.

experimental three years to advise it on applied candidates referred to it by sectional committees.[36]

Given the extent to which these developments were driven by public perception of the Society's attitude to technology, it is striking that for five months nothing was done to publicise them. The first public announcement was not until Florey's Anniversary Address on 30 November 1964 – and even then only after Florey had asked the other Officers whether they had any objection to his mentioning it.[37] He described the increase as the Society

deliberately proclaiming its special interest in the marvellous developments 'of modern technology and in those who are responsible for them ... We also hope that our interest in these matters will go some way to convince our technological colleagues that we are interested in them and all they mean to the nation.[38]

The *Daily Telegraph* hailed what it described a 'revolutionary change ... to give recognition to industrial scientists and technologists'.[39] The *Sheffield Telegraph*, however, influenced by Meredith Thring,[40] was less impressed, complaining: 'Though the Royal Society has at long last decided to admit more applied scientists, it has done so with more condescension than enthusiasm, to prevent the creation of a rival society which would certainly have greater influence than itself.'[41]

(ii)   Establish a 'C' side?

The Society's founding Charter provided for the existence of two Secretaries. From 1827, it became the practice that one of the Secretaries should be from the physical sciences and the other from the life sciences, embedding an 'A' side and 'B' side in the Society's affairs. When the Society's journal *Philosophical Transactions* reached unwieldy proportions, in 1887, it was analogously divided into A and B series, and the same

---

[36] CM 18 June 1964, minute 15. On sectional committees, see Annex. In 1977, the President, Alex Todd, decided that the special committee for applied science candidates, and a similar committee for general candidates, were no longer necessary and stood them down, their missions apparently accomplished. Todd to Ronald Keay, 8 June 1977: TODD Acc 1021, Box 33; CM 21 July 1977, minute 16 and appendix A; RS Circular 10(77).

[37] Memo from Martin to the Officers, 2 November 1964: HF/1/17/6/1.

[38] Howard Florey, 'Address at the Anniversary Meeting, 30 November 1964', *Proceedings of the Royal Society of London. Series B, Biological sciences*, 161 (1965), 448.

[39] *Daily Telegraph*, 1 December 1964; copy at HF/1/17/1/2.

[40] Thring, Professor of Fuel Technology at Sheffield and not an FRS, was one of those who had been independently exploring the options for an elite technology body and who had kept in close contact with Sutherland since the *Guardian* letter appeared. See M.W. Thring, 'The efficient development of new ideas in industry', *The Guardian*, 26 November 1963; Thring to Sutherland, 2 December 1963: Sutherland E27; and Thring to Sutherland, 9 December 1963: Sutherland E29.

[41] *Sheffield Telegraph*, 14 December 1964.

development occurred with *Proceedings* in 1904. Engineering was included in the A side but was not specially prominent within it. So it was natural that those seeking greater visibility for technology within the Society should think in terms of establishing a 'C' side, equal in rank and status to the A and B sides and with its own Secretary working alongside the Physical and Biological Secretaries, to provide a base and a champion for their ambitions.

Cyril Hinshelwood was one of those for whom the creation of a C side was the most obvious response to Gordon Sutherland's *Guardian* letter,[42] and the idea gained some currency. Sutherland himself supported it, suggesting to Alex Todd that the most likely alternative to a new organisation was the Royal Society creating 'a section C as large as A and B and of equal status at a very early date'.[43] He also told Florey that 'if a section C of about 40 new members were created at once with about 5–10 elections pa subsequently, I agree the need for a new Society would largely disappear'.[44]

But the idea was too radical for the bulk of the Fellowship. Their commitment to technology did not go that far. An informal meeting of Fellows in January 1964 showed no support for a C side.[45] Neither did the Fleck Committee, which saw it as a potential 'society within the Society' (though the existence of A and B sides apparently did not pose such a threat).[46] Harold Hartley alerted Sutherland to how the wind was blowing: 'Hinshelwood's idea of a C section ... is meeting strong opposition and any action is likely to be on an inadequate scale. Hinshelwood's solution would I believe have been the best for technology and the RS but it is not surprising to me that this lack of sympathy and generosity on the part of the academics will kill it.'[47] The idea was revived by Harold Barlow in 1973, but again turned down.[48]

(iii)   Technology medals

If a full-fledged C side was beyond the art of the possible, a dedicated medal was less threatening.[49] Since 1825, two gold medals had been

---

[42] Hartley to Sutherland, 4 January 1964: Sutherland E25. See also Fleck to Hinton, 12 December 1963: Hinton papers H21; and Fleck to Sutherland, 31 December 1963: Sutherland E23.

[43] Sutherland to Todd, 4 January 1964: TODD 5/21. Todd replied on 21 January 1964: 'I shall be most interested to see whether, in fact, the new Society gets formed or whether the effect may be to make the Royal Society go in for technology seriously.'

[44] Sutherland to Florey, 5 January 1964: HF/1/17/6/1.

[45] David Martin to Florey, 20 January 1964: HF/1/17/2/56.   [46] TC/4(64), C/56(64).

[47] Hartley to Sutherland, 22 February 1964: Sutherland E25.

[48] H.M. Barlow to Hodgkin, 5 November 1973; OM 17 January 1974, minute 3(b).

[49] The Society already had one technology medal in its gift, for pure or applied chemistry or engineering, sponsored by the Leverhulme Trust in 1960 to commemorate the tercentenary.

awarded annually by the Sovereign, on the recommendation of the Royal Society Council, for the most important contributions in the 'two great divisions of natural knowledge' – the A and B sides. In February 1965, prompted by the DES Permanent Secretary Maurice Dean and by Patrick Blackett, Florey secured the Queen's permission to add a third medal to the collection, 'for distinguished contributions in the applied sciences'.[50] He then got Council to agree that the Royal C Medal could equally well be awarded for applied work in the biological as the physical sciences.[51] In his final Anniversary Address that year Florey announced, a touch defensively: 'We are particularly delighted to have been enabled by the Medal further to proclaim that the Society wishes to continue to foster the applied sciences.'[52]

Another new medal followed the next year, when Mullard Ltd offered to sponsor an annual prize for an outstanding contribution to the advancement of science, engineering or technology leading directly to national prosperity in the UK. Mullard's aim was, explicitly, to 'encourage a transfer of attention from pure scientific pursuits toward manufacturing industry', including 'agriculture and other biological production'.[53] Blackett, then active in the Labour Government's Ministry of Technology, felt that the Society in recent decades had been 'somewhat remiss in not recognising more engineers and applied scientists by admission to the Fellowship', and was delighted to be able to announce the creation of the Mullard award in his first Anniversary Address as Royal Society President.[54]

There was almost another addition to the Society's armoury for esteeming applied science in January 1968, when the MacRobert trustees approached the Society about setting up an award 'for work within the fields of Applied Science and Technology'.[55] The Officers were keen to explore the possibilities, but at Christopher Hinton's prompting the trustees eventually decided to give the job to the Council of Engineering Institutions (CEI, the successor body to the EIJC) 'in conjunction with the Royal Society'. The Royal Society Council agreed to nominate three people to the main CEI selection committee but baulked at the 'in

---

[50] CM 4 March 1965, minute 26.   [51] CM 17 June 1965, minute 14.

[52] Howard Florey, 'Address at the Anniversary Meeting, 30 November 1965', *Proceedings of the Royal Society of London. Series B, Biological sciences*, 163 (1966), 432.

[53] CM 10 November 1966, minute 19; CM 13 July 1967, minute 24.

[54] P.M.S. Blackett, 'Address at the Anniversary Meeting, 30 November 1966', *Proc R Soc Lond A*, 296 (1967), v–xiv. The award is still made.

[55] OM/12(68). 'I hope you would agree that a project of this nature is in the national interest at this time.' See also Heughan (MacRobert Trust) to CEI, 9 January 1968; Finch to Heughan, 29 January 1968; and Hinton to Drucquer [who succeeded Wynne-Edwards as Chairman of the CEI in 1967], 6 February 1968: Hinton papers C38.

conjunction with' phrase, so that the formal rubric of the MacRobert Award referred only to the CEI.[56] The Society's Officers also rejected the suggestion that Fellows who had won the Award be identified as such in the Royal Society Yearbook.[57] Such were the nuances of managing the business of esteem.

A further prize – the Esso Award for Conservation of Energy, which the Royal Society Council approved after first debating whether Esso was a suitable sponsor for it – was added to the repertoire in 1974.[58]

(iv)   Technology lectures

A key figure in the background throughout this was the Duke of Edinburgh, because of his active interest in the applications of science and because of his likely role as patron of any new technology organisation seeking parity with the Royal Society. He had been elected to the Society in 1951, and his adviser Harold Hartley was anxious that the talk of the Society's apparent lack of concern about technology, and the possibility of a new body to fill the gap, should not create a rift between the Duke and the Society. So Hartley suggested to Florey that he invite the Duke to give a lecture to the Society. This grew into a proposal for a series of prestige evening lectures 'to advance the esteem of technology',[59] which found warm support within the Society. In the event, however, it took over two years to arrange the first lecture, and only seven had been given by the end of 1975 – none of them by the Duke.

Had the Society done enough?

The Society had shown itself willing enough to do things within its established comfort zone, though a bit wary at the edges; prepared when pushed hard to open up the election process a bit, provided that that did not impede the election of 'pure' scientists; but definitely not willing to go anywhere as radical as a C side that might give applied science parity with the whole of physical or biological science. It was all quite modest in relation to the perceived need. The Society's deeds

---

[56] OM 18 January 1968, minute 3(i); OM 15 February 1968, minute 2(d); OM 9 May 1968, minute 2(d); and CM 18 July 1968, minute 25 and Appendix A.

[57] Feilden to Martin, 22 December 1971; OM 13 January 1972, minute 3(b). The only non-RS prize mentioned in Yearbook entries is the Nobel Prize.

[58] CM 11 October 1973, minute 29; CM 30 November 1973, minute 9; CM 14 February 1974, minute 11. This attracted the satirical attention of *Private Eye*: Alan Hodgkin, 'Address at the Anniversary Meeting, 30 November 1974', *Proceedings of the Royal Society of London. Series B, Biological sciences*, 188 (1975), 110.

[59] Hartley to Florey, 8 March 1964: HF/1/17/2/56; Martin to Hartley, 26 October 1964: HF/1/17/2/53.

betrayed its continuing ambivalence even as its public rhetoric signalled new-found enthusiasm for the applications of science.

The 1965 round of elections to the Fellowship, the first with 32 places at stake and with Florey's promise of benefits for applied scientists, was expected to show some shift in direction. As soon as the outcome was known, Hartley commented grumpily to Sutherland: 'The list shows that the academics have hogged the additional places.'[60] Sutherland thought that the 1966 election round, however, was a 'great improvement' so far as recognition of applied scientists was concerned. But it was evident to him that it was going to take a very long time for the Society to have within it 'a really representative body of applied scientists and technologists'.[61] Hinton was not impressed: 'The Royal Society's deeds do not seem to me to match up with its expressed intentions ... it seems that the old pattern of elections remains little changed.'[62] Gordon Sutherland thought it was time for further action. So did Meredith Thring.

Sutherland and Thring were both present at a lunch with the Duke of Edinburgh on 27 April 1966. Thring took the opportunity to make his mark with the Duke.[63] The Duke was encouraging, though he commented that creating a 'Valhalla for engineering worthies' was unlikely to solve any problems.[64] And Sutherland warned Thring that while Harold Hartley was 'just as disappointed as we are about the very limited reaction of the Royal Society to the present need for recognition of applied scientists and engineers, I think he is doubtful whether one should open an immediate campaign for the promotion of a new society'.[65]

But the maverick Thring was determined, and pushed ahead with developing a blueprint for a Royal Society of Technology. The Duke, having first consulted Harold Hartley, was again mildly encouraging: 'I believe that eventually some such body as you propose is necessary and will emerge.'[66] In contrast, Patrick Blackett, as President of the Royal Society, told Thring dismissively that his plan 'would not be in the best

[60] Sutherland to Hartley, 25 January 1965; Hartley to Sutherland, 10 March 1965: Sutherland E25. Also Hartley to Hinton, 10 March 1965: Hinton papers, H23.
[61] Sutherland to Hartley, 25 April 1966: Sutherland E25.
[62] Hinton to Martin, 25 May 1966: Hinton papers, H23.
[63] Thring to Duke of Edinburgh, 27 April 1966, Duke of Edinburgh Library CEI Box 1; copy also at Sutherland E30. The Duke of Edinburgh later attributed the idea of a Royal Society of Engineers to Thring: HRH The Prince Philip, 'Promoting engineering', *Ingenia*, 41 (December 2009), 12–6.
[64] Duke of Edinburgh to Thring, 5 May 1966: Duke of Edinburgh Library CEI Box 1.
[65] Sutherland to Thring, 6 May 1966: Sutherland E30.
[66] Thring to Duke of Edinburgh, 7 June 1966; Duke of Edinburgh to Thring, copied to Wynne-Edwards and Hartley, 20 June 1966: Duke of Edinburgh Library CEI Box 1 and Sutherland E30.

interests of technology in the United Kingdom'.[67] The Officers did, though, recognise that they needed to reinforce, and secure greater publicity for, their efforts to engage with the applications of science.

There were only two organisations through which a new academy for engineering could be launched with obvious legitimacy: the Council of Engineering Institutions and the Royal Society. A successful launch would need the active support of one of these two and the acquiescence, at least, of the other. The Royal Society corporately was not going to put itself out to support Thring's initiative, even if he had the support of some individual Fellows. Thring recognised that he had to work with the CEI.

The Royal Society and the CEI saw themselves more as colleagues than competitors. Thring went to lobby the CEI Chairman, Robert Wynne-Edwards, on 5 July 1966. The CEI thereupon established an 'exploratory working party' chaired by Oliver Humphreys to consider 'possible alternatives to the proposal by Professor Thring for a Royal Society of Technologists'. Because of the latter's intended function of raising the status of technology, the Humphreys working party became known as the 'accolade committee' of the CEI. Coincidentally, Blackett also met with Wynne-Edwards on 5 July – to discuss a possible Wates Prize for engineering design[68] – and Wynne-Edwards invited the Royal Society to be represented on the accolade committee. The Society chose James Lighthill (then Physical Secretary), Ashley Miles (Biological Secretary), and the engineers Bob Feilden and Stanley Hooker for this assignment.[69]

The Royal Society leadership remained adamant that 'creation of a Royal Society of Technologists would be unsatisfactory for the technologists and could be damaging to the Royal Society'.[70] It was even willing at that stage to contemplate electing a special 'bonus' of ten extra applied science Fellows in 1968 to avert such an outcome.[71] For the CEI, Wynne-Edwards took the same view as the Royal Society. The accolade committee duly rejected Thring's proposal on the grounds that 'any such body would tend to usurp the functions of CEI and its constituent institutions'. The business of bestowing accolades, even to technologists, could safely be left to the Royal Society, with its long-established status, and to the individual engineering institutions.[72] Thring could find no way past the gatekeepers.

---

[67] Blackett to Thring, 24 May 1966: Sutherland E30.
[68] OM 14 July 1966, minute 2(a)(ii). The Wates donation was eventually used to help launch the Society's European Science Exchange Programme (see Chapter 9).
[69] OM 14 July 1966, minute 2(f).   [70] OM/108(66).
[71] OM 10 November 1966, minute 2(f). The bonus did not materialise.
[72] See handwritten timeline of events by John Coales: SC 168/1/3/10. Also OM 30 November 1966, minute 2(b).

To strengthen its hand, the Royal Society established an Industrial Activities Committee (IAC) in 1969. It was chaired initially by James Lighthill, with strong industrial representation in its membership and a dedicated secretariat. Its formal role was to advise Council on any matters related to scientific, engineering and technological work in industry, and it was excluded from anything to do with elections to the Fellowship.[73] Lighthill told the members at their first meeting in March 1970 that the creation of the IAC underlined the importance that the Society attached to its industrially oriented activities.[74] Its main work became organising a series of seven discussion meetings, 'Technology in the 1980s', though it also involved itself in discussions about the running of the NPL and about the Society's response to the 1971 Rothschild report (see Chapter 5). Lighthill was content that this pattern fitted the IAC's role and the Society's ambitions.[75] That was sufficient in the early 1970s; more would be needed as the decade progressed.

Meanwhile, the CEI for its part concentrated on maintaining good relations with the Royal Society and collaborating as opportunity arose, and on establishing itself within the engineering profession. At a meeting on 8 April 1970 the then CEI Vice-Chairman Eric Mensforth 'paid tribute to the attitude of the Royal Society in the past few years which was proving to be very helpful in the development of the CEI'. He was, Blackett recalled a week later, 'highly appreciative of what the Society had done in recent years in engineering and technology'.[76] At a special meeting of Royal Society Officers on 19 May 1970, Blackett elaborated: 'Sir Eric had been most appreciative of what the Society had done for the engineering professions. He thought it had put an end to the desire for a national engineering academy, although a few individuals still talked of it.' James Lighthill agreed that 'only a very small minority of engineers wanted an academy'.[77] The idea had not entirely gone away, but, with neither Royal Society nor CEI officially in favour, it was certainly not imminent. The Royal Society seemed in 1970 to have weathered the storm unleashed in 1963, its territory intact.

---

[73] OM 12 June 1969, minute 3(j); OM 9 October 1969, minute 3(m); CM 6 November 1969, minute 22 and CM 18 December 1969, minute 12. Lighthill had been Physical Secretary during 1965–8.

[74] IA/4(70), C/72(70).

[75] OM 14 January 1971: minute 2(g). Hodgkin later said that he had 'derived much satisfaction from attending the excellent meetings' organised by the IAC. Alan Hodgkin, 'Address at the Anniversary Meeting, 1 December 1975', *Proceedings of the Royal Society of London. Series B, Biological sciences*, 192 (1976), 377.

[76] OM/27(70); OM 16 April 1970, minute 2(a).   [77] OM/52(70).

The Fellowship of Engineering

It was a temporary respite. The arithmetic simply did not stack up. The industrial engineer Theo Williamson, a member of Lighthill's IAC, wrote to the President Alan Hodgkin in 1972 to highlight 'just how unrepresentative the present Fellowship is of the distribution of scientists in society'.[78] There were at least as many professional engineers as scientists in the country as a whole, if not twice as many, yet only 10 per cent of the Society's 700 Fellows were engineers (and a further 10–15 per cent were applied scientists). Moreover, only 6 per cent of the Fellows worked in industry, even though industry employed 80 per cent of all engineers and technologists and 60 per cent of all scientists. The Society's Council continued to review the matter and in 1974 appointed a group of senior individuals to examine options.[79] That group quickly recommended a substantial increase in annual elections, from thirty-two to forty, with effect from the 1976 round of elections. This time, Council accepted the recommendation immediately, noting that it was 'unlikely to involve a lowered standard' and that it would increase the pool of Fellows available to do the Society's work.[80] The requisite SGM, on 14 May 1975, agreed the increase, primarily in recognition of the marked growth of scientific activity since the increase to twenty-five in 1946.[81]

That was a step forward, but no one was seriously suggesting that engineers should constitute half the Fellowship of the Royal Society, and the Society had made clear that it would not offer the prospect even of the degree of parity implicit in the creation of a C side. If engineers wanted a means of delivering an accolade function and speaking with a unified and authoritative voice about engineering matters, they would have to establish their own organisation, and they would have to do so through the CEI to which virtually all the professional engineering institutions adhered. And engineers did indeed want that. Their colleagues in Sweden and America had already shown it was possible to have separate academies for engineering and for science. Now it was the UK's turn. Meredith Thring and others had found that head-on campaigning would not bring about the desired change. It was time for the Duke of

---

[78] D.T.N. Williamson to Alan Hodgkin, 8 May 1972: RMA345; also OM/55(72). Cf D.T.N. Williamson to Alan Hodgkin, 12 January 1972 [OM/11(72)], commenting on the 1971 Rothschild report and urging the Society to ' heed the writing on the wall and take the initiative by starting to play a wider role ... the scientific voice should be more frequently and purposefully heard in national decision-making ... This would ensure that we could not again be placed in the false position of appearing to defend the status quo regardless.'

[79] CM 18 May 1972, minute 9; CM 13 June 1974, minute 14.

[80] CM 16 January 1975, minute 8.

[81] CM 6 March 1975, appendix C; CM 15 May 1975, minute 10.

Edinburgh's softly softly approach, building consensus within the CEI community without overtly threatening the Royal Society with competition for territory or status.

The Duke of Edinburgh, as President of the CEI, constantly pushed his colleagues to find a way forward.[82] He organised dinners, effected introductions, badgered the CEI Board, floated ideas. One of these was for a CEI Senate, essentially focused on making the CEI a more coherent and influential voice for engineering at the national level. In the end, a Senate was established in 1971 but never met. However, the experience proved useful in highlighting many of the issues involved in creating such a grouping. It became clear that the key stumbling block was how to create a grouping that would be acceptable to the professional institutions, and whose individual members would carry unequivocal personal authority as outstanding engineers.

One of the keys to breaking the impasse was John Coales.[83] He was the 1971–2 President of the Institution of Electrical Engineers and had been active on key CEI committees for several years. He was also a Fellow of the Royal Society, elected in 1970, and had been involved in its debates about engineering and applied science. So – unlike Meredith Thring – he was an insider in both organisations.

Coales recognised that an elite body needed to have unchallengeable authority from the outset and that this would not necessarily be delivered by populating it with institutional presidents who in some cases owed their positions to skills other than outstanding track records as engineers. He identified about seventy Royal Society Fellows who were prominent as chartered engineers, and saw that if they formed the nucleus of the proposed body its status would be assured. One disadvantage of such an approach was that it might be too academically oriented for CEI tastes, so Coales also looked at the possibility of basing initial membership on the honorary fellows and medallists of the engineering institutions; but these, he decided, were of too variable a quality to serve his purpose.

During 1973, Coales tried out his idea on his old friend Alan Hodgkin, whose support as President of the Royal Society would be crucial, and on a number of other Fellows including Gordon Sutherland. He found them sympathetic. When Coales became Chairman of the CEI in early 1975,

---

[82] Interview with the Duke of Edinburgh.

[83] John Coales, *The history of the foundation of the Fellowship of Engineering* (unpublished Fellowship of Engineering paper F-91.12, 1991), paras 24–43. Coales here claims the credit, but his account was read in detail by a previous CEI Chairman Eric Mensforth, by the Secretary Michael Leonard and by Denis Rooke in his capacity as President of the Fellowship of Engineering, and includes many comments from them, so may be assumed to be reasonably reliable on attribution.

he seized the opportunity to get the organisation for elite engineers off the ground. He managed to build sufficient consensus round his model to make it work: the FRS engineers were complemented by outstanding individuals selected by a small committee from names put forward by each of the CEI member institutions.

Throughout his year as CEI Chairman, Coales made sure – as had his predecessors – to avoid treading on the Society's toes: its leadership was kept fully abreast of developments and received copies of the most significant documents. His personal friendship with Alan Hodgkin helped to secure the Society's goodwill, and the Society's Foreign Secretary Kingsley Dunham (who was also a CEI insider) used his position very effectively to build support for the CEI initiative within the Society.[84] The name Fellowship of Engineering for the new body was chosen deliberately as something that would not appear to compete with the Royal Society, and the full name CEI Fellowship of Engineering positioned it as a subset of an existing body rather than something wholly new.[85] By the time of the critical meeting of the CEI Board in October 1975, Coales was able to give explicit reassurances to his CEI colleagues that the Royal Society would collaborate closely with the proposed Fellowship.

The presidency of the proposed Fellowship of Engineering was linked to the presidency of the CEI. The man selected to succeed the Duke of Edinburgh in the latter role was Christopher Hinton, and he therefore became the inaugural President of the Fellowship. He had been elected to the Royal Society in 1954, had served as President of the Institution of Mechanical Engineers in 1966 and was the first Chairman of the Central Electricity Generating Board. He had initially been sceptical about Gordon Sutherland's proposal for a Royal Society of Technologists, commenting tartly: 'I do not believe that school-leavers take up a scientific career because the Royal Society exists.' He also thought that while academic scientists had enough spare time to do the work of the Royal Society on a volunteer basis, the country's top practical engineers would be too busy doing real work to attend to the needs of a new academy for technology.[86] However, five years later he was willing to air the suggestion that the CEI 'reconsider the proposal to form a National Academy of Engineers', while expressing satisfaction that 'close relationships have

---

[84] Kingsley Dunham, a geologist, had been President of the Institution of Mining and Metallurgy and hence a member of several key CEI committees. Coales recalled that David Martin, however, was ill disposed to the whole initiative: see transcript of meeting on 23 February 1990, SC 168/1/3/10.

[85] Duke of Edinburgh to Denis Rooke, 16 February 1991: Duke of Edinburgh CEI papers, box 2. The Fellowship of Engineering became the Royal Academy of Engineering when it secured a royal charter in 1992.

[86] Hinton to Sutherland, 11 December 1963; Hinton papers H21 and Sutherland E26.

been established with the Royal Society',[87] and he had been closely involved in developments at the CEI ever since.

The Fellowship of Engineering was formally launched by the Duke of Edinburgh at Buckingham Palace on 11 June 1976.[88] He reminded his audience that one aspiration of the Fellowship was to give engineers greater status in the social hierarchy in the way the Royal Society did for science.[89] He urged the Fellowship and the Royal Society to work together to create better relations between their respective disciplines.

The Royal Society and the Fellowship of Engineering

At the urging of James Lighthill and Sam Edwards (then chairman of the SRC, where he had established an Engineering Board), the Royal Society Council appointed a small group under the Physical Secretary Harrie Massey in January 1975 to look at its arrangements for dealing with 'the many interdisciplinary affairs related to engineering'.[90] Massey felt that the Society was still 'not very clear on the whole question of its attitude to "non-pure" science', and needed to try harder in 'the whole area of raising the status of engineering as it had clearly been instrumental in doing for pure science'. This should not be left to others, despite the progress then being made towards the creation of the Fellowship of Engineering. He therefore recommended the creation of a committee to examine the place of engineering and technology[91] in the Royal Society. Council appointed this further committee, with Alan Cottrell as chairman, in July 1975. The membership included John Coales, Angus Paton (Coales' predecessor as CEI Chairman) and Hugh Ford, so the committee was fully au fait with the various CEI initiatives as it worked to define the Royal Society's long-term role in this area.[92]

---

[87] Draft paper on future activities of CEI: Hinton papers C11.

[88] The founder Fellows included Meredith Thring and Fellows of the Royal Society such as Kingsley Dunham, Bob Feilden, Hugh Ford, Stanley Hooker, Angus Paton and Ned Warner. But they did not include Gordon Sutherland, because he was not a chartered engineer. He died in June 1980, before arrangements could be made for him to be elected an honorary Fellow.

[89] John Coales, *Foundation of the Fellowship of Engineering*, appendix K.

[90] CM 16 January 1975, minute 19(xii).

[91] This time the phrase was not taken to include the applications of biological science, since the Society's handling of that was not thought to need review. But Cottrell's committee did include one B-side Fellow, Charles Pereira (Chief Scientist at the Ministry of Agriculture).

[92] Minutes of the 22 April 1975 meeting of Massey's group: OM/50(75); OM 15 May 1975, minute 3(a); David Martin to Alan Cottrell, 19 May 1975: RMA345; CM 10 July 1975, minute 9.

The Officers asked Cottrell to begin by considering whether the Society should even attempt to speak for engineering and technology as it did for science: on this occasion, they did not take that as axiomatic. At its first meeting in November 1975, the Cottrell committee explicitly affirmed that it was indeed appropriate for the Society to interest itself in engineering as a discipline, while leaving matters concerning the engineering profession as such to the CEI. The nascent Fellowship of Engineering did not affect that. No organisation could match the Society's influence when 'making pronouncements of national concern': it was uniquely placed to make balanced, non-partisan and authoritative judgements and, therefore, in Cottrell's view, had a moral obligation to do so.

Cottrell did, however, acknowledge that the Society was hampered by 'the present small number of engineering and industrial Fellows and the high average age at which they were elected'.[93] An analysis by Angus Paton and John Coales reinforced the latter message. In the total Fellowship of then 790, there were 81 chartered engineers, half of them academics and only 18 of them aged 60 or less. This was just not enough to enable the Society to speak authoritatively across the full range of engineering issues: the voice of science lacked the capacity also to be the voice of engineering. The scale of expertise needed could be provided only by the CEI's proposed elite body, working closely with the Royal Society and including its engineering Fellows. The CEI initiative in their view was therefore 'highly desirable and fully justified'.[94]

John Coales told the Cottrell Committee, and separately the Officers, on 14 January 1976 that the CEI had achieved consensus around its Fellowship of Engineering proposal and that the initiative would definitely go ahead. The Society's Council told the CEI formally a week later that it 'has learned with satisfaction of the initiative of the CEI in proposing the formation of the Fellowship of the CEI and looks forward to cooperation with it in the future'. Coales made the most of this slightly muted enthusiasm, telling Todd how delighted he was and looking forward to 'a very happy and fruitful collaboration'.[95] The Duke of Edinburgh, too, was much relieved at the Society's officially positive line, whatever tensions may have simmered beneath the surface.[96] When Coales, in virtually his last act as CEI Chairman, sent letters of

---

[93] OM 9 October 1975, minute 3(h); minutes of 12 November 1975 meeting of the Engineering and Technology Committee, E&T/3(75) and C/217(75); CM 1 December 1976, minute 3(h).

[94] E&T/1(76).

[95] CM 15 January 1976, minute 17; Coales to Todd, 13 February 1976: OM/45(76); *Nature*, 259 (19 February 1976), 522.

[96] Duke of Edinburgh interview. On the simmering tensions, then and later, see interviews with Michael Atiyah and Julia Higgins.

invitation to the founding cohort of Fellows of the CEI Fellowship of Engineering, he was able to state prominently that he had the Royal Society's explicit backing. This was a key element in legitimising the new body, not least for those founder Fellows who were also Fellows of the Society.[97]

Even so, it was not all plain sailing. One particular bit of mischief was the suggestion put about by the Institutions of Electrical and Mechanical Engineering that the Royal Society would now stop electing engineers. The Cottrell Committee urged the Society to refute the suggestion publicly, and Todd issued a circular to all Fellows explaining the Society's involvement in discussions leading up the creation of the Fellowship and making very clear that it would not stop electing engineers.[98] Hinton for the CEI stressed his strong agreement with Todd's approach.[99]

Coales' 'happy and fruitful collaboration' included the establishment of joint study groups and lectures and, of course, meetings between the respective Officers. That was sufficient for the purpose. Hinton told his new Fellows in an inaugural newsletter about 'friendly discussions with the Royal Society which is anxious to help our Fellowship and to collaborate with it', and Todd welcomed the Fellowship in his 1976 Anniversary Address – 'not as a replacement for, but as a complement to, the Royal Society's role in engineering'. Coales himself recorded that 'From the start there was a generous welcome from the Royal Society, for instance with invitations to serve on joint committees.'[100] There were also occasional spats between the two bodies as the Royal Society got used to not being the only body at its level concerned with the useful arts, and the Fellowship of Engineering got used to the Society continuing to deal with engineering.[101] But a modus vivendi evolved, with regular meetings between the two sets of Officers and formal and informal collaborations as need and opportunity arose.

---

[97] Note also M.V. Wilkes to David Martin, 12 February 1976; Rawcliffe to Martin, 13 February 1976; OM/45(76).

[98] David Martin to Maurice Wilkes, 18 February 1976: OM/45(76); OM 4 March 1976, minute 2 (c); RS Circular 3(76), 23 March 1976.

[99] Hinton to Leonard, 13 April 1976: Hinton papers, C19.

[100] Newsletter, 26 October 1976: Royal Academy of Engineering archives, F-76.78; Alexander Todd, 'Address at the Anniversary Meeting, 30 November 1976', *Proceedings of the Royal Society of London. Series A, Mathematical and physical sciences*, 352 (1977), 457; John Coales, *Foundation of the Fellowship of Engineering*, para 41.

[101] See, for example, correspondence between Christopher Hinton, William Hawthorne and others, January–February 1981: HATN 8/24/3; OM/21(81), OM/30(81), OM/80 (81); and interviews with Fiona Steele and Peter Warren. Relationships became easier when Todd and Hinton were succeeded by the more emollient Andrew Huxley and Robert Caldecote as respective Presidents.

The Society's own role in engineering, separate from its relations with the new Fellowship of Engineering, still had to be clarified. The Cottrell Committee finalised its report on 8 July 1976, a month after the Fellowship was launched. It was then thoroughly discussed, by Council on 7 October and 16 December and by the Officers at a special meeting on 4 November.[102] One of the factors constraining how far the Society could go was its wish to secure a workable relationship with the Fellowship of Engineering. Council, led by Todd, was keen to avoid clashing openly with the Fellowship – which meant, ironically, resisting any suggestion for a major increase in the number of engineers elected into the Society. In the spirit of cooperation, Todd met with Hinton before final decisions were made on the Cottrell proposals, and the CEI leadership (including John Coales and Angus Paton) invited the Royal Society Officers to dinner at the Athenaeum[103] on 19 January 1977 for a further briefing session. The dinner reinforced the message that the Fellowship rather than the CEI would now be the natural partner for the RS, and highlighted the importance of cooperation between the two.

Echoing Harrie Massey's 1975 comment about the Society's ambivalence over 'non-pure science', Hugh Ford hammered home to the Society's Officers shortly after the Athenaeum dinner the crucial importance of taking forward at least one of Cottrell's main recommendations:

The Royal Society was too late in waking up to the opportunities and it now has to think very carefully how it identifies an effective and useful role for itself, now that the Fellowship of Engineering is well launched. The most important thing is to set up immediately a top level, fully serviced Committee to complement the Executive Committee of the Fellowship. We must have a strong say in engineering or else surrender it entirely to the Fellowship.[104]

As a way of reinforcing the Royal Society's commitment to engineering, technology and industry, Ford resurrected the idea of a C-side Secretary as the focal point for that area of the Society's work.[105] The Cottrell

---

[102] CM 7 October 1976, minute 15; CM 16 December 1976, minute 8; OM/167(76). See also Angus Paton to David Martin, 15 June 1976, and Ronald Keay ms notes on the 16 December Council meeting: RMA31A.

[103] There was a long-standing relationship between the Royal Society and the Athenaeum: Humphry Davy, while President of the Society, was one of the founders of the Athenaeum. When the Society launched its Yearbook in 1897, it included membership of the Athenaeum among the sparse personal information it provided about each Fellow. That practice stopped in 1978, after Mary Cartwright complained to the Officers about the attention given to an organisation from which she, as a woman, was excluded (OM/110(77)). In 1960, 28 per cent of Royal Society Fellows were members of the Athenaeum, and they constituted about 8 per cent of the Athenaeum's membership. The Athenaeum held complete runs of the Society's journals.

[104] Hugh Ford to Angus Paton, 14 February 1977: OM/44(77).

[105] OM/50(75); E&T/5(76). See also Hugh Ford interview.

Committee was supportive, but also cautious about delineating the workload of a C-side Secretary and about the division of labour with the A and B Secretaries. It suggested, for a three-year trial period, that a member of Council be identified as Vice-President responsible for applied science.[106] Council itself was more cautious still, with the result that Angus Paton was made a Royal Society Vice-President in December 1976 but given the applied science remit only informally, and without the right to attend Officers' meetings.[107]

Council was also half-hearted about Cottrell's other recommendations, endorsing his aspirations but diluting or discarding his prescriptions. For example, there would be no additional special measures to increase the number of engineers in the Society. However, as Hugh Ford had urged, the Industrial Activities Committee was reconstituted as the Engineering, Technology and Industries (ETI) Committee, properly resourced and charged with advising Council on 'any matters concerned with engineering, technology and applied sciences in industry'. It would include individuals active in the Fellowship of Engineering in its membership. Angus Paton, recently appointed Chairman of IAC, became the first Chairman of the ETI Committee.[108]

That provided some of the infrastructure and signalled the Society's determination to remain active in engineering, and in applied science more generally. But it still had to work out, in practice, just what it could do that would play to its strengths and have significant impact while not damaging its relations with the new Fellowship of Engineering. In that sense it had not made much progress since 1963. The following decades saw several attempts to respond to that challenge, summarised in the following section.

Promoting the applications of science

The Society's first published corporate plan, in 1986, included a strategic objective to 'encourage scientific research and to promote its application'. Similar phrases featured in later plans. With impetus from the ETI Committee and its successor, the Technology Activities Committee, the Society set about highlighting the importance of basic science for technology and wealth creation, and exploring the policy implications – as did other organisations at that time. It organised a steady stream of lectures

---

[106] C/62(76).
[107] CM 1 April 1976, minute 12; CM 7 October 1976, minute 15; CM 16 December 1976, minute 8.
[108] CM 31 March 1977, minute 6. See also paper by Ronald Keay, 'Future of the Industrial Activities Committee': RMA31A.

and discussion meetings on these themes during the 1980s and 1990s. It explored the constraints on R&D in small manufacturing companies through a study of the agricultural engineering industry. It gave evidence to inquiries on engineering R&D, on the Science Base and innovation, on technology foresight and on similar matters, and carried out policy studies on topics such as energy policy, environmental pollution, industrial research associations, university/business links and intellectual property. It fostered relations not only with the Fellowship of Engineering but also with such bodies as the Confederation of British Industry. Between 1986 and 1994 the Society devoted one of its annual soirées, and the accompanying exhibition, to an exploration of the relations between science and industry – building subconsciously on a precedent set 50 years previously.[109] Overall, the Society now saw the applications of science as encompassing not only engineering but also much else relevant to the wealth-creating sectors of the economy.

The Society also took practical steps to promote the applications of science. Most notably, in 1980 it implemented an ETI proposal developed by Bob Feilden for a scheme of industry fellowships to allow academic scientists and engineers to spend up to two years in industry and industrial scientists and engineers correspondingly to spend time in academe. The scheme was funded jointly with the SRC. Geoffrey Allen had succeeded Sam Edwards as chairman of SRC in 1977, the year after he was elected to the Society, with a brief from Shirley Williams to develop industry–academe relations. Industry fellowships were one of a number of initiatives he supported, in close consultation with the Society's President Alex Todd.[110] Subsequently, additional funding was raised from other Research Councils and from industry, and by 2013 nearly 50 industry fellows were being supported, with plans for further expansion. The 200 industry fellows appointed since 1980, assembled into a virtual network, would become valuable allies in the Society's renewed engagement with its founders' practical ideals.

This experience encouraged the Society to broaden its approach. From the early 2000s, it collaborated with Imperial College Business School to offer a very popular course, 'Innovation and the business of science', to URFs and other research fellows to help them work more effectively with industry. The implicit message was that practical application could be the concern of any scientist, irrespective of their starting point.

---

[109] Interviews with Geoffrey Allen, Eric Ash, Peter Cooper, Sam Edwards and John Horlock. CM 7 March 1985, minute 10; Sally M. Horrocks, 'The Royal Society and industrial R&D', S38–9.

[110] OM/81(79); CM 18 October 1979, minute 43; CM 14 February 1980, minute 4.

The receipt of a major legacy from the entrepreneur Brian Mercer in 1999[111] created a timely opportunity for a fresh initiative by the Society. This took the form of a scheme of major (£250,000) and minor (£30,000) awards to allow researchers to test the technical and economic feasibility of their ideas at a relatively early stage in the development process when commercial funding was typically hard to obtain. In 2008 the Society launched a complementary Enterprise Fund with the aim of raising private donations and making equity investments in emerging science-based businesses, such that any ensuing profits would be recycled back to the Fund for further investment.[112]

The Society also invested considerable effort reviewing how its accolade function was operating, to ensure that those primarily involved with the applications of science were not inadvertently overlooked as potential candidates for election to the Fellowship. In 2013, it expanded from forty-four to fifty-two the maximum number of Fellows it elected annually, mainly in order to increase the number of Fellows with industrial and other experience related to application. But it did not leave the outcome to chance. Learning from previous initiatives, it established 'temporary nominating groups' to help identify potential candidates of the highest quality in key areas, this being seen as the main weakness in the system. The move was reminiscent of the Applied Science Candidates Committee that Florey established at a corresponding stage in 1964.

Another lesson from earlier experience was to invest significant effort in ensuring that key outside groups were aware of new initiatives the Society was undertaking. In December 2003 it held the first of what became an annual series of events – quickly dubbed 'Labs to riches' – showcasing recent award-winners to influential audiences drawn from the worlds of business and finance as well as science and technology.

During the 1990s and 2000s the Society's leadership continued to mull over how most effectively to translate rhetoric into specific and sustainable activity – motivated both by the obvious importance of the issue and by the threat to its own reputation, expressed many years earlier by Bill Penney and others, should it fail to make its mark as an influential champion of the applications of science. Aaron Klug used his 1999 Anniversary Address to argue that the Society's concern with this was a defining characteristic of its long history and not a sudden embracing of a new fashion. But the feeling persisted that the Society was

[111] See interview with Hugh Ford for how the legacy was brokered.
[112] Christopher Snowden, 'Technological innovation in industry and the role of the Royal Society', *Notes and records of the Royal Society of London*, 64 (2010), S62.

underperforming on this part of its remit, much as it had done in the period before the Fellowship of Engineering was established.

In 2013, the Society held a Year of Science and Industry, both to animate and reinforce its own work in this area and to draw public and political attention to excellence in UK industrial science. The Year covered a wide range of events. It confirmed the Society in its commitment to the applications of science, and prompted further reflection on how it could be most effective in this role. The upshot was establishment of a strong permanent committee of Council on science, industry and translation, led by two Fellows, Simon Campbell and Hermann Hauser, with extensive experience in the pharmaceutical industry and venture capital, respectively. The emphasis was on the long term, and the committee's first task was to develop a five-year programme of work to embed the applications of science more fully into the Society's culture.

There was now no ambivalence about the Society's aspirations to engage creatively with the applications of science, and there was a determination to test out how, with its particular characteristics and resources, it could best take its commitment forward. The Society recognised that technical innovation drew on knowledge from a very wide range of scientific disciplines, and that the term 'industry' was equally broad. Its ability to draw together individuals from highly diverse backgrounds, and to facilitate mutual communication, would be a correspondingly crucial asset. So, too, would its traditional skills of catalysing clearly focused funding schemes, advising on public policy and honouring outstanding achievement. Hinshelwood and Florey would have understood what their successors were attempting, and would have endorsed their ambition.

5   Defending the Science Base

> The primary purpose of scientific research is the pursuit of knowledge in its own right, in the well-founded expectation that knowledge brings rewards. It is not necessary to define the rewards in advance in order to give value to this pursuit of knowledge.[1]

As the independent voice of science in the UK, the Royal Society has been giving advice about science, to the government and others, for much of its existence. At the beginning of the twentieth century the President, William Huggins, devoted one of his Anniversary Addresses to the 'very large amount of work' undertaken by the Society in providing 'assistance freely given, at their request, to different Departments of the Government on questions which require expert scientific knowledge'.[2] He regarded this work as 'subordinate' to 'the strict prosecution of pure natural science as the primary purpose of the Society', and observed that it placed a considerable burden on the time and goodwill of the Fellows. But the Society, said Huggins, was regarded by the government as 'the acknowledged national scientific body, whose advice is of the highest authority on all scientific questions, and the more to be trusted on account of the Society's financial independence'. So it had no choice but to be involved in giving expert advice.

Some of the advice instanced by Huggins was about people: recommending suitable individuals for particular positions. But much of it was of a technical character: the India Office wanted to know how best to organise research work in India; the Colonial Office asked about tsetse fly disease and malarial fevers, and the Foreign Office about sleeping sickness; the Board of Trade needed advice on metric and imperial weights and measures; the Colonial Office asked about volcanoes after an eruption in the West Indies caused 30,000 deaths. Some of the responses drew on existing expertise, some required new research. All played to the

[1] Royal Society, *The future of the Science Base* (September 1992), 3.
[2] Huggins' 1904 Anniversary Address was reprinted in William Huggins, *The Royal Society, or science in the state and in the school* (Methuen, 1906), 61–91.

Society's commitment to public service alongside its 'strict prosecution of pure natural science'.

Such dealings with government were not entirely straightforward, even within the Fellowship. As Huggins pointed out, the effort involved significant opportunity cost for those giving their expertise. The notion that policy advice should be a primary objective of the Society was a key element of the controversy over the election of a new President in 1945: putting its scientific authority to work in this way was all very well, but possibly not at the cost of diminishing that authority. Yet the Society valued its close relations with government and its insider status. Howard Florey, a signatory of the 1945 petition, spent much of his 1960–5 presidency defending the Society's advisory niche against the proliferation of competing machinery. In 1962 he remarked wistfully to science minister Hailsham: 'Fifty years ago if the Government wanted advice, it went to the Royal Society for it, but . . . [now] it went more and more to its own advisory bodies.'[3] By 1964 he conceded that the Society could not compete directly with the 'efflorescence of Government-sponsored Councils, Committees, Boards and Authorities' (see Chapter 2), but he determined that it would not abandon its responsibilities to speak for science.[4] For Florey as for Huggins, the Society had an obligation to harness its scientific status to the task of influencing policy.

So over the half century from 1960, policy advice became core business for the Society, transformed in both scale and character. Increasingly, advice was given at the Society's own instigation as well as in response to formal requests. Increasingly, the advice was made public and the process of preparing it was conducted publicly. Increasingly, too, it was given not only on how to run science (the subject of this chapter) and on technical matters that the government needed to understand (as in Huggins' 1904 Anniversary Address), but also on matters that were far more controversial in the public domain and that attracted the attention of a far more diverse set of people (Chapter 6).

The policy advice business

I started work at the Royal Society in November 1981 as the first member of permanent staff to work full time on science policy. The decision to create such a post was the culmination of several decades of mounting conviction that the Society needed to raise its game in the business of advising on policy matters, and the start of several decades of growth in its advisory activities.

[3] OM/37(62).   [4] OM/73(64) and Howard Florey, 'Anniversary Address 1964', 449–51.

In the mid 1960s, to complement the existing machinery of personal contact with individuals in key policy-making or policy-shaping roles, responses to formal inquiries and occasional full-scale reports, Howard Florey had initiated the idea of study groups to monitor socially important issues with scientific dimensions. Such groups, comprising Fellows and other experts but with no staff support, allowed the Society proactively to set some of the policy agenda rather than simply wait to be asked for its views on topics selected by others. Over the following twenty years, thirteen study groups were appointed on an eclectic mix of topics, and produced some influential reports.[5]

Apart from the study groups, most of the Society's formal policy advice work during the 1970s – in the sense of written material approved by Council – consisted of letters to ministers and submissions to departmental or parliamentary enquiries. Todd was keen to see the Society develop this work, and harangued his fellow Officers on the need for the Society to be more influential in public life. When a new staff post of Deputy Executive Secretary was created in 1977, the specification called explicitly for policy experience.[6] Todd kept up the pressure in his final two Anniversary Addresses, arguing that the Society had a duty to exercise influence through its formal and informal connections with key bodies and individuals and through marshalling the scientific evidence relevant to major policy issues.

Immediately after relinquishing the Royal Society presidency at the end of 1980, Todd procured another pulpit as a leading member of a House of Lords inquiry into the provision and coordination of scientific advice to government. The Royal Society gave evidence to the inquiry, indicating, a touch diffidently, its willingness to ramp up its own advisory work if the demand were there. The diffidence stemmed in part from the resource implications of such a move: any significant increase in advice work implied additional staff and therefore additional costs. The inquiry's overall conclusion – to call for a strengthening of the scientific dimension in government as a whole – made that ramping up more likely.[7]

---

[5] Study groups were appointed on: population (1965–70), ritualisation (1965), postgraduate training in science (1965–9), non-verbal communication (1966–70), marine pollution (1967–75), planetary studies (1969–73), human biology in urban environments (1972–4), atmospheric pollution (1975–7), long-term toxic effects (1974–7), safety in research (1978–81), risk (1979–83), nitrogen cycle (1979–84) and environmental geochemistry and health (1979–83).

[6] Interviews with John Mason and Peter Warren.

[7] House of Lords, *Science and government* (Select Committee on Science and Technology, 1981), vol II, 86–7 and 97–8. See also interview with Peter Warren, and Arnold Burgen to Ronald Keay, 2 January 1981: HATN 8/24/3. William Hawthorne was a member of the group that prepared the Royal Society's evidence, and his papers include a full set of the associated correspondence.

What made it inevitable was the Thatcher Government's announce-ment in March 1981 of a major cut in the UGC block grants to univer-sities over the following three years (see Chapter 3). The Society immediately saw that it would have to monitor and respond to the con-sequences. The recruitment of specialised policy staff was put in hand, deepening the Society's long-term commitment to the policy advice business.[8]

The upshot was a marked increase in policy activity across a range of topics, mainly but not only related to the Science Base. There was also a change in style. The extra staff capacity made it possible to undertake quantitative analyses of particular issues. For example, in 1983 the Society published a report estimating the impact of demographic changes on future demand for undergraduate places, which undermined the government's own facile predictions that were being used to justify cuts in provision.[9] A second formative project came early the following year, when the ABRC commissioned the Society to undertake a quantitative assessment of the health of basic science in the UK (see below). This broke new ground for the Society in two ways: it involved a contract under which the ABRC covered the full costs of the project, and it introduced the Society to the emerging and somewhat fraught practice of biblio-metric analysis.

These experiences prompted the Society to form the existing policy staff into a distinct Policy Studies Unit from early 1985 and to double their number to six, funded from a mix of industrial donations, commis-sions and the Society's private funds.[10] The aim was to reinforce the Society's effectiveness in the policy advice business by enabling it to add systematic data to expert argument, especially in matters related to policy for the Science Base, where much of its immediate focus was expected to be.[11] Following discussion with the Fellowship of Engineering, it was agreed in 1986 to run the Unit as a joint undertaking called the Science and Engineering Policy Studies Unit (SEPSU). SEPSU produced ten major self-initiated reports on a range of policy

---

[8] OM 21–22 March 1981, minute 15.

[9] Royal Society, *Demographic trends and future university candidates* (April 1983); 'Royal Society warning on cuts', *Nature*, 303 (1983), 194; 'Government predictions awry', *Nature*, 306 (1983), 634; 'Where UK academics go next', *Nature*, 308 (1984), 481–2; 'Grants body answered', *Nature*, 308 (1984), 488; John S. Rowlinson and Norman H. Robinson, *The record*, 34.

[10] There is a fashion in these things. The ABRC set aside an annual budget from 1983–4 to commission policy studies, and appointed its own team of three professional staff to focus on policy work in 1986.

[11] OM/109(84); CM 8 November 1984, minute 14; OM/41(85); Royal Society, *Corporate plan 1986–1996*, 9, 15–16.

topics, from measurements of the brain drain to analyses of international collaboration and a review of experience in quality management. It also carried out investigations under contract to a considerable variety of bodies in the UK and Brussels. However, as the market for such work gradually became more competitive, too many of the available contracted projects were of a character that did not require the combined weight of two national academies. SEPSU was closed in 1993, and the Society focused its policy activities on its own priorities.

The Society's first published corporate plan, in 1986, signalled 'the importance it attached to using its independence and its scientific authority to advise policy-makers about science. Subsequent corporate plans and other reviews of strategy reiterated the message with increasing enthusiasm.[12] In 1990 Council set up a substantive Science Policy Section to work alongside SEPSU. Close observers such as former Chief Scientific Adviser Robin Nicholson reinforced the Society's ambitions, urging it to take a more proactive role on strategic scientific issues.[13] So it continued to experiment with different approaches. In 1992 it combined with its analogue bodies in engineering, humanities and medicine to form the National Academies Policy Advisory Group (NAPAG) to tackle policy issues requiring a wider range of expertise.[14] This initiative was encouraged by the new science minister, William Waldegrave, if not by the various permanent secretaries, and led to the production of four major reports.[15] After a few years, NAPAG was superseded by more ad hoc collaborative arrangements between the academies.

From the later 1990s, the range and complexity of the policy issues demanding the Society's attention seemed to increase. The new Labour Government, stimulated by its Chief Scientific Adviser Bob May, seemed receptive to independent advice.[16] The Society gradually expanded its resources for offering that advice. In 2004 members of the Society's

---

[12] For example, a review of strategy in October 1992: 'Science advice was at the heart of the Society's primary role', defined as 'that of an independent academy, speaking on behalf of UK science': OM/94(92). Or, from a 2007 fundraising brochure: 'It is now more important than ever that policymakers and the public are effectively informed by timely, definitive, and unbiased advice.'

[13] In conversation with Bill Stewart (his successor as Chief Scientific Adviser), David Phillips and others: meeting of 14 February 1991, D.C. Phillips papers, MS Eng. c.5504, O.82.

[14] Interviews with Walter Bodmer, Aaron Klug and Martin Rees.

[15] *Intellectual property* (March 1995); *Energy and environment in the 21st century* (July 1995); *Research capability of the university system* (April 1996); *Regulation of medical devices* (March 1998). On the permanent secretaries' generic resistance to open, independent advice, see interviews with Michael Atiyah and Bill Stewart.

[16] Robert May's *The use of scientific advice in policy making* (Office of Science and Technology, 1997) outlines his approach.

secretariat dealing with international relations were moved into the policy advice team to strengthen its capacity to address international issues, creating a section of over twenty staff dedicated to the Society's role of providing expert independent advice on scientific matters. In 2008, the section was renamed the Science Policy Centre to give it greater prominence, and a high-powered advisory group was instituted to help Council determine the Centre's priorities.

Organisation and funding of the Science Base

Closest to the Society's heart as a focus for its efforts to influence policy was the well-being of the Science Base. The 'Science Base' in this context meant scientific and technological research carried out in higher education institutions (mainly universities) and in the research institutes and facilities supported by the Research Councils and by bodies such as the medical research charities.[17] Science carried out in the laboratories of government departments or in industry was also of great concern, but the bulk of the Fellowship was drawn from the Science Base and it was there that the Society carried most weight.

The Society's view of how to get the most from the Science Base revolved consistently around identifying individual excellence and backing it with resources, opportunities and scope for creativity. So, for example, the Society championed the dual-support system for funding academic research; it advocated selective distribution of funds; it generally resisted top-down initiatives favouring newly fashionable areas of research; and it pushed, carefully, at the boundaries of public acceptability when new advances opened up new scientific possibilities. It understood the political reality that public funding had to generate public benefit, and argued its view as to how that benefit would best be secured. It also understood the political reality that, especially in the biological sciences, research methods had to have broad ethical agreement. It engaged strongly in associated public debate and took initiatives to institute voluntary, scientist-led, regulatory regimes, as a prelude, or sometimes as an alternative, to more formal regulation reinforced by legislation. There were many opportunities for the Society to promote its views on the Science Base in this period.

The publication of the Dainton and Rothschild reports in 1971 was a milestone in British postwar science policy. They essentially dealt with the overarching question, accentuated as the period of rapid budgetary expansion came to an end, of how best to organise the Science Base so as

---

[17] This was the definition used in *The future of the Science Base*.

to yield maximum practical benefit for the UK. Rothschild was also prompted by a proposal, formulated confidentially by an internal Civil Service committee, to move the Agricultural Research Council from DES to the agriculture ministry.[18] This had been preceded by discussions at the prime ministerial level about how far ministers should seek to influence the details of research spending and whether it was time to discard the iconic 'Haldane principle',[19] which interventionist-minded ministers saw as preventing them from focusing the Science Budget directly on useful research.[20]

The headline proposal in the Dainton report (effectively the work of the CSP[21]) was the creation of what eventually became the Advisory Board for the Research Councils (ABRC) to guide the distribution and use of the Science Budget. Rothschild's headline proposal, on the other hand, was to transfer money from three Research Councils (Agriculture, Medicine and Natural Environment) to their relevant lead government departments and, under the so-called 'customer/contractor principle', to allow the latter to commission research from the former. The two reports were seen as pitching scientific independence against administrative accountability.[22] Crisp and controversial, it was the Rothschild report that attracted most of the attention.

---

[18] The report of the committee of inquiry into food and agricultural research, chaired by Paul Osmond and submitted to the Prime Minister in May 1970, is at TNA PREM 13/3238. Note also CM 17 December 1970, minute 26.

[19] The 'Haldane principle' is a reference to the 1918 Report of the Machinery of Government Committee (Cd 9230), which argued that research other than that conducted by a government department for its own needs should be the responsibility of a minister who was 'immune from any suspicion of being biased by administrative considerations against the application of the results of research'. The phrase gained in the telling in later years, and came to mean that ministers should adopt a hands-off approach to research paid for by their departments. See, for example, Zuckerman to Trend, 30 October 1968: TNA CAB 168/218; and Philip Gummett, Scientists in Whitehall (Manchester University Press, 1980), 25.

[20] See exchanges between Solly Zuckerman, William Armstrong, Harold Wilson and others between October 1968 and August 1969: TNA CAB 168/218 and PREM 13/3238. Zuckerman, as government Chief Scientific Adviser, wanted to prohibit Harrie Massey, then Chairman of CSP, from raising the profile of the Haldane principle and reasserting the independence of the Science Base from ministerial interference. Zuckerman was overruled on tactical grounds by the Prime Minister, on the advice of William Armstrong (Permanent Secretary in the Civil Service Department) and Burke Trend (Cabinet Secretary). The Osmond inquiry was then set up as a more subtle approach to countering the CSP and strengthening ministerial control of research. It set the scene for Rothschild, who regarded Haldane as irrelevant.

[21] Dainton succeeded Massey as Chairman of CSP in 1970.

[22] Roger Williams, 'Some political aspects of the Rothschild affair', Science studies, 3 (1973), 31–46. See also Philip Gummett, Scientists in Whitehall, 195–202, and Tom Wilkie, British science and politics since 1945 (Blackwell, 1991), 78–89.

All this was of intense concern to the Royal Society, and prompted it to articulate its views on the Science Base. Patrick Blackett and Alan Hodgkin were members of the CSP during their respective terms as Royal Society President, and to that extent were involved in the preparation of Dainton's report – though constrained by official confidentiality in how they could use their inside knowledge. As soon as the Society was formally notified of the Rothschild inquiry, it established a strong committee.[23] The committee's first action, at the Prime Minister's prompting, was to meet with Rothschild. It was an exchange of views more than a meeting of minds, but served to spell out the issues at stake.[24]

The Dainton and Rothschild reports were published together on 24 November 1971 in a Green Paper with a request for comments within two months.[25] Alan Hodgkin was sufficiently agitated by Rothschild that he used his Anniversary Address the following week to kick off an unaccustomed consultation with the Fellowship about the Society's response, commenting that 'any country which has a reasonably satisfactory method of supporting science should think very carefully before dismantling it'.[26] The Society itself was not directly affected by Rothschild, in the sense that there was no proposal to shift any of its public funding to a 'customer' department, but its mission to represent the views of practising scientists in high-level debates about policy meant that it was energetically engaged. Over 100 Fellows sent in comments to Hodgkin, and 170 Fellows (about 25 per cent of all Fellows in the UK, a very high proportion) dissected the Dainton and Rothschild reports at a special meeting in early January 1972, appreciating both the importance of the issues and the unfamiliar experience of being asked by the President for their input to the Society's response.[27]

The Society was entirely supportive of the principle that 'the welfare of the nation, in short and long term, must be the major factor in determining the allocation of substantial public funds for science and technology', and told Rothschild so in the early stages of his review.[28] It was not a question of 'pure' science versus 'applied' science: it was about whether

---

[23] CM 17 December 1970, minute 26; CM 14 January 1971, minute 14; CM 1 April 1971, minute 6; CM 13 May 1971, minute 13.

[24] Paper RD/1(71) (minutes of 25 May 1971 meeting with Rothschild), in Royal Society Committee on the Government Research and Development Study, 7/3/4/6; CM 15 July 1971, minute 11.

[25] *A framework for government research and development* (Cmnd 4814, 1971).

[26] Alan Hodgkin, 'Address at the Anniversary Meeting, 30 November 1971', *Proceedings of the Royal Society of London. Series B, Biological sciences*, 180 (1972), ix–x; CM 30 November 1971, minute 6.

[27] OM/19(72), OM/23(72); Roger Williams, 'Rothschild affair', 43.

[28] CM 15 July 1971, appendix C, repeated in the Society's formal response to Cmnd 4814 the following year.

these were helpful categories for organising the nation's scientific effort, and about how to optimise national performance across the spectrum. Theo Williamson, a member of the Society's Industrial Activities Committee (see Chapter 4), briefed Hodgkin about that committee's 'round condemnation' of Rothschild's proposals. But he applauded the Green Paper insofar as it represented 'the official recognition in the UK of a world trend to couple science more effectively to the social and economic needs of the community'.[29] Williamson urged Hodgkin to be constructive about that objective, even as he deprecated Rothschild's proposed means. Hodgkin readily took his advice.

The Society's formal response to the Green Paper criticised Rothschild as 'a misleading oversimplification', prepared in undue haste with inadequate consultation.[30] 'Oversimplification' applied both to Rothschild's assumptions about how customer/contractor relations really worked in industry and to his binary pure/applied analysis: the Society greatly preferred Dainton's threefold taxonomy of pure/strategic/applied. It extolled the track records of the three Research Councils at the heart of Rothschild's reforming proposals. Some of their work was indeed applied and could, possibly, be funded directly by relevant departments, but it was a much smaller proportion than Rothschild had arbitrarily determined. On the other hand, the Society warmly welcomed Rothschild's idea of placing chief scientific advisers at senior levels in all relevant government departments and ensuring they were able to exercise genuine influence.

The whole debate was conducted in public to an unprecedented extent for such a relatively arcane topic,[31] and the Society made sure that its views were circulated to the press as well as to all interested parties. Mrs Thatcher, who as Secretary of State at the DES was the Cabinet minister with the most to lose from Rothschild, was broadly sympathetic to the Society's approach.[32] In the end she was successful in defending the Science Budget to the extent of securing a significantly smaller transfer to customer departments than Rothschild had proposed. Hodgkin made conciliatory noises in his 1972 Anniversary Address. Given that the

---

[29] D.T.N. Williamson to Alan Hodgkin, 12 January 1972: OM/11(72).

[30] Royal Society, *Memorandum by the Council on the consultative document (Cmnd 4814)* 'A framework for government research and development' (February 1972).

[31] Rothschild complained later that the scientific community, with few exceptions, had been hostile to his report, had largely misunderstood it, and had mounted a 'major, well-organised public campaign' against the recommendations dealing with the Research Councils. Rothschild to Douglas Allen (Treasury Permanent Secretary), 13 January 1972: TNA T 224/2469.

[32] As she told Hodgkin and others on 3 March 1972: OM/33(72). Also Jon Agar, 'Thatcher, scientist', *Notes and records of the Royal Society of London*, 65 (2011), 215–32.

government appeared to have made up its mind well in advance, the outcome of the consultation was, he thought, not as bad as it might have been. He pledged the Society to continue being 'a strong voice for the importance of pure science, both for national prosperity and as one of the great intellectual activities of this country'.[33]

The Society's public defence of the Science Base and of the importance of fundamental research involved both rhetoric and practical action. Both were stepped up very considerably in the wake of the 1981 cuts in university funding. The practical action is described in Chapter 3. One thread running through the rhetoric was a strong commitment to the dual-support system, under which the core infrastructure for academic research was funded by the UGC via block grants to each university and additional project support came from the Research Councils. This provided a crucial element of flexibility, especially at the early stages of research. The balance between these two funding streams was disrupted by the 1981 cuts and by endless government fretting over exactly how the UGC block grants were used by universities. The Society repeatedly attacked government claims that it had safeguarded scientific research by safeguarding the Science Budget (essentially, the money going to the Research Councils) when, at the same time, it was eroding the UGC budget from which the greater portion of university funding for research was derived. A survey of Fellows documented the consequences of the erosion, and the Society made sure it received due publicity.[34]

The 1981 cuts were widely perceived in practice to have fallen most heavily on the research function of universities. The Society in November 1982 therefore called for the research element in block grants to be explicitly earmarked in order to safeguard the dual-support system.[35] A year later, the UGC under its new chairman, Peter Swinnerton-Dyer, and with support from the DES, took up this theme in the context of funding research more selectively[36] – a policy that the Society strongly and publicly endorsed. Under Ted Parkes, the UGC had opposed earmarking of

---

[33] Alan Hodgkin, 'Address at the Anniversary Meeting, 30 November 1972', *Proceedings of the Royal Society of London. Series B, Biological sciences*, 183 (1973), 9–11. Three years later, Hodgkin concluded from detailed analysis of the new arrangements in practice that the customer/contractor approach was 'self-defeating': Alan Hodgkin, 'Anniversary Address 1975', 381–8.

[34] Andrew Huxley, 'Address at the Anniversary Meeting, 30 November 1981', *Proceedings of the Royal Society of London. Series B, Biological sciences*, 214 (1982), 149–50; *Nature*, 300 (9 December 1982), 473; Jon Turney, 'Science suffering says FRS survey'.

[35] See, for example, the Society's comments on the 1982 ABRC/UGC report *Support of university scientific research* (the Merrison report).

[36] Richard Bird (Deputy Secretary at DES) to Peter Swinnerton-Dyer, 20 October 1983: D.C. Phillips papers, MS Eng.c.5501, O.58; UGC Circular Letter 16/83, 1 November 1983; interview with Peter Swinnerton-Dyer.

elements within the block grant on the grounds that it diminished the autonomy of individual universities, but Swinnerton-Dyer took a different line. He publicised the general formula used by the UGC to calculate the block grant allocated to each university, and announced the UGC's wish to associate a discriminatory quality factor with the term in the formula representing core funding for research. This ushered in, in 1986, the first of the Research Assessment Exercises that were to become an elaborate and unloved feature of British academic life. In 1987 the ABRC took the logic a step further with its controversial RTX proposal, that universities be categorised (and funded) according to whether they carried out teaching and substantial research across the full range of fields, or whether they offered teaching to MSc level only without advanced research facilities, or whether they offered a broad range of teaching and some, more specialised, research. However, that was pushing selectivity too far, though it took the government well over a year to reject it. The Royal Society thought it inappropriate to apply RTX to whole universities, but could see a case for applying it to individual university departments – which in effect is what the Research Assessment Exercise did.[37]

David Phillips stepped down as Biological Secretary of the Royal Society in April 1983 in order to succeed Alec Merrison as Chairman of ABRC. The Secretary of State at DES, Keith Joseph, signalled to Phillips that he, and probably the Prime Minister, would be willing to go out on a limb in support of the Science Base generally, in exchange for dispassionate evidence about problems and opportunities and increased flexibility in a range of management practices.[38] Encouraged by this, Phillips quickly invited Ron Mason to document how government departments had cut back the amount of research they were commissioning from Research Councils to the lowest level since the Rothschild transfer was effected.[39] Then, early in 1984, Phillips began expressing concern about the state of basic science as a whole. The familiar mantra, that Britain was world class at basic research but lagged in the business of exploiting its

[37] ABRC, *A strategy for the Science Base*, 1987; CM 30 November 1987, minute 5. See also David Phillips, 'A strategy for science in the UK', *Science and public policy*, 15 (1988), 3–12, for his subsequent reflections, and interviews with John Ashworth and John Kingman. A Royal Society report under Ron Oxburgh calling for academic geophysics to be concentrated in a small number of centres of excellence was in some ways a precursor of this: Royal Society, *Support of geophysics in the United Kingdom* (June 1985).

[38] Meeting between Joseph and Phillips 13 July 1983 and various associated exchanges: D. C. Phillips papers, MS Eng.c.5528, O.237 and MS Eng.c.5517, O.177 and O.178.

[39] ABRC, *A study of commissioned research* (Mason report, 1983); Clive Booth, 'A war of independence', *Times higher education supplement* (9 December 1983), 16. Mason had just completed a term as Chief Scientific Adviser to the Ministry of Defence.

results, might be wrong: perhaps Britain was lagging in basic research as well. He invited the Royal Society to study whether the evidence objectively supported the perception of a 'decline in basic science'.[40]

The study was carried out by SEPSU. It produced consistent evidence of a decline in the UK's global position in physics as a whole and in most sub-fields of physics, but a considerably stronger performance in genetics.[41] Another SEPSU study, on the movement of skilled researchers in and out of the UK, provided clear quantitative evidence about a phenomenon that was shown to be more nuanced and less dramatic than implied by the plentiful pessimistic rhetoric then current.[42] But there was enough anecdote and enough circumstantial evidence in the mid 1980s to fuel a sense of British science in decline[43] – so much so that on 13 January 1986 a campaigning group, Save British Science, was launched, via a half-page advertisement in *The Times* paid for by 1,500 individual contributors, to lobby for increased funding for basic research. The initiative secured the support of 100 Royal Society Fellows and an equal number of MPs. The Royal Society itself, sympathetic to the cause but institutionally uncomfortable with that style of public campaigning, and upstaged in public profile, kept a careful distance.[44]

The government policy of greater selectivity, between institutions and, to some extent, between fields of research, implied a need for greater central control. Selectivity would not happen spontaneously. By the end of 1986 this was well recognised within government. The Chief Scientific Adviser, John Fairclough, wondered about enhancing the role of ABRC and giving it some executive responsibility. The DES Permanent

---

[40] David Phillips to Brian Flowers, 12 January 1984: D.C. Phillips papers, MS Eng.c.5528, O.237; Phillips to Huxley, 22 February 1984: OM/35(84); CM 12 April 1984, minute 22; CM 14 June 1984, minute 9; Andrew Huxley, 'Address at the Anniversary Meeting, 30 November 1984', *Proceedings of the Royal Society of London. Series B, Biological sciences*, 223 (1985), 414–5.

[41] P.M.D. Collins, D.M. Hicks & S. Wyatt, *Evaluation of national performance in basic research* (ABRC Science Policy Studies No1, 1986); paper ABRC(FL)(86)32. See also D.C. Smith, P.M.D. Collins et al., 'National performance in basic research', *Nature*, 323 (1986), 681–4.

[42] SEPSU, *Migration of scientists and engineers to and from the UK* (SEPSU Policy Study No 1: Royal Society, 1987); P.M.D. Collins et al., 'Flows of researchers to and from the UK', *Nature*, 328 (1987), 27–8; David Fishlock, 'A trickle not a flood', *Financial Times*, (30 June 1987), 1 and 21.

[43] John Irvine, Ben Martin et al., 'Charting the decline in British science', *Nature*, 316 (15 August 1985), 587–90; Ben Martin, John Irvine et al., 'The continuing decline of British science', *Nature*, 330 (12 November 1987), 123–6. There is, of course, a fine tradition of declinist literature going back to Charles Babbage's 1830 *Reflections on the decline of science in England*.

[44] Booklet produced to mark the 20th anniversary of SBS; OM 6 March 1986, minute 3(w). The Society's relatively low profile in public debate was still troubling Council two years later: CM 17 December 1987, minute 9, CM 14 January 1988, minute 13.

Secretary, David Hancock, wanted to use ministerial power to force the UGC, ABRC and Research Councils to decide explicitly which research fields and research institutions would, and would not, receive major funding.[45] One move in that direction was to morph the UGC into the Universities Funding Council in 1988, with a senior industrialist (Henry Chilver) as Chairman and a significantly more top-down approach in its relations with universities. The ABRC was also reconstituted, in 1990, as a 'smaller, more authoritative, body with a specific remit to promote greater harmonisation of the Research Councils' activities'.[46] It became more tightly bound to the machinery of government, and the Royal Society lost its treasured right to nominate an independent assessor to it.[47] ABRC itself was dissolved with the implementation of the 1993 White Paper (see below).

The Royal Society President, George Porter (Figure 5.1), was all in favour of the best scientists getting most resources, but like many scientists he was distinctly wary of top-down direction about what science should be done, and therefore did not want to concentrate executive power in a few hands. Initiatives such as Interdisciplinary Research Centres (IRCs) or funding programmes in specific fields appeared to be politically necessary devices for demonstrating 'relevance' and thus safeguarding the budget. But they also looked dirigiste, and were liable to be funded at the expense of responsive-mode budgets, so the Royal Society generally set its face against them unless they came with new money and thus no opportunity cost for other activities.[48] The Society believed strongly that, so far as basic science was concerned, the best way to secure national benefit was to give individual researchers as much freedom as possible to set their own agendas.[49] Porter used his 1987 Anniversary Address to beat the drum for responsive-mode funding, where money was allocated in response to proposals from researchers based solely on their own ideas rather than externally mandated priorities: 'What is needed is a

[45] D.C. Phillips papers, MS Eng.c.5528, O.224.

[46] David Phillips to John MacGregor (Secretary of State, DES), 15 November 1989: copy at RMA1077.

[47] Correspondence between George Porter, John MacGregor (Secretary of State at DES), David Phillips and others, January–March 1990: RMA1077.

[48] To the vexation of John Fairclough. See Roger Elliott (Physical Secretary of the Royal Society) to Fairclough, 15 March 1988, Elliott to David Phillips, 25 April 1988: D.C. Phillips papers, MS Eng.c.5519, O.193, and interview with Elliott. By 1989 the ABRC had stopped prioritising IRCs in its spending strategy.

[49] But the Society did make some exceptions. For example, in 2006 it argued that responsive-mode funding could not deliver adequately on nanoscience and on that occasion it called for more directed funding. Royal Society and Royal Academy of Engineering, 'Nanoscience and nanotechnologies: opportunities and uncertainties': two-year review of progress on government actions (October 2006).

Figure 5.1 George Porter. © The Royal Society

clear commitment from the Government to support . . . without direction
from above, the individual effort of those who seek to improve natural
knowledge.' This led to some anguished exchanges with John Fairclough,
one of the main champions of IRCs, eventually mollified to some extent

over lunch at the Athenaeum between the two of them and David Phillips.[50]

In May 1988 Phillips met with the Royal Society Council for a very full discussion of issues facing the Science Base.[51] These included related concerns such as Porter's view that Research Councils tended to favour their own in-house institutes over supporting work in universities, and the hazards of funding 'big science' from the same overall budget as responsive-mode grants.[52] The challenge for science policy was to achieve a workable balance between securing maximum funding for scientific research from both public and other sources (which required alertness to the objectives of the funders), maximising national benefit from that funding, dealing with the seemingly inexorable rise in the cost of doing research, and at the same time maximising scope for individual creativity. The Society's Council recognised the complexities of such a balancing act, and appreciated Phillips' frankness, but it continued to push the importance of responsive-mode funding when it thought that the balance had tilted too far away from it.[53] Experience with its newly reinstated Research Grant Scheme (Chapter 3) provided useful quantitative ammunition.

Two years later, in September 1990, the Society was again badgering David Phillips about responsive-mode funding. Phillips tried to offer reassurance that the package of funding mechanisms then in place was in fact a reasonable balance. The Society's commitment to responsive funding was coupled with a long-established aversion to identifying fields of high priority, for fear they might become the focus of top-down initiatives,[54] and at this meeting George Porter stayed true to form by not engaging with an ABRC move to identify the most exciting and potentially rewarding areas of science. ABRC members expressed some sympathy for his position.[55]

---

[50] John Fairclough to George Porter, 10 December 1987; Porter to Fairclough, 17 December 1987: David Phillips papers, MS Eng. c.5519, O.193.

[51] CM 11 February 1988, minute 14; special Council meeting, 3 May 1988.

[52] Project overruns and foreign exchange fluctuations in areas like astronomy and particle physics tended to threaten flexible parts of the budget, and on one occasion consumed the whole of the ABRC flexibility margin.

[53] For example, in a submission to the UFC in October 1989 about research policy. Also interview with Tony Epstein.

[54] See, for example, Royal Society, *Encouragement* (1960), where a list of important fields needing a stimulus is prefaced with the disclaimer that it is not a priority list; and the Society's very wary dealings with an ABRC study on 'growth points in science' (CM 10 February 1983, minute18) and with Charles Reece's study for ACARD of 'promising areas of science' (CM 1 March 1984, minute 16).

[55] Note of 13 September 1990 meeting: RMA1077; ABRC minutes, 24 October 1990. Also CM 4 October 1990, minute 35, and interviews with Michael Atiyah and Bob May.

At the end of Porter's presidential term, the Society established what it called its 'Science Inquiry', investigating factors likely to affect the well-being of UK science over the coming decade and considering their policy implications.[56] The Inquiry was led by the incoming President, Michael Atiyah. It became an outlet for the frustrations of the previous decade and a focus for hopes of improvement. Over 300 individuals and organisations submitted written evidence to it, and all five Research Councils and the UFC also gave oral evidence. The ensuing report, *The future of the Science Base*, set out the Society's vision of science as a human activity of intrinsic value, generating knowledge about the natural world that brought both tangible and intangible benefits. Science was central to wealth creation in advanced industrial nations; it was also central to meeting the increasing pressures for improved quality of life. Such benefits would accrue to the UK only if the UK was actively engaged in research, and thus tied into the key scientific networks in Europe and throughout the world. Research was full of surprises: management structures had to be able to accom-modate the unexpected. The dual-support system therefore needed strengthening, with a significant shift to funding more independent researchers of the URF type. Towards the basic end of the research spectrum, grant funding should so far as possible be allocated on a responsive-mode basis.

The future of the Science Base was launched at a crowded press confer-ence on 1 October 1992. The Major Government had by then embarked on its own review of science policy, which would lead to the most sig-nificant restructuring of British science since the Rothschild and Dainton reports. The Society made sure that William Waldegrave, Major's science minister,[57] had an early copy of its report to help shape his thinking.

The government's White Paper *Realising our potential* followed in May 1993. It, too, spoke of wealth creation and the quality of life. It was not content to leave these to chance, though it recognised the dilemma involved:

The Government does not believe that it is good enough simply to trust to the automatic emergence of applicable results which industry then uses. Nor, on the other hand, does it believe that scientists should or could be told, from above, what to work on in order to generate relevant and industrially applicable results.[58]

---

[56] CM 4 October 1990, minutes 13 and 34; CM 13 December 1990, minute 11; CM 14 February 1991, minute 15; *Nature*, 349 (17 January 1991), 183; interviews with John Enderby and Bob May.

[57] The full title of his post, which carried a seat in Cabinet, was Chancellor of the Duchy of Lancaster and Minister of Public Service and Science, but science was his main focus. Mrs Thatcher had previously spoken for science herself at Cabinet level.

[58] *Realising our potential* (Cm 2250, 1993), 2. See also interviews with Bill Stewart, William Waldegrave, Peter Warren and Helen Williams.

But the whole thrust of the White Paper was to put the 'user community' (broadly analogous to Rothschild's 'customer') centre stage, for example in the management of Research Councils and in a new prominence given to engineering, technology and innovation. Quoting Michael Atiyah, the White Paper acknowledged the importance of nurturing the most creative individuals and providing for responsive-mode funding, though always within the limits of financial constraint. There was a cautiously friendly welcome, or at least acceptance, for the White Paper's approach.[59]

As in its dealings with Rothschild, the Royal Society endorsed the principle that scientific research funded by public money should serve publicly determined objectives, and it appreciated the clear statement of those objectives in the White Paper.[60] But it was moved to comment again on the process through which public benefit was most likely to be secured. As well as highly trained scientists for employment throughout the economy, what basic research produced was surprises, and it had to be organised accordingly. Moreover, basic research underpinned all areas of science, not just those furthest from obvious application. A long-term perspective was essential. The Society was encouraged by the White Paper's efforts to deal with science across the whole of Whitehall, and by its recognition of the need for an international perspective in research.

In the years that followed, the Society continued to promote the case for the Science Base and to oppose measures likely to weaken its ability to contribute to national well-being. For example, largely in order to protect the basic research that they carried out, the Society strongly and in the main successfully opposed repeated government attempts to privatise public sector research establishments during the mid 1990s.[61] It worked with government's review of the future of higher education in the early 2000s, and on numerous other inquiries into aspects of how the Science Base was working. And in 2010, it produced an influential report, chaired by Martin Taylor, on the long-term future of the Science Base called *The scientific century: securing our future prosperity*.

In that report the Society stressed, again, the importance for the future competitiveness of the UK of maintaining public investment in the Science Base. Support for excellent individuals was central to this: the report called for a higher proportion of Research Council expenditure to be assigned to responsive-mode ('investigator-led') funding, and for an increase in longer postdoctoral fellowships. The report also attached high importance to

[59] *Nature*, 363 (3 June 1993), 381–2.
[60] Peter Collins, 'Realising our potential', *Science and public affairs* (Summer 1993), 5–7.
[61] Aaron Klug, 'Address at the Anniversary Meeting, 29 November 1996', *Notes and records of the Royal Society of London*, 51 (1997), 124–6; interviews with Michael Atiyah, Aaron Klug, Bill Stewart and William Waldegrave.

strengthening the UK's international scientific connections, and to strengthening the position of departmental chief scientific advisers. This restatement of the Society's long-held views on the Science Base, its arguments quotably expressed and underpinned by extensive factual evidence, proved timely. Widely reported, and buttressed by numerous private exchanges with key government figures, it helped protect science from the worst of the cuts in public funding that the new government imposed after the May 2010 General Election.[62] The Chancellor of the Exchequer, George Osborne, chose the Society as the venue for a major speech in November 2012 in which he stressed the value of science as a driver of the economy and his personal commitment to securing its long-term funding. The Prime Minister, Tony Blair, had made similar points in a speech at the Society ten years previously. In April 2013, the Presidents of the Academy of Medical Sciences, British Academy, Royal Academy of Engineering and Royal Society issued a joint statement, *Fuelling prosperity*, to reinforce the political momentum behind sustained support for research in the effort to 'transform society, revitalise the economy and improve our health and wellbeing'.

Public acceptability of research methods

Defending the Science Base involved more than dealing with Whitehall infighting, making the case for public funding and arguing about how funds should be used, important as it was to get such matters right. It also involved dealing with public concern about how science was done. In the late 1960s and early 1970s, scientists and non-scientists alike became increasingly alert to ethical issues raised by scientific research. Social responsibility became the watchword. The British Society for Social Responsibility in Science was launched at (though not by) the Royal Society in April 1969.[63] In its early days it included a good number of Fellows among its members as scientists sought to give a lead in openly establishing and implementing ethical guidelines.

[62] Richard van Noorden, 'Royal Society sets out case for investment in research', *Nature*, 464 (9 March 2010), 155; Geoff Brumfiel, 'UK scientists celebrate budget reprieve', *Nature*, 467 (27 October 2010), 1017; Martin Rees, 'Address at the Anniversary Meeting, 30 November 2010', *Notes and records of the Royal Society of London*, 65 (2011), 200–3; interview with Martin Taylor.

[63] 'More about social responsibility', *Nature*, 221 (29 March 1969), 1190; 'Public and private responsibility', *Nature*, 222 (26 April 1969), 320; 'Welcome to the BSSRS!', *New scientist*, 42 (17 April 1969), 101; Hilary Rose and Steven Rose, 'Knowledge and power', *New scientist*, 42 (17 April 1969), 108–9; Struther Arnott, T.W.B. Kibble and Tim Shallice, 'Maurice Hugh Frederick Wilkins', *Biographical memoirs of Fellows of the Royal Society*, 52 (2006), 474; Hannah Gay, *The Silwood circle* (Imperial College Press, 2013), 13.

One focus for public concern was genetic engineering. The ABRC appointed a working group in autumn 1974, with advice from the Society about suitable members, to assess potential benefits and hazards. Its work was completed by the end of the year, and published by the DES to stimulate debate. The Society remarked on the 'thoroughly responsible concern among scientists themselves in the matter', and endorsed ABRC's proposal for guidelines on good practice provided they were drafted by knowledgeable experts and applied flexibly. The Society subsequently corresponded with a DES group appointed to prepare the guidelines, judging its final draft to be a workable 'combination of scientific freedom and public safety'.[64] Later it objected strongly to regulatory proposals from the Health and Safety Commission that it regarded as unduly sweeping and restrictive.[65]

Interest in, and concern about, genetic engineering were widespread. The new European Science Foundation set up a working group, with Royal Society representation, to study it. More famously, a group of geneticists and others met at Asilomar, on the Californian coast, in February 1975, to draw up guidelines for the safe use of recombinant DNA technology. In California as in London, voluntary self-regulation was seen as preferable to government-imposed regulation. The intersection of voluntary regulation and government financial support was the challenge for the Spinks biotechnology committee in 1978. The topic was given additional prominence by a long-planned and, in the event, well-timed scientific meeting at the Society on new horizons in industrial microbiology. The Spinks report, co-sponsored by the Society, shaped the development of biotechnology in the UK during the 1980s.[66]

The use of live animals in research was another example of scientific process where public opinion was divided and where the Society contributed actively to the debate. Its objective, again, was to secure a regulatory regime that would command sufficient public support for research to continue.

The Society was strongly represented on the Advisory Committee for implementation of the 1876 Cruelty to Animals Act, and the Society's President was one of the few individuals with designated power to

---

[64] CM 10 October 1974, minute 32; CM 13 February 1975, minute 17; CM 17 April 1975, minute 13 and Appendix A; CM 12 February 1976, minute 6; CM 4 March 1976, minute 14. The virologist David Tyrrell, then on Council, was particularly active in these initiatives.

[65] CM 7 October 1976, minute 12; CM 11 November 1976, minute 14.

[66] Robert Bud, *The uses of life: a history of biotechnology* (Cambridge University Press, 1993), 158–60, 175–8; Robert Bud, 'From applied microbiology to biotechnology: science, medicine and industrial renewal', *Notes and records of the Royal Society of London*, 64 (2010), S17–29; OM 30 November 1978, minute 3(c); interviews with John Ashworth and Peter Warren.

authorise licences for animal experiments.[67] When Lord Halsbury, President of the Research Defence Society (RDS), introduced an ultimately unsuccessful bill in 1979 to update the 1876 Act, the Society found itself uneasy about the RDS's forthright approach, and wary of getting trapped on what was clearly an emotive issue on which Fellows held a variety of opinions.[68] It therefore established its own committee to deal with the Halsbury Bill, with analogous European legislation, and with other policy issues to do with animal experiments including ethical concerns. This became a standing committee in 1983, chaired initially by Andrew Huxley (Figure 5.2). One of its early tasks was to produce guidelines on the care of laboratory animals in collaboration with the Universities Federation for Animal Welfare, which were subsequently adopted in the Home Office code of practice.[69]

Although the Society had not been secretive about its position on animal experiments and the advice it was giving policy-makers, it had not particularly gone out of its way to attract public attention on the matter. But in October 1983 the Home Office minister David Mellor urged scientists to do more to make the case in public for animal experiments, and Andrew Huxley used his Anniversary Address the following month to do just that.[70] Opposition to animal experiments continued, of course, becoming in some quarters more personal and more violent.[71] From 1992 the Society actively supported the work of the Boyd Group, formed that year to bring together a wide spectrum of opinion for reasoned discussion. The Society's Animal Experiments Committee monitored legislative and other developments, and prepared submissions to government and parliamentary inquiries as needed. In 1999, the Committee drafted guidelines for research involving animals, which everyone supported financially by the Society or publishing in its journals would be required to follow. In 2002 the Society published a formal statement of its position on the use of animals in research, justifying the practice in terms

---

[67] By 1979, the Society regarded this as no longer appropriate (CM 30 November 1979, minute 6); the role was eventually removed in the 1986 Animals (Scientific Procedures) Act.

[68] OM 6 November 1975, minute 2(i), and correspondence at D.C. Phillips papers, MS Eng. c.5474, L.14.

[69] CM 17 May 1979, minute 16; CM19 July 1979, minute 17; CM 18 October 1979, minute 18; CM 8 October 1981, minute 8; CM 14 July 1983, minute 6; CM 12 June 1986, minute 11; John S. Rowlinson and Norman H. Robinson, *The record*, 30; interview with Andrew Huxley.

[70] CM 13 October 1983, minute 17; Andrew Huxley, 'Address at the Anniversary Meeting, 30 November 1983', *Proceedings of the Royal Society of London. Series B, Biological sciences*, 220 (1984), 385–94. Also interviews with Robert Hinde and Andrew Huxley.

[71] Interviews with Patrick Bateson, Colin Blakemore, Walter Bodmer, David Sainsbury and Martin Taylor.

Figure 5.2 Andrew Huxley at the staff Christmas party after collecting his OM, December 1983. © Peter Collins

of benefits derived and stressing the need to take all possible measures to minimise suffering. It also published guidance for scientists, setting out in measured detail the case for animal experiments and describing the regulatory environment within which such work took place. Following oral evidence from the Biological Secretary Patrick Bateson and the former Foreign Secretary Anne McLaren, the Society's position was reflected in an influential 2002 report by the House of Lords Select Committee on Animals in Scientific Procedures. And in 2006 the Society collaborated with the Academy of Medical Sciences, the MRC and the Wellcome Trust in a report, led by David Weatherall, on the use of non-human primates in research, which sought to give a balanced analysis of a particularly contentious area of animal research.

A third controversial research practice that attracted sustained Royal Society intervention was in vitro fertilisation (IVF) and associated technologies. This was important scientifically, clinically and commercially, and depended on public acceptance. The Society was involved in advising government about the implications of the science, supporting the development of regulations to keep research practice demonstrably within ethical bounds, and engaging with public concern.

The trigger was the Warnock Inquiry on human fertilisation and embryology, set up in 1982 to develop principles for the regulation of IVF and embryology. The Royal Society submitted evidence to the Inquiry, setting out the relevant elements of human embryology and explaining some potential research objectives. It took a relatively liberal position on the ethics, arguing for an absolute distinction between IVF embryos intended for reproductive purposes and those destined for research use and, for the latter, flexibility to keep the embryos for longer than the implantation stage suggested by the MRC. The Warnock Report accepted the Society's case for research and, by a majority, proposed to allow research on a developing human early embryo until just before the appearance of the primitive streak, that is to fourteen days from fertilisation.[72] In later exchanges with Mary Warnock and with government departments working on the implementation of her proposals, the Society argued vigorously that limits to research on embryos should not be specified in inflexible legislation but should be left to a licensing authority able to adapt to scientific and social developments.[73]

---

[72] This outcome owed a good deal to Anne McLaren, later the Society's Foreign Secretary: Rose Morgan, *The genome revolution* (Greenwood Press, 2006), 77.

[73] CM 3 March 1983, minute 15; 'Embryology needs rules, not new laws', *Nature*, 302 (28 April 1983), 735–6; Tom Beardsley, 'Human embryo experiments: societies urge a softer line', *Nature*, 302 (28 April 1983), 739; CM 8 November 1984, minute 11; CM 17 January 1985, minute 10. Interviews with Brian Heap, Aaron Klug, David Sainsbury.

The Society's contributions to the IVF debate were led throughout by the embryologist Richard Gardner, and in summer 1987, as key parliamentary debates loomed, he urged the Society to take a more prominent role in persuading the public about the case for IVF and related research.[74] The Society therefore prepared a simple summary fact sheet about the science and sent it to the press and all parliamentarians; it held a seminar for science writers and journalists; it invited leading parliamentarians and others to a policy forum and dinner; and it held private meetings with individual ministers. Between autumn 1989 and spring 1990, ahead of a series of further parliamentary debates, the Society undertook a sustained programme of similar lobbying and briefing initiatives with the aim of defending the legal right to carry out embryo research under carefully defined conditions. There appeared to be a real chance that such research could be banned, but in the end both Houses of Parliament voted in favour of research up to the 14-day limit, and in favour of the Human Fertilisation and Embryology Authority to regulate it.[75]

The birth of a healthy lamb called Dolly, derived through IVF from the nucleus of an adult mammary cell, was formally announced by Ian Wilmut on 27 February 1997.[76] This was in some ways the culmination of work first published by John Gurdon in 1962, for which he finally shared a Nobel Prize in 2012.[77] It was a matter of intense scientific interest, since it demonstrated that it was possible to re-programme a highly differentiated cell and use it to produce a viable clone. It was a matter of intense commercial interest, for the same reason. And it was a matter of intense public interest, since it opened up all sorts of ethical issues, especially if the technique could be made to work with humans. Debate started immediately, with various interest groups staking out their grounds, political leaders calling for moratoriums on research into cloning and scientists arguing against overhasty action.[78] The Royal Society, like other organisations, set up a committee to think it through carefully, and in early 1998 published a widely disseminated briefing, *Whither*

[74] OM 14 May 1987, minute 3(l); CM 11 June 1987, minute 13; CM 9 July 1987, minute 15.

[75] CM 5 October 1989, minute 10; CM 2 November 1989, minute 10; CM 14 December 1989, minute 3; CM 8 February 1990, minute 15; CM 17 May 1990, minute 17; CM 13 December 1990, minute 10.

[76] Ian Wilmut et al., 'Viable offspring derived from fetal and adult mammalian cells', *Nature*, 385 (27 February 1997), 810–3. The story had been leaked the previous weekend, so the furore had already started.

[77] Ruth Williams, 'Sir John Gurdon: godfather of cloning', *Journal of cell biology*, 181 (2008), 178–9.

[78] Ehsan Masood, 'Cloning technique reveals legal loophole', *Nature*, 385 (27 February 1997), 757; Declan Butler and Meredith Wadman, 'Calls for cloning ban sell science short', *Nature*, 386 (6 March 1997), 8–9.

cloning?, prepared by a group chaired by Brian Heap. This explained the science of cloning and the scientific significance of Dolly, and discussed potential therapeutic benefits that might be secured in the long term. It echoed the general consensus against reproductive cloning of humans[79] but argued that human cloning for research purposes raised no new ethical issues not already covered under the 1990 Human Fertilisation and Embryology Act.

The Society was keen to ensure that the possibility of therapeutic cloning was not lost in a general ban on reproductive cloning of humans. Therapeutic cloning implied the production of embryonic stem cells[80] with nuclei taken from adult cells, or the extraction of stem cells from embryos remaining after courses of IVF treatment, and then keeping them in culture for future use. The Society set out the case for allowing such work in evidence to the government's Chief Medical Officer, Liam Donaldson, in late 1999, highlighting both research and medical aspects. Donaldson agreed. The government started drafting legislation, and promised a free vote on the issue. The failure at the end of October 2000 of a private member's Bill that supported therapeutic cloning raised the stakes. The Royal Society combined with the MRC and the Nuffield Council on Bioethics to arrange for a briefing meeting in Parliament and the wide distribution of briefing material ahead of a crucial House of Commons debate in December 2000. In the end, the Commons voted by a 2:1 majority in support of research related to therapeutic cloning. The intervention of individuals suffering from conditions that might eventually be alleviated through stem cell research proved particularly influential. The House of Lords followed suit the next month after a repeat briefing.[81]

The debate then moved to a larger stage. The Society wanted a global ban on reproductive cloning, and it wanted to keep research options open. It sent briefing material to relevant Members of the European Parliament in March 2001 when a committee report hostile to both therapeutic cloning and stem cell research was due to be debated. A subsequent debate in the full European Parliament endorsed the Society's line. The UN then set up a committee to examine a possible international convention against reproductive cloning of humans. Its agenda was hijacked, and it soon found itself looking at a possible ban on therapeutic cloning as well. The Society convened a multinational

---

[79] Reproductive cloning was banned in the UK under the Human Reproductive Cloning Act 2001.

[80] Relatively undifferentiated cells able to give rise to more specialised progeny.

[81] Robert May, 'Address of the President at the Anniversary Meeting, 30 November 2001', *Notes and records of the Royal Society of London*, 56 (2002), 124–5.

group under Richard Gardner in 2003 to draft a short statement arguing the cases for a reproductive ban and against a therapeutic ban. This was sent to the Executive Committee of the InterAcademy Panel (IAP; see Chapter 10), which agreed it after long debate.[82] It was eventually endorsed by 63 national science academies and sent as an official IAP statement to each member of the UN committee just before it held a crucial meeting in November 2003. By the narrowest of margins, that committee voted to delay decision for two years, thus fending off pressure to ban cloning for therapeutic or research purposes. This was, apparently, the result more of geopolitical considerations than of scientific logic, but the outcome suited the scientific lobby. In matters of public concern, scientific logic is rarely the whole story.

[82] Robert May, 'Address of the President at the Anniversary Meeting, 30 November 2004: global problems and global science', *Notes and records of the Royal Society of London*, 59 (2005), 109–10. The Society was a leading member of the IAP Executive Committee. One of the most tricky issues in finalising the IAP statement turned out to be the logic for a ban on reproductive cloning. This was finally agreed to be a matter of risk to both mother and foetus, which was deemed to be so great as to be unethical. However, it was also agreed that if in future that risk was reduced to an acceptable level, there would still be 'strong ethical, social and economic objections'.

6   Doing science publicly

> Scientists must learn to communicate with the public, be willing to do
> so, and indeed consider it their duty to do so.[1]

The Royal Society's work on policy advice drew it increasingly into the
public sphere as it engaged with topics that were publicly controversial.
The Society's concern with science and mathematics education also
broadened its range of engagement beyond the bounds of professional
science. It was motivated here both by the need for technical skills in the
workplace and by a wish for science to be more securely embedded in the
nation's culture. It recognised that the scientific enterprise needed
public support, most obviously because of its demands on the public
purse and its ever-growing societal impacts. In a landmark 1985 report,
the Society elevated communication with the public to the status of
professional duty for scientists capable of doing it well. From the late
1990s this included public engagement, covering dialogue and debate as
well as provision of information. By the turn of the century, the public
dimensions of science were entrenched in the Society's vision of its core
business, and its target audiences included many who were not career
scientists.

Public understanding of science

The Royal Society's work on school education during the 1950s and
1960s was focused initially on teachers and particularly on the need for
more properly qualified teachers in science and mathematics. In 1957 it
launched a scheme jointly with the Science Masters' Association, a precursor of the Association for Science Education (ASE), to help teachers
study for higher degrees by carrying out scientific research in schools.[2]
The scheme paid for equipment and for the travel costs involved in

---

[1] Royal Society, *The public understanding of science* (September 1985), 6.
[2] David Layton, *Interpreters of science: a history of the Association for Science Education* (John
Murray/ASE, 1984), 258. The ASE was formed in 1963 from five teachers' organisations.

facilitating collaboration between teachers and relevant university departments. By 1960 the scheme was supporting 66 research projects, and it soon became the custom that pupils should be actively involved in them, thus broadening the scheme's objectives.[3] From 1962 the Society collaborated with specialised learned societies in a series of education committees addressing issues in the teaching of individual sciences. Having shied away from doing so in 1957, the Society finally established its own wide-ranging education committee in 1969. One of the prompts for this was the success of a pamphlet published that year by the Society called *Metric units in primary schools*, of which over 80,000 copies were sold.[4] The first member of Royal Society staff working full-time on education, Don Harlow, an experienced secondary school physics teacher and a senior figure in the ASE,[5] was recruited the same year.

Education activities continued to grow through the 1970s, with a continuing emphasis on issues around recruitment and training of science and mathematics teachers and continuing concern over the low numbers of pupils choosing science subjects at A level. The Society appointed a group under Harry Pitt in 1981 to review secondary science education. Its report, published the following year, called for all pupils to study a broad science curriculum for 20 per cent of their time up to age sixteen, and for a less specialised approach to sixth-form studies. It also highlighted the need for 'much greater awareness and enlightenment about science and its role in society'[6] as the prerequisite for a sensible public debate on science education. It therefore recommended a review of how public understanding of science might be enhanced, with a strong emphasis on the role of the media.[7]

Neither Florey nor Blackett nor Hodgkin had included public understanding of science among the strategic objectives that they envisaged for the Royal Society during their presidential terms. It was, they felt, more an issue for the British Association for the Advancement of Science, the

[3] *Nature*, 189 (4 February 1961), 358. For a good number of years, the Scientific Research in Schools scheme produced one of the exhibits at the annual soirée. The scheme morphed into the Partnership Grants scheme in 1995 with a substantially greater budget, initially via industrial sponsorship.

[4] CM 20 June 1957, minute 28(x); CM 18 December 1969, minute 20; OM/108(69); John S. Rowlinson and Norman H. Robinson, *The record*, 18.

[5] David Layton, *Interpreters of science*, 105, 261–2. Among other activities, Harlow had been closely involved in the ASE's response to the discussions about raising the public esteem of engineering and technology in the early 1960s (see Chapter 4).

[6] Not a new theme for the Society. Its 1960 *Encouragement* report (see Chapter 2) had commented: 'Those who are not destined to follow science as a profession should at least ... be disabused of any idea that ignorance of science is a mark of social or intellectual superiority ... The image of the scientist in the public mind still needs rectification.'

[7] Royal Society, *Science education 11–18 in England and Wales* (November 1982).

newly launched weekly magazine *New scientist*, the museums or the broadcast media. This was made explicit by the Treasurer Alex Fleck in 1963.[8] But the Society cared sufficiently about public understanding that in 1971 it helped the British Association secure an annual grant from the Science Budget for its work of 'publicising science and providing an opportunity for broad discussion of scientific affairs', and for over 30 years it acted as the channel for that grant.[9] It had a roughly similar arrangement with the Royal Institution until 2008.

When Harry Pitt's review of secondary education was being set up, the Society was also contemplating a study on 'the whole issue of science and public understanding'.[10] Five years earlier it had taken steps to make its own activities more publicly accessible by establishing a Press Officer role on its staff – albeit by adding the function to an existing member of staff, Peter Cooper, and restricting it to a coordinating function, taking just 20 per cent of his time.[11] It had also distributed a booklet summarising its various funding schemes, and had launched a newsletter aimed initially at Fellows but soon sent also to the press and other interested parties.[12] More radical than publicising itself, however, was the notion that the Society should get involved in publicising science. In March 1981 the Officers overturned Alex Fleck's position on this and accepted a proposal from the Deputy Executive Secretary Peter Warren for a study on the 'public presentation, appreciation and understanding of science'. Underpinning the proposal were the notion that modern democracy required a scientifically literate population, and the wish to 'minimise adverse responses to science'.[13] The study would provide context for any new initiatives the Society might undertake in this area, and would supply a rationale for its support for the public understanding activities of the British Association and Royal Institution. The Society harboured doubts as

[8] See Chapter 4.

[9] CM 17 June 1971, minute 13; CSP meeting, 24 September 1971: TNA ED 215/59; CM 14 October 1971, minute 17; CM 11 November 1971, minute 11; CM 30 November 1971, minute 8. From 2004, the Society ceased to act as intermediary and the British Association secured its grant directly from government.

[10] CM 2 April 1981, minute 6.

[11] OM 4 March 1976, minute 3(b); OM/40(76) and OM/70(76); interview with Peter Cooper. The role was initially labelled 'Information Coordinator'.

[12] Such advertising had previously been inhibited by fears that it might generate unmanageable numbers of applicants. *RS News* was launched in January 1980; for the initial (generally positive) comments, see OM/44(80). An earlier proposal for a news bulletin for Fellows about the Society's activities, partly in imitation of the NAS, had been rejected in 1964: RMA1369 and CM 17 December 1964, minute 27.

[13] OM 2 March 1981, minute 3(c); OM/33(81); Peter Warren interview. Warren had written a similar paper during his time in the Cabinet Office a few years previously.

to whether either of those bodies had 'found the best way of doing this important job', and it wanted to examine the issue.[14]

So Harry Pitt's recommendation fell on prepared soil, and the Society's review of public understanding of science was launched in early 1983. With a membership ranging from a leading schoolteacher to the investment director of the Prudential Assurance Company, it was chaired by the geneticist Walter Bodmer. Bodmer was active in both British Association and Royal Institution and, as Research Director (later Director-General) of the Imperial Cancer Research Fund, funded by charitable donations, was well aware of the significance, and complexity, of public understanding. His report was published in September 1985.[15] It targeted sensible recommendations at various external audiences, but it also did three more influential things: it set out a clear case for why public understanding mattered; it told the scientific community that active engagement with the public was not evidence of a faltering research career but rather a respectable activity for successful researchers, indeed a 'duty'; and it told the Royal Society itself that it should make improving public understanding one of its major activities instead of leaving it to others. Council's acceptance of the Bodmer report was a commitment to doing science publicly.

So when the Society produced its first corporate plan a few months after publishing the Bodmer report, one of the eight objectives it set itself was promoting 'science education, awareness and understanding'. To embed this objective and carry forward the Bodmer programme, the Society, jointly with the British Association and Royal Institution, established the Committee on Public Understanding of Science (COPUS). The fact that George Porter headed all three bodies about that time provided irresistible logic for this collaboration, and he served as the first chair of COPUS.[16] COPUS established a media fellowships scheme to improve scientists' understanding of the media; it worked with television producers[17] and interactive science centres; it instituted prizes for the best science books

---

[14] David Phillips (as Biological Secretary) to Walter Bodmer, 30 March 1981: D.C. Phillips papers, MS Eng. c.5478, L.32. Bodmer, in reply, thought the BA and RI did well with younger audiences.

[15] Royal Society, *Public understanding*; Walter Bodmer, 'Public understanding of science: the BA, the Royal Society and COPUS', *Notes and records of the Royal Society of London*, 64 (September 2010), S151–61; P.M.D. Collins and W.F. Bodmer, 'The public understanding of science', *Studies of Science Education*, 13 (1986), 96–104. Unusually for a Royal Society report, it quickly became known by the name of its Chairman. Interviews with Walter Bodmer, Peter Cooper and Peter Warren.

[16] Peter Briggs, *The BA at the end of the 20th century: a personal account of 22 years from 1980 to 2002* (British Association archives), 23–5.

[17] COPUS reported that its dealings with the BBC had, by 1993, led to a 'change of heart' and that the BBC 'now acknowledged the importance of science'. CM 16 December 1993, minute 17.

aimed at adult, and at young, readers; it ran courses with the Women's Institute, the Girl Guides Association, and senior civil servants; and it took many other initiatives. Particularly, it set up a grant scheme with funding from the Society's PGA to support practical projects in public understanding, distributing almost £100,000 annually by 1992.[18]

Beyond the Society, the Bodmer report brought a new urgency to the whole business of communicating science to the public.[19] The Economic and Social Research Council picked up one of its recommendations and funded a programme of sociological research into public understanding; the Science Museum experimented with 'hands-on' exhibits; chairs in public understanding of science appeared in Imperial College (John Durant, 1989) and Oxford (Richard Dawkins, 1995); the Institute of Physics and the Science Museum launched a new journal, *Public understanding of science*, in 1992; the Science Museum, UCL and Birkbeck started a long-running seminar on public understanding in 1993; and the government, at COPUS's prompting, launched an annual national Science, Engineering and Technology Week in 1994. Interest in public understanding of science moved into the professional mainstream. William Waldegrave's 1993 White Paper made it compulsory: the restructured Research Councils included stimulating public interest and promoting public understanding in their formal objectives and mission statements.

After 1985, the Society did much of its public understanding work through COPUS, but it also acted separately. As Bodmer had recommended, it instituted the Faraday Award in 1986 to encourage research scientists to promote public understanding and to underline the respectability of such activity. It ran press briefings with the Association of British Science Writers. It put on 'lectures for the public', and expanded public access to the annual exhibition so that the latter became a landmark in the calendar.[20] It started running courses to help its more outgoing research fellows to get involved in public communication, and some, such as Brian

---

[18] The government established a similar scheme in 1993, as promised in the Waldegrave White Paper, and soon handed it over to the Royal Society, more than doubling the budget of the COPUS grants scheme.

[19] Richard Jones, 'Introduction', in David Bennett and Richard Jennings, eds., *Successful science communication* (Cambridge University Press, 2011), 2; Simon Lock, 'Deficits and dialogues: science communication and the public understanding of science in the UK', in David Bennett and Richard Jennings, eds., *Successful science communication*, 18–20.

[20] In the postwar years, pupils from selected schools were sometimes invited to see the scientific exhibits that embellished the Society's grand soirées (e.g. CM 8 April 1948, appendix D). From the mid 1980s, the exhibition was opened up to more and more visitors and became correspondingly more ambitious in scope. In 2011 the Summer Science Exhibition, as it was by then known, ran for a week and attracted 14,000 visitors from all backgrounds. Interview with Colin Blakemore.

Cox, Athene Donald, Lucie Green and Marcus du Sautoy, subsequently became household names for their communication work.

The Bodmer report's narrative about why better public understanding mattered linked it with national prosperity, public and private decision-making, and enriching the life of the individual. 'Understanding' embraced not just scientific facts but also the methods and limitations of science and its social implications. The underpinning idea was that better understanding went with better decision-making, '*not* because the "right" decisions would then be made, but because decisions made in the light of an adequate understanding of the issues are likely to be better than decisions made in the absence of such understanding'.[21] In both public and private life, people were increasingly faced with decisions requiring an appreciation of scientific and technical factors and an ability to evaluate, for their own situation, the advice on offer. Some degree of familiarity with science was therefore a universal need.

There was certainly a strand of thinking within the Royal Society and the wider scientific community, and indeed within government,[22] that improved public understanding would somehow smooth the path for science. But this was not pivotal to the work of the Bodmer Committee – though some commentators imputed such an assumption to the Committee. On the contrary, Bodmer called for research into just these issues. In 1998 COPUS commissioned a survey of relevant sociological research since his report, and that survey set out very clearly the complexities and context-specific character of the public understanding of science. That survey also highlighted a 1995 finding that 'the more scientifically informed are more discriminating in their judgements', which chimed in well with Bodmer's approach.[23]

In view of the proliferation of initiatives and organisations by then active in public understanding, COPUS moved towards more of a coordinating and enabling role in the late 1990s, while its three sponsoring bodies (and many other organisations) continued to develop their individual programmes in what was then more widely known as science communication.

---

[21] Royal Society, *Public understanding*, 9; emphasis in original.

[22] In 1993, the government committed itself to fund activities 'designed to increase public understanding and appreciation of science and technology': *Realising our potential*, 67.

[23] COPUS, *To know science is to love it?* (Royal Society, 1998). See also John Ziman, 'Public understanding of science', *Science, technology & human values*, 16 (1991), 99–105; Brian Wynne, 'Knowledges in context', *Science, technology & human values*, 16 (1991), 111–21; Alan Irwin and Brian Wynne, eds., *Misunderstanding science?* (Cambridge University Press, 1996); Robert May, 'Address at the Anniversary Meeting, 29 November 2002', *Notes and records of the Royal Society of London*, 57 (2003), 119; Laura Bowater and Kay Yeoman, *Science communication* (Wiley-Blackwell, 2013), 13–19.

However, quite what needed coordinating was less and less clear, and differences between COPUS's three sponsoring bodies over both style and substance began to emerge. COPUS was eventually closed in late 2002, amid some controversy among the sponsoring bodies and COPUS itself over its future programme.[24] By that time its original mission, to put public understanding firmly on the agenda of the scientific community, had arguably been accomplished. The Royal Society continued building its own programme of public activities. By 2006 its corporate objectives referred to inspiring 'an interest in the joy, wonder and fulfilment of scientific discovery'; in 2012 it was a commitment to 'ensuring that everyone has the opportunity to appreciate the value of and engage with science'. In 2015, the Society appointed Brian Cox as its first professor for public engagement in science. It was a long way from the attitude fifty years previously of leaving it all to the British Association et al.

In parallel with this, the Society continued to develop its formal education work and to do so through a mutually reinforcing combination of policy advice and practical initiatives. As with public understanding, this required the Society to operate mostly in collaboration with others.[25] It worked with the Joint Mathematical Council in establishing and developing the Advisory Committee on Mathematics Education (ACME) in 2002, and in 2006 it persuaded six other scientific organisations to join with it in forming the Science Community Partnership for Supporting Education (SCORE). These two bodies, with secretariats based at the Royal Society, provided unified voices for their respective communities in seeking proactively to influence education policy in ways that would not have been possible for their members acting individually, and proved effective on the national stage. Impact was significantly enhanced by the appointment of Professor Michael Reiss, on a two-year part-time secondment from the Institute of Education, as the Society's Director of Education.[26]

The Society continued to pursue its own education initiatives alongside the ACME and SCORE work, culminating in 2014 in publication of a major and well-received report, led by Martin Taylor and Julia Higgins, that set out a high-level vision for the future of science and mathematics education.[27] This was of a piece with Harry Pitt's 1981–2

[24] David Adam, 'British science champion quits post', Nature, 417 (6 June 2002), 577 and 420 (12 December 2002), 598; Peter Briggs, The BA, 112–4; interviews with Eric Ash, Colin Blakemore, Walter Bodmer and Peter Cooper.

[25] Interviews with Peter Cooper, John Horlock and Martin Rees.

[26] Interviews with David Boak, Brian Follett, Julia Higgins, David Sainsbury and Martin Taylor.

[27] Royal Society, Vision for science and mathematics education (June 2014).

review. It stressed the importance of raising the general level of mathematical and scientific knowledge and confidence in the population in order to help individuals make informed choices, empower them to shape scientific and technological developments, and equip them to work in an advanced economy. It called for a broad and balanced education for all through to age 18, within a baccalaureate-style framework including science and mathematics at all stages. And it highlighted the need to train and nurture many more inspiring teachers and to increase the status of the teaching profession. The Royal Society, in its work on education as in its work on science communication, was sending a strong message that everyone must have real opportunity to engage with science.

Scientific dialogue and public controversy

The public encounter with science was about more than comprehension of formal scientific knowledge or even of scientific process. The encounter on occasion also embraced the context and credibility of all involved and, hence, assumptions about social organisation.[28] This could generate robust exchanges from competing perspectives, especially when the science was developing rapidly, was subject to arguments among the experts, and had potentially controversial implications for society. As the Royal Society moved increasingly to address the scientific aspects of more publicly controversial matters, it therefore had to acquire new skills in the art of doing science publicly. Dialogue and engagement came to the fore, alongside continued and increased efforts to communicate the findings and processes of science. The Society learned from experience, as did everyone else.

Sometimes the Society's scientific reputation and independence allowed it to play a useful honest broker role in public controversy. An instance in the 1980s was environmental pollution – specifically, acid pollution of freshwater lakes in Scandinavia. This was controversial in Scandinavia, obviously; and it was controversial in Britain, whose coal-fuelled power stations were suspected of being the source of the nitrogen and sulphur oxides concerned. The newly appointed chairman of the Central Electricity Generating Board, Walter Marshall, was being pushed by his Scandinavian colleagues to do something about it. Elected to the Royal Society twelve years previously, he turned to the Society in 1983 for help in establishing whether British power stations really were to blame.

[28] Jane Gregory and Simon Lock, 'The evolution of "public understanding of science": public engagement as a tool of science policy in the UK', *Sociology compass*, 2 (2008), 1252–65.

The upshot was an extensive research programme, led by the Society's Treasurer John Mason (who had just retired from the post of Director-General of the Meteorological Office) and funded by the CEGB and the National Coal Board, to an eventual total of £5.6 M, under terms that guaranteed the Society's scientific independence.

The Society secured the collaboration of the national science academies of Norway and Sweden at the outset of the programme. Interim results in 1986 and 1987 were sufficiently clear and sufficiently authoritative to prompt a major investment by the CEGB in flue gas desulphurisation and to influence the government's response to the European Union's Large Plants Directive. The final results of the research programme were presented at a scientific conference hosted by the Society in March 1990, with the Prime Ministers of all three countries making speeches at the accompanying dinner. There was extensive coverage in the national press, and the core scientific conclusions were generally accepted. With Mrs Thatcher by now convinced that acid rain as described was indeed real, it was a good example of what the Norwegian Prime Minister Jan Syse called 'the principle that political discussions must be founded on scientific evidence'.[29]

The Society played a similar honest broker role, but with rather less impact, in accepting a contract from UK Nirex Ltd in 1993–4 to advise it about the scientific aspects of a possible deep repository for long-lived radioactive waste. Council took the view that it had a public responsibility in the matter provided it could guarantee demonstrable scientific independence. A group led by Alan Muir Wood subsequently produced both a full technical report and a widely circulated summary version.[30] However, on this occasion, politics trumped the science in shaping what happened next: moments before Parliament was dissolved for the start of the 1997 general election, the outgoing Environment Secretary John Gummer rejected Nirex's planning application for the repository.

Michael Atiyah, who became President on 30 November 1990, was not shy of engaging with political issues.[31] As he was preparing to take office, Council described the Society as 'the conscience of the science community'

[29] Richard Southwood, 'Surface Waters Acidification Programme', *Science & public affairs*, 5 (1990), 74–95; OM 21 April 1983, minute 3(c); CM 16 June 1983, minute 15; CM 13 October 1983, minute 19; RMA396 and 1403; national newspapers around 23 March 1990; Richard Stevenson, 'Acid rain: swapping fishermen's tales', *Chemistry in Britain* (May 1990), 397; John Campbell, *Margaret Thatcher: the iron lady* (Jonathan Cape, 2003), 643; interview with John Mason.

[30] CM 19 May 1993, minute 14; Royal Society, *Disposal of radioactive wastes in deep repositories* (November 1994); David Oldroyd, *Earth, water, ice and fire* (Geological Society memoir 25, 2002), 274.

[31] Interview with Michael Atiyah.

and beefed up the policy advice function. Atiyah was a long-standing and active member of the Pugwash Conferences on Science and World Affairs,[32] and was one of those who had persuaded the Society in 1982 to take an interest in defence matters, at a time of heightened concern about nuclear weapons. The Society's interest was somewhat tentative, and constrained by a strong wish to stay away from politics. But inspired by an analogous group at the US National Academy of Sciences, and after further consultation with the Fellowship, Council agreed in 1988 to establish a temporary group on Scientific Aspects of International Security (SAIS), the apolitical injunction still in force. SAIS was put on a permanent footing the day Atiyah became President.[33] It broadened its interests beyond nuclear to include biological and chemical weapons, and prepared a series of reports that attracted public attention. These responded in the main to developments in international treaty negotiations and to issues around combating terrorism.

Of greater public controversy in the defence field was the use of depleted uranium on the battlefield – particularly the Gulf in 1991 and Kosovo in 1999. Depleted uranium, a by-product of uranium enrichment in the manufacture of nuclear fuel rods, is a toxic metal, weakly radioactive and very dense, and was used in the tips of armour-piercing shells. Such shells were first deployed in the Gulf. Subsequent reports of soldiers with Gulf War syndrome and of deformed patients in Iraqi hospitals gradually focused public attention on the health and environmental impacts of depleted uranium. SAIS launched a very thorough independent study of depleted uranium at the end of 1999. The study was conducted as openly as practicable, with considerable media coverage from the outset and with all interested parties invited to submit evidence. The Society published two technical reports dealing respectively with health and environmental impacts, in 2001 and 2002, and separate summaries aimed at general audiences.[34] When the first report was published, the Society organised a public meeting to discuss its findings,

---

[32] On Pugwash, see for example R.A. Hinde and J.L. Finney, 'Joseph Rotblat, 1908–2005', *Biographical memoirs of Fellows of the Royal Society*, 53 (2007), 309–26. Pugwash is a small town on the Nova Scotia coast where the first conference was held in 1957, at the instigation of Joseph Rotblat and Bertrand Russell. Presidents of the Pugwash Conferences organisation have included a number of Royal Society Fellows: John Cockcroft, Howard Florey, Dorothy Hodgkin, Joseph Rotblat, Michael Atiyah and M. S. Swaminathan.

[33] CM 14 October 1982, minutes 28, 29; CM 14 July 1988, minute 49; CM 3 November 1988, minute 45; CM 30 November 1990, minute 46. Martin Rees was another active participant, as was Anne McLaren (Foreign Secretary during Atiyah's presidency). SAIS was eventually absorbed into the Science Policy Centre's advisory group in 2008.

[34] Royal Society, *The health hazards of depleted uranium munitions* (Part I, May 2001; Part II, March 2002); Robert May, 'Anniversary Address, 2001', 126; Peter Moszynski, 'Royal

at which the Chairman of the working group, Brian Spratt, presented the report, and representatives of two campaigning groups and the Ministry of Defence responded. The audience included Gulf War veterans and others with personal stakes in the issue. The Society's account of the science was generally accepted, though those campaigning on low-level radiation and on Gulf War syndrome remained unconvinced.

An outbreak of bovine spongiform encephalopathy (BSE), a fatal neurological disease in cattle, came to light about 1986. A government committee chaired by Dick Southwood in 1988–9 made recommendations about how to bring the outbreak under control, which in due course were successfully implemented. The committee also examined the chances of BSE being transmitted to humans, and eventually assessed them as 'remote'. The political process then turned that assessment into public assurance that there was 'no conceivable risk' from eating British beef, though debate on the matter was continuing among scientists. The Royal Society organised a briefing for science journalists on BSE in 1990, and held a two-day conference on the science of prion diseases (such as BSE) in 1993. BSE itself was traced back to scrapie, the prion protein disease in sheep, which was known not to be harmful to humans. The government was keen to restore the flagging fortunes of the British beef industry and constantly stressed that there was no evidence that BSE could be transmitted to humans in the form of the human prion disease CJD. Then, on 20 March 1996, the Health Secretary Stephen Dorrell told Parliament that ten young people had been diagnosed with a new and fatal variant of CJD most likely linked to exposure to BSE in the period before the Southwood recommendations had been implemented. The press was filled with apocalyptic warnings of the impending epidemic, and the blame game began immediately – to the further erosion of public confidence.[35] With unaccustomed speed, the Royal Society sought to provide a basis for public discussion by issuing, on 2 April 1996, an authoritative summary of what was known scientifically about BSE and CJD and where the uncertainties lay.[36] Updates followed in July 1996, July 1997 and May 2001 as significant new scientific results became available. The Society also leaned on a

Society warns of risks from depleted uranium', *British Medical Journal*, 326 (3 May 2003), 952.

[35] *Nature*, 365 (9 September 1993), 93; *Nature*, 380 (28 March 1996), 271, 273–4; *Nature*, 381 (30 May 1996), 351, 353–4; *Nature*, 382 (29 August 1996), 755–6; Hannah Gay, *Silwood circle*. For a different angle on prion disease, see John Collinge, 'Lessons of kuru research: background to recent studies with some personal reflections', *Philosophical transactions of the Royal Society of London. Series B, Biological sciences*, 363 (2008), 3689–96.

[36] Royal Society, *BSE and CJD – the facts to date* (2 April 1996). This briefing underpinned discussion in, for example, the House of Lords on 17 April 1996.

reluctant agriculture ministry to release data needed by independent experts for epidemiological calculations.[37]

The Blair Government appointed Nicholas Phillips in 1998 to investigate what had gone wrong with BSE under the previous Government. Phillips' very lengthy report, the result of a very lengthy process, highlighted some important lessons for policy advice in areas of strong public interest. These lessons included the need for greater openness and, particularly, the importance of acknowledging areas of uncertainty and not rushing to precipitate conclusions for the sake of appearing decisive. Public confidence was better secured by putting factual information, with clear signals about its limitations, into the public domain, than by secrecy and patrician reassurances. Given reliable information, individuals would make their own judgement about risks and benefits in their own specific contexts. This approach was also reflected in, for example, the Chief Scientific Adviser's 1997 guidelines on the science advice process, and the establishment of the Food Standards Agency under John Krebs in 2001.[38]

Aaron Klug (Figure 6.1), who succeeded Michael Atiyah as President in late 1995, shared his willingness to engage with the scientific aspects of politically difficult issues.[39] He recognised the need for better public understanding of science, and also for the Society itself substantially to strengthen its Press Office and to interact more effectively with the media. The Press Office, enlarged and invigorated, was to make a very important contribution to the Society's policy influence in the coming years. But even that was not sufficient in circumstances where debate about the science was interwoven with debate, explicit and implicit, about quite different matters. That rapidly became clear with the next controversy to engage the Society: the furore over genetically modified (GM) crops that erupted in 1998–9.[40]

The fuse for this was the decision by a number of supermarkets in the spring of 1998, in response to what they perceived as public opinion, to stop selling foods with GM components, despite the initial market success of a GM tomato purée. This, and the growing quantity of unsubstantiated information in circulation, prompted the Royal Society to prepare a report detailing what was and was not known scientifically

---

[37] Declan Butler, 'BSE researchers bemoan "ministry secrecy"', *Nature*, 383 (10 October 1996), 467–8; Robert Matthews, 'Now scientists must bridge the credibility gap', *Daily Telegraph* (26 October 2000).

[38] Robert May, *Scientific advice in policy making*; Lord Phillips et al., 'Lessons from the BSE Inquiry', *FST journal*, 17 (July 2001), 3–8; interviews with Peter Lachmann, Bob May, Bill Stewart and William Waldegrave.

[39] Interview with Brian Heap.

[40] The Society had held a scientific discussion meeting on GM as early as 1992, but at that stage GM was not publicly controversial in the way it later became.

Figure 6.1 Aaron Klug. © Anne Purkiss

about GM plants.[41] The report, published in September 1998 under the

[41] Royal Society, *Genetically modified plants for food use* (September 1998). On the Royal Society and GM more generally, see interviews with Michael Atiyah, Patrick Bateson,

chairmanship of Peter Lachmann, recognised that consumer confidence would rightly be a major influence on whether GM technology could be successfully deployed, and called for a cross-government regulatory body to oversee use of GM. This was accepted and implemented. During the following months, the Society's report became a key source document in government circles about the science of GM.[42]

But shortly before the Society's GM report was published, on 10 August 1998, Árpád Pusztai from the Rowett Research Institute gave a television interview in which he claimed experimental evidence that GM potatoes containing a lectin gene from a snowdrop could have harmful effects on rats. This, so to speak, lit the fuse, though it smouldered for some months before really flaring up. In February 1999 a group of twenty-three scientists came out publicly in support of Pusztai,[43] and the anti-GM polemic escalated. The Royal Society, vexed that un-refereed and unpublished research was being used to shape public policy, issued a series of press statements trying to counter some of the polemic. Then, with government encouragement, it decided to submit such of Pusztai's work as it could access to a thorough peer review, in consultation with Pusztai himself. On 18 May it held a press conference to publicise the conclusion of that review, that Pusztai's work was seriously flawed and that 'no conclusions should be drawn from it'.[44]

This attracted massive media attention. It also provoked further invective, to a degree unparalleled in the Society's experience of giving scientific advice. The Society was accused by anti-GM campaigners of being in the pocket of big business,[45] of running 'a "rebuttal unit" to push a

David Boak, Walter Bodmer, John Enderby, Brian Heap, Julia Higgins, Aaron Klug, Peter Lachmann, Robert May, Noreen Murray and David Sainsbury.

[42] Aaron Klug, 'Address at the Anniversary Meeting, 30 November 1998', *Notes and records of the Royal Society of London*, 53 (1999), 162.

[43] 'Top researchers back suspended lab whistleblower', *Guardian*, 12 February 1999. At this point the *Daily Mail* launched into its 'Frankenstein foods' frenzy. The *Daily Telegraph*, in a more balanced piece three days later headlined 'Vested interests cloud search for truth', showed the twenty-three to be mostly colleagues of Pusztai or environmental activists – 'a shadowy environmental coalition ... mostly obscure figures', the *Telegraph* editor Charles Moore told the House of Commons Select Committee on 23 May. The Prime Minister Tony Blair assured *Telegraph* readers on 20 February that, contrary to rumour, he was indeed acting on the Royal Society's September 1998 recommendations.

[44] Royal Society, *Review of data on possible toxicity of GM potatoes* (June 1999); Natasha Loder, 'Royal Society: GM food hazard claim is "flawed"', *Nature*, 399 (20 May 1999), 188. The press conference attracted fifty journalists and seven television crews, and generated worldwide publicity.

[45] Stanley Ewen, 'Health risks of genetically modified foods', *Lancet*, 354 (21 August 1999), 684; but note Patrick Bateson, 'Genetically modified potatoes', *Lancet*, 354 (16 October 1999), 1382, detailing the Society's methodology in its review.

pro-biotech line and counter opposing scientists and environmental groups',[46] of 'breath-taking impertinence' in presuming to undertake its review.[47] Its science, however, was not seriously challenged. The Society responded to some comments and ignored others. It continued giving scientific advice to the various government and parliamentary inquiries and public consultations that examined particular aspects of genetic modification in the following months and years, and continued with its unprecedentedly intense interactions with the media. So did other scientific organisations, and individual experts such as Derek Burke who had just completed an eight-year stint chairing the Advisory Committee on Novel Foods and Processes.

In its September 1998 report, the Society had put GM in the context of feeding the rapidly growing global population. In 1999 it pursued the international dimension of the issue by convening a group of seven science academies to develop a statement on the role of GM in global agriculture, which was published simultaneously in all the countries concerned just ahead of the July 2000 G8 summit.[48] It also contributed to debates about regulation at EU level. The Society updated its 1998 report in 2002, and again in 2009 when a detailed study led by David Baulcombe concluded that GM had a significant role in achieving global food security in the long term.[49]

'Policy advice is a hazardous business', Aaron Klug commented ruefully at the end of 1999. But, like William Huggins nearly a century earlier, Klug affirmed that 'the Society's special position ... means that we have a duty to get involved', adding 'we need to think further about the role of the expert in society'.[50] Reflecting over the summer of 1999 on its bruising GM experience, the Society recognised the need for a fresh approach to what was vaguely termed 'science in society'. It had established a Science in Society Committee in 1995 under the chairmanship of Derek Roberts, tasked with organising a series of public discussions about the interactions of public policy, technological change and scientific

[46] Laurie Flynn and Michael Gillard, 'Pro-GM food scientist "threatened editor"', *Guardian*, 1 November 1999.

[47] Richard Horton, 'Health risks of genetically modified foods', *Lancet*, 353 (29 May 1999), 1811.

[48] Royal Society et al., *Transgenic plants and world agriculture* (July 2000); *Nature*, 399 (24 June 1999), 715 and 721. The national science academies of Brazil, China, India, Mexico and the United States, and the Third World Academy of Sciences, as well as the Royal Society, co-authored this document.

[49] Royal Society, *Genetically modified plants for food use and human health – an update* (February 2002); Royal Society, *Reaping the benefits* (October 2009).

[50] Aaron Klug, 'Address at the Anniversary Meeting, 30 November 1999', *Notes and records of the Royal Society of London*, 54 (2000), 101, 104.

knowledge.[51] The speakers included social scientists, journalists, politicians and industrialists, and the discussions were widely disseminated. The GM experience, and an inquiry by the House of Lords S&T Select Committee into science and society, showed the need for something more ambitious.

So, alongside its programme of activities in science communication, the Royal Society determined in late 1999 to develop a programme of activities to respond to the increasing public demand for greater involvement in the conduct of science. It revamped its Science in Society Committee, appointed Paul Nurse (who, coincidentally, had succeeded Walter Bodmer at the Imperial Cancer Research Fund) to chair it, and charged it to 'develop a widespread, innovative and effective system of dialogue with society, to involve society more positively in influencing and sharing responsibility for policy on scientific matters, to embrace a culture of openness in decision-making, to take account of the values and attitudes of the public, and to enable the Society to promote national science policy'.

This challenging undertaking was much facilitated by a donation of £1 M over five years from the medical scientist Ralph Kohn through the Kohn Foundation. This donation funded a programme that included a scheme to pair individual research scientists and MPs so that each could experience something of the other's world;[52] a series of structured national dialogues, with regional meetings, televised debates and published discussions, on such topics as trust in science, genetic testing, pharmacogenetics, ICT and health care, and information security; and events with schoolchildren around the social and ethical impacts of science and technology. The Society also published guidelines to help scientists to judge whether their work might be of public interest and deal with the media accordingly. At the outset the Society commissioned an external review of best practice in conducting public dialogue, and throughout the programme it worked closely with social scientists to ensure that it learnt from experience. It collaborated with numerous partners in delivering all this, thus considerably enhancing the effectiveness of the programme and extending the Society's range of contacts. When the Kohn donation came to an end in 2006, public funding allowed science-in-society activities to continue, recognition of the progress made up to that point.[53]

---

[51] The published discussions covered such topics as science, policy and risk; science, technology and social responsibility; and science and the law.

[52] This was later extended to MEPs, through the European Academies Science Advisory Council, and was imitated in other countries.

[53] Royal Society, *Science in society: report* (July 2004); Royal Society, *Science in society: the impact and legacy of the five year Kohn Foundation funded programme* (September 2006);

This science-in-society programme was conceived in the context of bolstering science's 'licence to practise'[54] and of making the Society's policy advice more influential in areas of actual or potential public interest and public controversy. The cultural shift was soon manifested in the depleted-uranium project, with its careful interactions with veterans' groups. It was also manifested in another major policy project at that time, when the Blair Government commissioned the Society, after the 2001 foot and mouth outbreak, to advise on the scientific aspects of future strategy for dealing with infectious diseases in livestock.[55] The foot and mouth outbreak had had immense public impact, with personal liveli-hoods jeopardised, parts of the country closed off, and television screens filled nightly with scenes of mass slaughter and its aftermath as 6 million sheep, cattle and pigs were culled. The Society therefore worked in, and with, the spotlight of public interest. Its infectious diseases working group, chaired by Brian Follett, included a farmer, two vets, the deputy chair of the Food Standards Agency, a European Commission official and a representative of a group campaigning for better food and farming, as well as research scientists with relevant expertise. It consulted widely on the detailed framing of its remit. It solicited 400 submissions of evidence, many from non-scientists, and published them alongside its own report. It held focus groups on specific issues. It held open meetings in three areas of the country particularly affected by the outbreak. It established a dedicated website and issued three progress reports during the 11 months the project was under way. Its resultant report[56] gained wide public acceptance and proved effective at both national and EU levels in shaping future policy.

Scientific dialogue in areas of public controversy was difficult in entrenched adversarial situations, as with GM. It was a little easier – though certainly not without its own particular complexities[57] – in areas where full-scale public controversy had not yet broken out. At the turn of the century, nanotechnology was entering the early stages of public controversy, pushed there mainly by writers of popular fiction, some prominent individuals and

Martin Rees, 'Address at the Anniversary Meeting, 30 November 2006', *Notes and records of the Royal Society of London*, 61 (2007), 80–1; interviews with David Boak, Ralph Kohn and David Sainsbury.

[54] Aaron Klug, 'Address at the Anniversary Meeting, 30 November 2000', *Notes and records of the Royal Society of London*, 55 (2001), 172.

[55] 'Royal Society has "free hand" to run inquiry', *Times higher education supplement* (17 August 2001).

[56] Royal Society, *Infectious diseases in livestock* (July 2002); Robert May, 'Anniversary Address, 2001', 125; interview with Brian Follett.

[57] Tee Rogers-Hayden and Nick Pidgeon, 'Moving engagement "upstream"? Nanotechnologies and the Royal Society and Royal Academy of Engineering's inquiry', *Public understanding of science*, 16 (2007), 345–64.

a few campaigning groups in Canada and elsewhere.[58] Keen to pre-empt if possible a GM-style polarisation of public opinion, the Blair Government announced in June 2003 that it was commissioning the Royal Society and the Royal Academy of Engineering to review the current state of nanotechnology and consider possible future developments and their societal implications.

In a notable example of 'upstream public engagement', the two academies appointed a working group under Ann Dowling that included the heads of Forum for the Future and the National Consumer Council, a senior social scientist and an eminent philosopher among its members. The group consulted publicly on exactly which issues it should address. It invited individuals and organisations to register their interest in the project and contribute via an interactive website. It commissioned, and published, a survey of public attitudes to nanotechnology. It held workshops with scientists and engineers, with civil society organisations, with health and environmental experts, with regulators and with industrialists. Workshop summaries and other evidence received were published on the website as the project progressed. The report, published in July 2004, made recommendations about research and application, but also about managing environmental impacts, about regulation more generally and about the need for sustained, properly funded, public dialogue. It attracted international media coverage, and positive reactions from an unusually wide spectrum of opinion as a result of its inclusive approach. The Society subsequently held a number of international workshops to stimulate the development of nanotechnology. In 2006, the Society and the Academy published a critical review of the government's progress in implementing their recommendations, calling, among other things, for greater efforts to ensure that public dialogue helped to shape policy-making.[59]

Climate change provided occasion for public controversy in which the battle lines were drawn differently from the debates just described. The Society's involvement with the science went back to at least the 1957–8 International Geophysical Year and the subsequent international experiments, organised through ICSU (see Chapter 10) with Royal Society support, that generated the data on which a science of the global climate could be based.[60] In 1989, as evidence for anthropogenic global

[58] Geoff Brumfiel, 'Nanotechnology: a little knowledge …', *Nature*, 424 (17 July 2003), 246–8.

[59] Royal Society and Royal Academy of Engineering, *Nanoscience and nanotechnologies: opportunities and uncertainties* (July 2004); Royal Society and Royal Academy of Engineering, *Nanoscience: two-year review of progress*; interviews with David Boak, John Enderby, Julia Higgins, Robert May and David Sainsbury.

[60] International Council for Science, *ICSU and climate change: 1962–2006 and beyond* (ICSU, 2006).

warming was mounting, the House of Lords Select Committee on Science and Technology undertook an inquiry on the subject. The Society's submission to that inquiry, prepared by a group under John Mason, argued for dedicated research funding and set out, both for the House of Lords and for the wider public, what was already known about the science and what the research priorities were, including research into social impacts.[61] The Society later produced a number of reports on energy policy linked to the objective of minimising CO_2 emissions, which attracted criticism from environmental campaigners when they advocated greater use of nuclear power for generating electricity, and support from them when they did not.[62] But by 2000 the focus of attention was increasingly on well-organised groups, many backed by vested interests from particular industrial sectors or ideologies, that were systematically trying to undermine the scientific consensus. The nature of the debate changed.

Climate change was an archetypically international policy issue, and the Intergovernmental Panel on Climate Change (IPCC), in which a number of Royal Society Fellows held key positions, set the pace on building global consensus around the science and around developing policy options for mitigation and adaptation. The Society regarded the IPCC as authoritative, and organised meetings to disseminate its periodic findings. When the American President George W. Bush announced on 29 March 2001 that the United States would not sign the Kyoto agreement on reducing emissions of greenhouse gases, claiming among other things that the science was too uncertain, the Society organised a group of seventeen national science academies to sign a statement endorsing the IPCC's scientific competence and stressing the need for urgent action. The statement appeared as an editorial in the American journal *Science* on 18 May 2001.[63] The signatories ensured that the statement was distributed to their national policy-makers as well as to participants in relevant international bodies.

Since vigorous efforts were being made to mislead public opinion about the science of climate change, the Royal Society invested considerable time in clarifying the boundaries between what was firmly established,

[61] Royal Society, *The greenhouse effect: the scientific basis for policy* (July 1989).
[62] Royal Society, *The greenhouse effect: the scientific basis for policy* (July 1989); Royal Society and Royal Academy of Engineering, *Nuclear energy – the future climate* (June 1999); Royal Society and Royal Academy of Engineering, *The role of the Renewables Directive in meeting Kyoto targets* (October 2000).
[63] Royal Society et al., *The science of climate change* (May 2001); *Science*, 292 (18 May 2001), 1261, 1275; Robert May, 'Anniversary Address, 2004', *Notes and records of the Royal Society of London*, 59 (2005), 102–5. For convenience, most of the signatories were drawn from the IAP Executive Committee.

what was tentative and what was unknown. A major meeting on this at the end of 2001, based on the third IPCC Assessment Report, generated a widely distributed document describing the current state of understanding.[64] The British Prime Minister Tony Blair, hosting the July 2005 G8 summit meeting, let it be known well in advance that he was making climate change a central element on the summit agenda. Faced with pre-emptive spoiling actions by sceptical lobby groups, the Society prepared a fresh guide for journalists about the science of climate change.[65] In reaction to some striking examples of distorted reporting in the media, the guide was later developed into a document aimed at the general public, and very widely distributed.[66] Meanwhile, the Society also led the drafting of a statement stressing that enough was already known scientifically about the phenomenon and the causes of climate change to mandate urgent remedial action, which was signed by the national science academies of all G8 countries, and of Brazil, China and India which had also been invited to the 2005 summit.[67] The statement was published a month before the summit began, and received global coverage. The final communiqué from the summit, however, was seen as disappointingly vague, and the event as a whole was overshadowed by terrorist bombings in London.

Public controversy continued, and the Royal Society produced a stream of detailed studies, with non-technical summaries, on such climate-related topics as carbon sinks, economic instruments for reducing CO_2 emissions, ocean acidification, carbon capture and storage, the impact of climate change on biodiversity and on agriculture, biofuels, and geoengineering – in each case summarising the current science and its implications for policy. It engaged with key events such as the publication of new IPCC assessments and meetings of the UN Framework Convention on Climate Change. It also provided briefings for parliamentarians and monitored developments in the media. In 2010 it published a new summary of the science that set out clearly what was widely agreed, what was subject to some continuing debate, and what was not well understood. And in 2014 it produced a further update, jointly with

---

[64] Royal Society, *Climate change: what we know and what we need to know* (August 2002).

[65] Robert May, 'Under-informed, over here', *Guardian* (27 January 2005), 10; Bob Ward, 'The Royal Society and the debate on climate change', in M.W. Bauer and M. Bucchi, eds., *Journalism, science and society: science communication between news and public relations* (Routledge, 2007), 159–72.

[66] Royal Society, *Guide to facts and fictions about climate change* (March 2005); Royal Society, *Climate change controversies: a simple guide* (June 2007, updated December 2008).

[67] Royal Society et al., *Global response to climate change* (June 2005); Robert May, 'Address at the Anniversary Meeting, 30 November 2005: Threats to tomorrow's world', *Notes and records of the Royal Society of London*, 60 (2006), 113–7.

the NAS. Scientific information, the two academies insisted, was 'a vital component of the evidence required for societies to make sensible policy decisions': it could not be marginalised as merely a matter of political opinion. The document rapidly gained global coverage. As 'two of the world's most august scientific institutions', according to one major newspaper, the Royal Society and the NAS were committed to their duty to provide such authoritative information to policy-makers and to the wider public – all the more because of the febrile nature of the public debate.[68] It was the same sense of duty that William Huggins had acknowledged in his Anniversary Address 110 years previously.

[68] Royal Society and National Academy of Sciences, *Climate change: evidence and causes* (February 2014); 'Proof positive', *Independent* (27 February 2014), 2.

7 Science and international politics

> Encourage the international proclivities of science and scientists in order
> to build bridges of understanding.[1]

There is an international dimension to the Royal Society because there is
an international dimension to how science is done. There is an international dimension to how science is done because some disciplines (such as
geophysics, meteorology, oceanography, epidemiology) depend on data
from several parts of the globe; and because some disciplines are too
expensive to be resourced by a single country; and because nearly all
disciplines are investigated by groups in several countries and progress
is accelerated if those groups can interact with each other, collaboratively
and competitively. But there is more to it than that.

The second half of the twentieth century was a period of the freezing
and thawing of the UK's relations with several parts of the world, of
Europe re-forging its identity in the aftermath of war, of the
Commonwealth forming itself in the post-colonial era, of global communications transforming the options for international collaboration. All
these developments helped to shape how science was done, and all were
to some extent shaped by science.

The Royal Society epitomised the global prestige of UK science.[2] With
its renowned history and its widespread networks, its independence
coupled with its close and influential contacts with the UK Government,
it was in a good position to engage with the scientific dimension of
international relations and, in the process, to promote the interests of
science. It did so wholeheartedly.

---

[1] Stanford Research Institute, *Possible non-military scientific developments and their potential
impact on foreign policy problems of the United States* (Senate Committee on Foreign
Relations, 1959) – copy at TNA CAB 124/2726.

[2] For comments on the Society's reputation outside the UK, see interviews with Brian Heap
and Tony Epstein, and CM 14 October 1993, minute 38. Nearly 200 scientific institutions from 45 countries sent congratulatory messages to mark the Society's tercentenary in
1960.

Science and foreign affairs

King Charles II, launching the Royal Society, stated his ambition 'to extend not only the boundaries of the Empire, but also the very arts and sciences' – thus giving a global geopolitical setting at the outset to the Society's mission to improve science and technology. In a phrase excessively quoted by those later giving talks on the Society's history, King Charles empowered the Society to 'enjoy mutual intelligence and affairs with all and all manner of strangers and foreigners . . . without any molestation, interruption, or disturbance whatsoever'. His 'indulgence' in granting this privilege was, of course, restricted to 'things philosophical, mathematical, or mechanical', but it ensured that an internationalist outlook could permeate the Society's activities.[3]

Three hundred years later, the Royal Society found itself operating in a world where there was a growing interaction between 'things philosophical, mathematical, or mechanical' and things political and geopolitical.[4] Governments throughout the developed world were increasingly recognising the relevance of science and technology to their international concerns. After the Second World War and the advent of the nuclear age, the discourse was gradually broadened to embrace wider impacts of science and technology beyond the defence context. Over a period of two decades, from about the mid 1950s to about the mid 1970s, it became common wisdom that aspects of international science and aspects of foreign policy were or could be interlinked to mutual advantage. By the same token, it became clear that science was too important to be left to the whim of consenting scientists. International science had to accommodate the exigencies of international politics, just as international politics had to accommodate the process and progress of science.

The Royal Society's leaders were more than willing to play the game in order to foster the international dimension to the conduct of science. The Society explicitly saw itself as 'the body historically responsible for advice to and action on behalf of the Government in all relations with foreign

---

[3] See the second, 1663, Charter.
[4] There is now a considerable literature on this overlap. See, for example, Warner R. Schilling, 'Science, technology, and foreign policy', *Journal of international affairs*, 13 (1959), 7–18; Jean-Jacques Salomon, 'International scientific policy', *Minerva*, 2 (1964), 411–34; Jean-Jacques Salomon, *Science and politics* (Macmillan, 1973), chapter 9; Brigitte Schroeder-Gudehus, 'Science, technology and foreign policy', in Ina Spiegel-Rösing and Derek de Solla Price, eds., *Science, technology and society* (Sage Publications, 1977), 473–506; Brigitte Schroeder-Gudehus, 'Nationalism and internationalism', in R. C. Olby, G.N. Cantor, J.R.R. Christie and M.J.S. Hodge, *Companion to the history of modern science* (Routledge, 1990), 909–19; and other work cited below. Note also Royal Society and American Association for the Advancement of Science, *New frontiers in science diplomacy* (January 2010).

science'.[5] It held several cards that made it useful to government in this context: its Fellows were networked with scientific colleagues throughout the world; it could speak with authority, from the perspective of the 'working scientist', about the scientific issues that government needed to understand; and, being 'owned' by no single part of the government machine, it was able to deal with all parts independently of the relentless turf battles within Whitehall. This was particularly important for its dealings with the Foreign Office and the Overseas Development Ministry. The Society was also independent of government as a whole, and hence was the natural body to represent the UK in international, non-governmental, scientific fora. And it was able to field very strong teams that could deal with the most senior figures in Government and the Civil Service on socially equal terms.

One milestone in the growing politicisation of international science was a report by the American physicist Lloyd Berkner, instigator of the International Geophysical Year, called *Science and foreign relations*. This was submitted to the State Department in 1950, and spelled out the diverse reasons why those responsible for American foreign policy needed to accommodate science in their thinking, both as a shaper of the context in which they worked and as an instrument for achieving their objectives.[6] The State Department moved quickly to implement Berkner's recommendations on such matters as adding scientific attachés to embassy staffs. The UK moved more gradually, adding over the next ten years Stockholm, Bonn, Moscow, New Delhi and Tokyo to Paris and Washington as capitals in which its diplomatic representation explicitly included non-military aspects of science. In the same period, the Office of the Minister for Science established the (ultimately ineffectual) Overseas Research Council (ORC);[7] DSIR strengthened its machinery for international liaison; ACSP set up a committee on Overseas Scientific Relations to advise on 'policy questions arising out of overseas scientific relations of Departments'; and the Foreign Office created a delightfully named Scientific Relations and Outer Space section – with four staff – to

---

[5] Henry Dale to Henry Tizard, 16 February 1945: HD/6/8/6/11.

[6] John Krige and Kai-Henrik Barth, 'Introduction: science, technology, and international affairs', *Osiris*, 21 (2006), 1–21; Ronald E. Doel, 'Scientists as policymakers, advisors, and intelligence agents: linking contemporary diplomatic history with the history of contemporary science', in Thomas Söderqvist, ed, *The historiography of contemporary science and technology* (Harwood Academic Publishers, 1997); 'International scientific liaison', *Nature*, 166 (15 July 1950), 81–2; C.E. Sunderlin, 'United States science offices abroad', *Nature*, 166 (15 July 1950), 87–8.

[7] TNA CAB 124/1485–92; OM/51(59); *New scientist* (6 August 1959). The ORC was upstaged by the creation of the Department of Technical Cooperation (forerunner of the Ministry of Overseas Development) in 1961, and was dissolved with the implementation of the Trend Report in 1964.

handle aspects of civil science. Such moves were seen internally as reflecting the 'increasing influence of scientific matters in questions of foreign
policy and commercial relations'.[8] The Royal Society helped to populate
this evolving advisory machinery.

An initiative that had more direct impact on the Society was another
American report, produced in September 1959 for the Senate Committee
on Foreign Relations, that highlighted how foreseeable developments in
civil science would create major headaches for American diplomacy.[9]
Examples included international competition arising from exploitation
of the oceans, polar regions and space; explosive population growth
attributable to improved public health and medicine; rising expectations
in the developing world fuelled by improved communications; increased
global demand for raw materials and energy; and the impact of new
synthetic products on (US-friendly) suppliers of their natural analogues.
The problems, concluded the report gloomily, would outweigh the capacity of future science to deliver compensating solutions. Possible policy
responses included increased spend on research coupled with increased
focus on relevant disciplines (including social science and psychology,
seen as specially germane to foreign policy), and encouraging 'the international proclivities of science and scientists in order to build bridges of
understanding' – for example, more international conferences, exchange
programmes and collaborative projects. The British science attaché in
Washington promptly sent copies of the report to DSIR, to the Minister
for Science, Lord Hailsham, and to the Foreign Office.

An opportunity for the Royal Society

The Foreign Office was then in – slightly leisurely – discussion at a very
senior level with the Treasury, DSIR and ACSP about the adequacy of the
UK's civil research effort in relation to the needs of foreign policy. The
Senate Committee report gave them plenty to chew on. However, beyond
agreeing the importance of the issue, they were unsure how to approach it.
It was one of those things whose importance was matched by its vagueness,
and which thus seemed to lack urgency. Eventually, in May 1960, the
Foreign Office Policy Steering Committee decided to park the issue with

[8] DSIR, 'Overseas Liaison Group and scientific attachés', May 1960: TNA T 218/646.
Foreign Office memo 27 June 1962: TNA CAB 124/2726. R.A. Butler to Howard
Florey, 24 July 1964: RMA742.
[9] Stanford Research Institute, *Possible non-military scientific developments*; see also 'United
Nations collaboration in the sociological sciences', *Nature*, 185 (27 February 1960), 561–
4, and 'Science and foreign policy in the United States', *Nature*, 185 (27 February 1960),
584–5.

the scientists.[10] The Deputy Secretary Paul Gore-Booth therefore wrote to David Martin, the Assistant Secretary (head of staff) at the Royal Society, seeking 'a really authoritative external view' on the matter because science was now 'so important that, like economics, its relationship to foreign policy needs special consideration'. He was then sent to India as High Commissioner. Martin made haste slowly, and it took the rest of the year for him to submit a paper to Gore-Booth's successor, Patrick Reilly.[11]

Martin's paper, echoing Cyril Hinshelwood's comments at the recent tercentenary (see Chapter 8), set out the classical picture of science as 'largely free from national interests ... one of the few common languages of mankind ... Friendly international relations among scientists tend to reduce international tension and should be encouraged.'[12] However, Martin also acknowledged that scientists were not always adept at recognising the political significance of their activities: the Royal Society and others involved in the IGY, for example, had been well aware of the impending launch of *Sputnik* (October 1957) but had missed its wider implications.[13] Nevertheless, he suggested, government needed early warning of potentially significant scientific developments, and a high-level committee of Foreign Office representatives and experienced scientists could facilitate this. Such a committee could, further, help the UK extract political capital from the prestige attached to its scientific leadership. Though of course Martin did not mention it, his proposed committee would also underline the Society's determination to manage its own interactions with any part of Whitehall and not to have to channel everything through Lord Hailsham and the Office of the Minister for Science or the DSIR.[14]

[10] Meeting of 10 May 1960: TNA FO 371/152116.

[11] Paul Gore-Booth to David Martin, 17 May 1960: TNA CAB 124/2726. Copies of Martin's paper *Scientific progress and foreign policy* are at OM/6(61), HF 1/17/1/10, and TNA CAB 124/2726.

[12] David Martin was passionate about international collaboration in science. See interview with Chris Argent.

[13] By way of an addendum to that story, the use of the Jodrell Bank telescope in tracking *Sputnik* is well known. But in 1960 Jodrell Bank became centre stage in government thinking about how to track the launch of intercontinental ballistic missiles, and thus brought a whole new appreciation of the value of basic research. Francis Graham-Smith and Bernard Lovell, 'Diversions of a radio telescope', *Notes and records of the Royal Society of London*, 62 (2008), 197–204.

[14] To illustrate the problem of Whitehall protocol: Alcon Copisarow, Director of the Forest Products Research Laboratory and previously science attaché in Paris, offered to help David Martin with his paper, and Martin convened a group of experienced Fellows to meet with him. When Copisarow's DSIR boss, Harry Melville, heard of this, he forbade Copisarow from talking directly with the Society and told him to channel all communication through DSIR. The briefing meeting was cancelled. TNA DSIR 17/827; also Martin to Tommy Thompson, 5 December 1961: HWT B180, and Martin to Patrick Blackett, 21 May 1965: RMA742.

Martin had spent the first half of 1960 finalising the Society's report for Hailsham on *The encouragement of scientific research in the UK* (see Chapter 2). Never one to miss a chance, he now used the platform of a discussion about science and foreign policy to launch a further bid for additional resources for the Society. Hence much of his response to Gore-Booth, and a follow-up paper five months later,[15] was devoted to arguing the case for the Society to expand its new schemes of visiting professorships and scientific exchanges within and beyond the Commonwealth, hosting more international meetings, disseminating UK scientific literature to other countries, and working more fully with international scientific bodies. Martin estimated that the Society's publicly funded international programme needed to be trebled, to £150,000 p.a., to accommodate these ambitions. This, he argued, would strengthen the international standing of UK science and thus render it a more effective instrument of foreign policy.

Martin's papers highlighted an awkward generic issue for the Society. It could speak to government with authority, and with independence, but it was still *parti pris* in the sense of being a potential beneficiary of its own recommendations when it came to implementing policy advice. It was not always disinterested, nor could it be. The Society needed to tread carefully lest its advice about the best interests of British science be interpreted as the best interests of the Royal Society, and discounted accordingly.

Reactions in Whitehall

In this instance, it looked as if David Martin might have overplayed his hand. Patrick Reilly and others were concerned at how the agenda was morphing from a general consideration of interactions between science and foreign affairs into a narrow focus on the Society's own funding ambitions. Richard Griffiths, head of the Arts and Science Division at the Treasury, was scathing: 'The Royal Society ... have come across the splendid idea that most of the Foreign Office's philosophical doubts can be resolved if the Royal Society were given more money ... The Foreign Office are not anxious to be used as an instrument for securing increased travel grants for the Royal Society.' To make matters worse, Griffiths also rejected the core notion of linking scientific and foreign policy objectives. He took a much narrower view, telling David Martin bluntly: 'It will be a sad day when scientists in general or the Royal Society in particular are given additional subventions in order specifically that they should act as

---

[15] 'The Royal Society and international scientific relations', OM/59(61); cost estimates in OM/68(61).

instruments of national foreign policy. The present grants to the Royal Society ... are justified by reference to the needs of British science and not of British prestige.'[16]

Martin's paper circulated among interested parties in Whitehall, and formed the basis of a meeting between the Society and the Foreign Office on 5 July 1961. The meeting rehearsed the impact of civil science on issues of concern to the Foreign Office, on much the same lines as the 1959 Senate Committee report.[17] The Society had been advised privately not to use the occasion to lobby for extra funding, but the Foreign Office proved sympathetic to the idea that the Society might provide one of the channels through which it accessed emerging science. Reilly himself wanted informal biannual meetings with the Society that might develop over time into a consultative committee of the type proposed by David Martin. His recent spell as Ambassador to Moscow had convinced him that the Foreign Office benefited considerably from the Society's international expertise, and he saw mutual advantage in maintaining close relations.[18]

However, at a time when groups on international scientific cooperation were proliferating through Whitehall, and when momentum was building towards the setting up of what became the Trend Enquiry into the organisation of civil science, the Society's historic consultative role in respect of science and foreign policy was under competitive pressure. Roger Quirk, a senior official in Hailsham's Office, partisanly advocated the new ACSP Committee on Overseas Scientific Relations, which now came under Hailsham and on which both the Royal Society and the Foreign Office were represented, as the natural vehicle for discussions about science and foreign policy. Both the Office of the Minister for Science (Quirk and Frank Turnbull) and the DSIR (Harry Melville)[19] were determined to prevent the Foreign Office bypassing established Whitehall channels in its dealings with science. In a detailed 1963 paper on how to strengthen the impact of science and technology on foreign relations, DSIR argued: that policy advice should be sought through existing structures, reinforced by a new post of Scientific Adviser within the Foreign Office; that scientific exchanges should be managed primarily by the British Council; and that the Royal Society should concentrate just

[16] Griffiths to Arnold France, 2 June 1961: TNA T 218/508; Griffiths to Martin, 2 June 1961: OM/88(61); and other exchanges at TNA CAB 124/2726, T 218/508, and FO 371/161233, some summarised in OM/94(61).

[17] Official Foreign Office minute of the meeting, and Roger Quirk minute to Frank Turnbull, both at TNA CAB 124/2726; also OM 13 July 1961, minute 2(f).

[18] Patrick Reilly, 'Science and foreign policy', memo dated 5 October 1961: TNA CAB 124/2726.

[19] See exchanges at TNA DSIR 17/827.

on representing the UK internationally in non-governmental scientific bodies such as ICSU.[20]

Despite Patrick Reilly's supportive leadership, there were also arguments inside the Foreign Office about how much science it really needed, whether diplomatic staff themselves should have first-hand knowledge of science,[21] whether the Office should have its own in-house chief scientific adviser, and how it should access external advice.[22] But the core thesis that the world of foreign policy had to pay serious attention to the world of civil science had entered mainstream thinking: the Foreign Office ran a summer school on science and foreign policy in 1963, with Howard Florey and other Fellows among the contributors.

The international dimension featured substantially in the Trend Enquiry. The Society's evidence to Trend stressed the case for it to have a strong role both in international projects and organisations and in international relations. Trend, however, took the DSIR line that the Society's international work should concentrate on promoting non-governmental international scientific cooperation. Reviewing their strategy in the post-Trend dispensation, the Society's Officers affirmed their own more ambitious vision. They committed to 'continue and intensify' their work on non-governmental international scientific relations, where the Society's independence was prerequisite. But they also committed to 'seek to advise the Foreign Office on science and foreign policy at every opportunity', as an alternative to the Foreign Office establishing too much of its own scientific capability.[23] And the Foreign Office happily continued to refer to the Royal Society as 'the traditional adviser of Her Majesty's Government on scientific matters'.[24]

In the post-Trend world, the new Department of Education and Science closed the international committees (populated mainly by active scientists) of the now-defunct DSIR and ACSP, and replaced the former with two new committees of its own on International Scientific Cooperation (CISC) and Overseas Scientific Relations (COSR), both dominated by civil servants. The Royal Society remained keen to stake

---

[20] TNA DSIR 17/827 and CAB 124/2727. On ICSU, see Chapter 10.

[21] At that time, only twenty-seven members of the Foreign Service held degrees in science subjects. 'Science and foreign affairs', *Nature*, 202 (13 June 1964), 1039–41.

[22] See, for example, numerous memos at TNA FO 371/161233, FO 371/161234 and FO 371/170982, especially papers by Patrick Reilly dated 16 November 1961 and by Ronald Hope-Jones dated 7 February 1963. The post of Chief Scientific Adviser to the Foreign Office was eventually created in 2009.

[23] OM/73(64). See also a minute of a discussion on science and foreign policy at the Foreign Office on 20 November 1963: TNA CAB 124/2727.

[24] Nicolas Cheetham (Foreign Office Assistant Secretary) to David Martin, 13 June 1963: OM/47(63).

its own claims to be involved in government discussions about international scientific relations, 'as it was important the views of the working scientists should be heard as well as those of the civil servants'. When in September 1965 the Society was eventually notified officially of the existence of COSR and CISC and told that it could ask for their papers and would be invited to attend meetings when the civil servants deemed there to be matters of direct interest, it was not impressed.

So the Society took a different route. The new Council for Scientific Policy, chaired by Harrie Massey, at first had no international committee of its own. Harrie Massey and Howard Florey therefore set up a joint CSP/Royal Society group to review governmental and non-governmental international scientific relations, and by the spring of 1966 that group had become the CSP Standing Committee on International Scientific Relations, with strong Royal Society representation. The Foreign Office was supportive, recognising that the Massey/Florey group was unusually alert to the political aspects of international science.[25] At the same time, the Royal Society beefed up its machinery for delivering a more strategic approach to its own work on international relations.

Encouraged by these moves, the Foreign Office set about initiating a sequel to the 5 July 1961 meeting with the Royal Society leadership. The logic was the same as before: the 'immense influence which scientific developments have' on both political and economic relations between nations, the lack of scientific expertise within the Foreign Office, and the expectation that the Royal Society – 'the supreme scientific body in this country' – would be able and willing to supply 'the best scientific advice available'.[26] The then President Patrick Blackett was particularly interested in helping the Foreign Office acquire earlier information and better understanding of the significance of developments in science and technology, and was keen to put the Society's expertise at its disposal. Support within the Foreign Office for such an approach was coupled with continuing preference for keeping it informal, and continuing sensitivities about not tripping over the official Whitehall channels of communication.

[25] OM 29 January 1965, minute 2; C Wigfull to Ronald Keay, 1 September 1965: CAB 124/2983 and OM/(69(65); OM/86(65), OM 30 November 1965, minute 2(a), and OM 13 January 1966, minute 2(c); internal Foreign Office memos by James McAdam Clark, 7 February 1966, and Ben Strachan, 17 March 1966: TNA FO 371/189399. On the CSP international committee, see also P.M.S. Blackett, 'Anniversary Address, 1966', *Proceedings of the Royal Society of London. Series A, Mathematical and physical sciences*, 296 (1967), xiii–xiv.

[26] J.A. Thomson (Head of Foreign Office Planning Department) to John Nicholls (Foreign Office Deputy Secretary), 5 August 1966. Also Thomson to John Rennie (Nicholls' successor), 14 March 1967. TNA FCO 49/38.

Regular meetings at ministerial level – complementing very extensive interactions at staff level – eventually resumed in April 1969. The saturation briefing prepared for the minister, Fred Mulley, for the first such meeting highlighted the 'importance which the FCO attach to the closest cooperation with the Royal Society and with the British scientific community; the extent to which we are dependent on scientists both in and outside government in the formulation of policy; and the increasing importance of scientific and technological considerations and subjects in the general framework of foreign policy'. Even allowing for the ritual courtesies of the occasion, this does signal an openness to external help with the scientific dimensions of foreign policy. The meeting confirmed the scope for useful interactions between the Society and the FCO, both in terms of helping each other interpret the changing world of international relations and in terms of specific, mutually beneficial, actions. The FCO, for example, agreed to lobby the Treasury on tax issues related to international exchanges, to lobby DES on funds for exchanges with Japan, and to set up meetings between the Society and other groups dealing with Western European relations. The Society, in its turn, invited the FCO to strengthen its participation in the CSP international committee, agreed to find relevant experts for various tasks within the FCO, and offered help with briefings for science attachés. More subtly, and minding its independence, the Society was able to sound out the FCO on UK relations with China, Cuba and Yugoslavia, on manoeuvrings in the EEC over scientific collaboration, and on future scientific initiatives with the Commonwealth.[27] The upshot was an agreement to intensify connections between the Society and the FCO, while keeping it all informal to avoid trouble with the rest of Whitehall.

Further ministerial meetings followed on a roughly annual basis for the next three decades, in effect fulfilling the functions that David Martin had prescribed in 1960 for the proposed high-level 'early warning' committee. Some meetings focused on specific issues like the 1972 UN conference in Stockholm on the human environment, while others ranged across the breadth of international relations. But there were always limits to the intimacy. The briefing for the FCO minister Lord Chalfont in 1970 warned him candidly:

During the past few months the Royal Society have been very cooperative and a good working relationship has been established with them. Their advice will be extremely useful ... but we should make no commitment about the way it will be fed into the governmental machine ... Further, the DES are concerned lest the

---

[27] TNA FCO 55/233. FCO's minutes of the meeting also at OM/48(69), and the Royal Society version at OM/46(69).

direct links between the Royal Society and their Department be supplanted by a dialogue between the RS and the FCO.[28]

Relations between the Society and the FCO continued to develop, to their mutual advantage in sometimes unexpected ways. In June 1972, the Society hosted an evening for all foreign science attachés in London, plus their ambassadors, and invited the UK Foreign Secretary, Alec Douglas-Home, to make a speech. Three months later, it sought help from the FCO to airfreight scientific material out of Uganda via diplomatic channels after Idi Amin's sudden decision to expel Asians from that country. When transmitting this request to the High Commission in Kampala, at a time of extreme pressure, the FCO in London justified adding to the Commission's burdens on the grounds that it was 'much indebted to the Royal Society for high level advice on a range of subjects of importance to HMG, which is freely and willingly made available by the Royal Society and its Fellows ... It behoves us to do anything we can to lend them a hand.'[29] The material duly arrived back in England. In 1974 the FCO Permanent Secretary, Thomas Brimelow, was guest of honour at the Society's annual dinner.[30] During the 1970s and early 1980s, the Central Policy Review Staff (CPRS), sitting at the heart of government, also tackled a number of international science issues and worked closely with the Society in that context.[31]

H.W. ('Tommy') Thompson, as the Society's Foreign Secretary from 1965 to 1971, periodically complained that the UK failed to exploit the prestige value of its scientific achievements on the international stage. From the mid 1970s, this became a more prominent theme in the Society's dealings with government, rather as Richard Griffiths in the Treasury had earlier feared. In February 1976, for example, the Society urged a CPRS review of the UK's overseas representation that: 'Although Britain's place in the world in terms of economic and military strength is now much declined, its place in science is now second to none ... an asset which could be used with increasing effectiveness.'[32] In October that

---

[28] TNA FCO 55/381. In 1975 the DES asked to send an observer to RS/FCO ministerial meetings: both the Society and the FCO rejected the request. OM 17 April 1975, minute 2(e).

[29] J.B. Ure to J.C.E. Hyde, 7 September 1972: TNA FCO 55/917.

[30] Tommy Thompson to Thomas Brimelow, 14 October 1974: HWT/B.186; Thomas Brimelow, 'Anniversary dinner 1974', *Notes and records of the Royal Society of London*, 30 (1975), 5–14.

[31] Interviews with John Ashworth and Robin Nicholson.

[32] OM/50(76). In 1971, the United States had nineteen science attachés in post, France had fifteen and the UK and Japan had seven. Washington was host to more foreign science attachés than any other capital, with twenty, followed by London (15), and Paris and Tokyo (14). J.W. Greenwood, 'The scientist-diplomat: a new hybrid role in foreign

year, the President Alex Todd reminded the FCO minister David Owen that 'The UK still held a powerful position in the scientific world which, unlike other sources of influence, had not diminished in recent years.'[33] The prestige factor was, of course, key to being able to corral science to foreign policy objectives in the first place: it was because other countries wanted access to UK science that the international dimension to the practice of science in the postwar era could have this rich complexity. And, though a small, niche player in budgetary terms, the Royal Society could play the prestige card effortlessly.

affairs', *Science forum*, 19 (1971), 14–8; Derek de Solla Price, 'The world network of scientific attachés', *Science forum*, 21 (1971), 34–5. On the Society's positive view of the value of attachés, see John Ashworth interview.

[33] FCO minute of the meeting at OM/153(76). See also FCO paper dated 9 December 1977 prepared for a ministerial meeting with the Royal Society – 'Scientific contacts laying the foundation for eventual economic and commercial benefits to the UK, with particular reference to the Middle East': RMA622; and interview with Bill Stewart.

8   Keeping the door open

The door is often open to scientists when it is closed to others.[1]

Launching the tercentenary celebrations in July 1960, the President, Cyril Hinshelwood (Figure 8.1), affirmed his belief that 'the flow of ideas and of goodwill through scientific channels has contributed, and can contribute, sometimes in adverse circumstances, towards the peace and harmony of all mankind'.[2] In the following decades there were plenty of adverse circumstances, plenty of opportunities for the Royal Society to discover whether it could live up to this high calling. Its aim, throughout, was the promotion of science. But the Society was also alert to the possibilities for tangential political progress – not to mention direct competitive benefit to the UK – and worked in quiet collaboration with the British Government to keep open the doors of communication between nations at times of international tension.

It was a role that was thrust upon the Society by the events of the period – not for the first time in its history. As an independent, non-government body of global repute, and seen as concerned much more with fundamental research than with economically or militarily significant technology, the Royal Society was sufficiently credible and sufficiently non-threatening to have some chance of success. So, for example, it participated in the UK's engagement with the Soviet Union during and after the Cold War; with China, notably in the wake of the Cultural Revolution and Tiananmen Square; with South Africa during and after the apartheid era; and with Argentina around the time of the Falklands War. The work could be controversial, within and beyond the Fellowship. Having to thread its path through the various adverse circumstances pushed the Society to clarify what it stood for and what types of initiative it could legitimately, and usefully, undertake.

---

[1] Tommy Thompson, 'International relations: a progress report by the Foreign Secretary, 1968': C/71(69).
[2] Cyril Hinshelwood, 'Tercentenary Address', 23.

Figure 8.1 Cyril Hinshelwood giving the opening address at the
tercentenary celebration, 19 July 1960. © The Royal Society

The Soviet Union

During the Second World War, the Royal Society and the Soviet
Academy of Sciences exchanged messages of fraternal solidarity, high-
lighting the contribution of scientific research to the struggle.[3] The Royal
Society's messages were printed in full in *Pravda* and *Izvestia*, and the
Soviet Academy elected the Society's President Henry Dale an honorary
member.[4] Catching the spirit of the moment, Dale himself commented to

---

[3] CM 17 July 1941, minute 20 and 28 May 1942, minute 5; HD 6/8/7/1; Henry Dale,
 'Address at the Anniversary Meeting, 1 December 1941', *Proceedings of the Royal Society of
 London. Series A, Mathematical and physical sciences*, 179 (1942), 233–60. See also
 'Correspondence with foreign academies, societies and other bodies', *Notes and records
 of the Royal Society of London*, 4 (1946), 69–74.
[4] Dale resigned from the Soviet Academy in 1949 in protest against its endorsement of
 Lysenkoism: W.S. Feldberg, 'Henry Hallett Dale, 1875–1968', *Biographical memoirs of
 Fellows of the Royal Society*, 16 (1970), 152–3.

the Fellowship in 1941 on the 'new and growing intimacy of collaboration with our colleagues of Soviet Russia'. However, as the war in Europe drew to a close, relations between the allies became more complicated and relations between their academies of science followed suit.

Making the most of a slim opportunity, the Soviet Academy of Sciences contrived to celebrate its 220th anniversary with a fortnight of speeches, scientific visits and general conviviality in June 1945, and sent invitations to 146 individual scientists and 54 learned societies across the globe. Henry Dale's instinct was to respond with a friendly message and leave it at that. When the scale of the occasion became apparent, Dale became more effusive in his apologies, telling the Soviet Ambassador that the celebration would underpin 'a progressive development of friendly collaboration between the scientists of our two countries, which we regard as a matter of the first importance to the future of the world'; but he still refused to attend. He had more urgent matters to deal with, including the controversy over his successor as President of the Royal Society (see Chapter 1). In the end, twenty-one British delegates did go, led by Robert Robinson, at that stage one of the Society's Vice-Presidents. The British Government was supportive to the extent of offering air transport at very short notice. What proved much more difficult, however, was securing exit visas from the UK. At the last minute, the British Government refused exit visas to, among others, J.D. Bernal, Patrick Blackett, Paul Dirac, Nevill Mott and Ronald Norrish, despite Dale's best efforts behind the scenes. It transpired in due course that the refusal was at the behest of General Groves, the highly security-conscious head of the Manhattan Project, and, given the timing (two months before Hiroshima), it would have been impossible to provide a full explanation to those affected. The incident caused considerable ill feeling for a while among sections of the British scientific community: it was clear that some doors were not always open, even to the most senior scientists.[5] Dale used his final Anniversary Address on 30 November 1945 to warn about the dangers of unwarranted secrecy undermining the progress of science.

Relations between the Royal Society and the Soviet Academy broke down completely in 1948 over the Lysenko affair, which reached a climax that year when the Soviet Academy declared Lysenko's erroneous ideas about genetics to be the only permissible approach to the

[5] See numerous exchanges at HD 6/8/7/1, including correspondence with the Editor of *The Times*; AVHL I 3/36 and I 3/67; A.V. Hill, 'Cancelled visit of British men of science to the Academy of Science of the USSR', *Nature*, 155 (1945), 753; A.V. Hill, *Memories and reflections*, chapter 101. A suitably enthusiastic account of the visit was published in *Notes and records of the Royal Society of London*, 4 (1946), 65–7.

subject.[6] Restoration of relations was gradual. In 1955, the Soviet Academy invited the Society at just two weeks notice to be represented at a conference in Moscow on the peaceful uses of atomic energy – a month before a major international conference in Geneva on the same subject. The Foreign Office sent mixed signals about the political desirability of attendance, and the Society eventually declined the invitation.[7] However, with encouragement from the Foreign Office,[8] the Society did agree to work with the British Council to set up an exchange scheme with the USSR so that scientists from each country could visit the other for varying periods of time to give lectures and carry out collaborative research. The Soviet Academy sent over a full-scale delegation in November 1955 to discuss the details. The next year the Society reciprocated with an equally heavyweight delegation to Moscow and Leningrad. A formal agreement was approved by both sides in December 1956, paving the way for the first formal agreement, three years later, between the British and Soviet Governments for cultural exchanges more generally.[9] On the back of this progress, the British Government established a scientific attaché post in Moscow in 1958, and the Soviet Ambassador Jacob Malik visited the Royal Society in 1959 to admit three Fellows – Paul Dirac, Cyril Hinshelwood (then President of the Society) and David Watson – to foreign membership of the Soviet Academy.[10]

[6] Jennifer Goodare, *Representing science in a divided world: the Royal Society and Cold War Britain* (PhD thesis, University of Manchester, 2013), 94; David Joravsky, *The Lysenko affair* (Harvard University Press, 1970).

[7] CM 16 June 1955, minute 14, and 14 July 1955, minute 15.

[8] John Deverill, *Royal Society experience in facilitating scientific interchange with the Soviet Union*, typescript September 1973: RMA622.

[9] CM 13 October 1955, minute 17; 15 December 1955, minute 6; 12 July 1956, minute 9; 13 December 1956; correspondence between the Royal Society and the British Council: OM/53(56); H.G. Thornton, 'A note on the visit to Russia of the Royal Society delegation in 1956', *Notes and records of the Royal Society of London*, 12 (1957), 230–6; John van Oudenaren, *Détente in Europe: the Soviet Union and the West since 1953* (Duke University Press, 1969), 289. The Soviet Academy had previously tried, unsuccessfully, to organise exchanges through scientists such as Bernal and Blackett. Alex Todd observed that the orthodox channels of the Royal Society provided a surer route to official approval: Alexander Todd, *A time to remember*, 131–5. Hinshelwood later told Hailsham that the November 1955 visit from the Soviet Academy 'may have made quite a useful contribution to the general unfreezing': Hinshelwood to Hailsham, 22 January 1960: RMA222.

[10] 'Visit of His Excellency the Soviet Ambassador, 19 November 1959', *Notes and records of the Royal Society of London*, 14 (1960), 160–2. Dirac had been elected in 1931, Hinshelwood in 1948 and Watson in 1932. Continuing the courtesies, the Society cabled congratulations to the Soviet Academy on Yuri Gagarin's pioneering space flight in April 1961, and Gagarin visited the Society later that year: CM 20 April 1961, minute 25. A somewhat warmer cable to the NAS greeted John Glenn's flight in February 1962: CM 12 April 1962, minute 24.

With strong support from the Foreign Office, the Royal Society quickly established similar exchange agreements with Warsaw Pact countries: Bulgaria (1960), Poland (1962), Hungary (1962), Romania (1963) and Czechoslovakia (1964).[11] The UK share of the funding came mostly from the British Council. Such initiatives were symbolically significant, and also elicited a certain amount of useful information about scientific activity in the various countries. Outgoing British scientists were required to write reports on their return, which were valuable in the planning of future visits. Edited versions of the reports were circulated to the British Council and the Foreign Office.

But the exchange schemes were necessarily small scale and, despite protestations of mutual goodwill, were beset by administrative inefficiencies. Additional challenges arose from unequal demand for travel to and from the UK,[12] and from the different expectations of the British and Soviet systems about the extent to which science could be centrally planned and controlled. Positions on both sides hardened in periods of heightened diplomatic tension. After the building of the Berlin Wall in August 1961, the UK and other NATO countries refused visas to East German representatives, including scientists wanting to attend international conferences. That made it almost impossible under ICSU rules against exclusion of bona fide scientists to hold international scientific conferences in the UK, and the Royal Society had to lobby the UK Government at the highest level to secure a relaxation of the visa ban.[13] Howard Florey led an extensive visit to the USSR in 1965 and saw the general difficulties at first hand.[14] His valedictory Anniversary Address later that year gave a downbeat assessment: 'We have to be content with relatively little. Agreements reached with the Soviet Union and other communist countries spell out in detail exchange arrangements with which we are all pleased, but real international working has scarcely been touched.'

Warsaw Pact troops invaded Czechoslovakia on 20 August 1968. It was a sharp reminder, if one were needed, of the Cold War context.

[11] Foreign Office officials started to worry whether they were overdoing it: 'Perhaps we should be a little more cautious in future in pressing the RS and others to develop contacts with Eastern Europe? We don't want them to get browned off.' Memos between Bob Brash, Richard Speaight and others, September–October 1966: TNA FO 924/1594.

[12] The Foreign Office estimated that in 1966–7 about 2,000 Soviet scientists visited Britain and about half that number of British scientists visited Russia, mostly for conferences rather than as official representatives. Briefing on Anglo-Soviet contacts dated 16 January 1968: HWT C.2.

[13] OM/80(62), OM/89(62), OM/100(62), OM/32(63), OM/56(63), C/97(64), OM/25 (66), HF 1/17/1/10, HF1/17/1/21, HWT B.522, RMA742, TNA FO 924/1594, TNA FO 371/189399.

[14] Interview with Terry Garrett.

The invasion occurred while an international geological congress was being hosted by the Czechoslovak Academy of Sciences. The troops occupied the Academy's buildings in Prague and brought the congress to a premature end. The reformist President of the Czechoslovak Academy, František Šorm, promptly wrote in protest to Mstislav Keldysh, President of the Soviet Academy, and, for his pains, was in due course removed from most of his posts. The President of the National Academy of Sciences in Washington, Frederick Seitz, who had visited Prague just before the invasion, also wrote to Keldysh to express his 'grave concern' that scientific relationships would be 'seriously jeopardised' and to urge Keldysh to help Czechoslovak scientists preserve the 'substantial degree of intellectual freedom' that they had already achieved. The Royal Society resisted pressure from within the Fellowship to make a public statement and/or break off relations with the Soviet Academy, on the grounds that no good would come of it, but voiced its 'strong disapproval' to the Soviet Ambassador and encouraged individual British scientists in touch with Soviet counterparts to make their feelings known. The Officers wrote collectively to Šorm to offer sympathy and support, and immediately expanded the exchange scheme as a way of keeping the door open to continued collaboration. They also quietly got on with helping refugee Czechoslovakian scientists to find suitable posts in the UK. Tommy Thompson visited Šorm and the Czechoslovak Academy in February 1969.[15] The Foreign Office minister, Fred Mulley, affirmed that the Society's actions had been much appreciated in Czechoslovakia.[16]

One door that did close, at least temporarily, as a direct result of the invasion of Czechoslovakia was the fledgling Anglo-Soviet Consultative Committee on Bilateral Relations. This had been initiated in 1967 by the British Government as a high-level but non-governmental body to give confidential advice to the British and Soviet Governments on aspects of bilateral relations where non-governmental initiatives could play a significant role. Such aspects included science, industrial development, culture, sport and tourism. Tommy Thompson represented science on the Committee. However, the invasion occurred before the Committee could hold its inaugural meeting. The British immediately cancelled the meeting as incompatible with the post-invasion climate of opinion in the UK. It was October 1969 before most of the British side felt ready to convene the full Committee, with Thompson arguing for a longer period of suspension because of the increasingly hostile behaviour of the Soviet

[15] OM 23 September 1968, minute 2; CM 10 October 1968, minute 35; CM 7 November 1968, minute 17; HWT B.355, B.356, B.359, B.360.
[16] OM/46(69).

Academy towards the Royal Society following his February visit to Prague. The Committee Chairman, Humphrey Trevelyan, recently Ambassador in Moscow, commented laconically: 'Given the Soviet concept of peaceful cooperation as a political and ideological struggle, periodical set-backs in Anglo-Soviet relations have to be expected.' In the event, the first meeting of the full Committee was held in April 1970. The fact that it happened at all was probably more significant in terms of détente than any specific outcomes that may have resulted. Plans for a follow-up meeting in late 1971 were put on hold indefinitely amid mutual recriminations and mutterings of administrative incompetence.[17]

In the early 1970s the Royal Society found itself grappling more and more with the challenge of mounting an effective response to the persecution of scientific colleagues in other countries. Individual Fellows were horrified at the stories emerging from the Soviet bloc, and were eager to express solidarity with persecuted colleagues. The Society was entirely content that they should do so. But framing an institutional as distinct from a personal response was not straightforward. To do nothing was to risk being perceived as indifferent or even complicit; but to take inappropriate action might make matters worse. There were plenty in the UK and elsewhere ready to urge boycotts of, at least, official invitations to visit countries that restricted the freedom of scientists.[18] But it was recognised, by the British side of the Anglo-Soviet Consultative Committee among others, that that would play into the hands of governments whose restrictive policies were directed against their own citizens rather than against foreigners and whose aim was to isolate their own citizens from foreign contact. The chemist Otto Wichterle in Prague, a staunch reformist being systematically hounded by the post-invasion government, despaired at calls in *Nature* to boycott a conference being organised in Czechoslovakia, lambasting their proponents for appeasing their own consciences while doing nothing practical to help.[19]

Tommy Thompson fully appreciated the dilemma.[20] The Society's representatives took every opportunity to tell their Soviet counterparts

---

[17] Papers and correspondence relating to the Consultative Committee are at HWT C.1–16. Note also OM/31(70) and OM 14 May 1970, minute 2(a), OM/69(70) and OM 11 June 1970, minute 2(e).

[18] For example, see Francis Crick, John Kendrew et al., 'International conferences', *Nature*, 224 (4 October 1969), 93–4; *Nature*, 225 (10 January 1970), 120.

[19] Undated letter from Wichterle; note also undated letter from Alexander Kasal to Ewart Jones giving news of what Šorm, Wichterle and others were up against: HWT B.526. Also David Hughes (British embassy in Prague) to Thompson, 22 December 1969, reporting that Wichterle had been banned from travelling outside Czechoslovakia: HWT B.527. And 'When to boycott', *Nature*, 226 (9 May 1970), 482–3.

[20] See, for example, his paper *Czechoslovakia, the USSR and the Royal Society*: OM/7(70).

how strongly they disapproved of the situation in Czechoslovakia and other parts of the Soviet bloc where the suppression of dissent was intensifying. A group of over 100 Fellows wrote to the Prime Minister about the treatment of dissident scientists, and Thompson himself, by then no longer the Society's Foreign Secretary, debated the issue on television with George Porter.[21] But the Society remained opposed to making a corporate public protest, and this policy was continued by Thompson's successor as Foreign Secretary, Kingsley Dunham. Dunham pointed out that maintaining relations with the Soviet Academy provided a channel through which the feelings of the British scientific community could be communicated directly to the highest echelons of Soviet science. But the policy was controversial among the Fellowship as a whole, sometimes publicly so.[22] The Society's refusal to follow the NAS example of formal public protest at the Soviet Academy's persecution of the physicist Andrei Sakharov added fuel to the fire.[23] One Fellow particularly uncomfortable with the Society's position, John Ziman, wrote to the entire Fellowship in February 1974 asking for comment. He received 175 responses – a very high response rate, 23 per cent, for this sort of initiative – showing roughly equal levels of support for the existing policy and for the Society taking a more publicly conspicuous approach. The President, Alan Hodgkin, met with Ziman to discuss his findings. Like the Fellowship as a whole, the Royal Society Council was itself divided on the issue and considered it at length several times in early 1974, eventually reaffirming the established preference for keeping the door open to engagement with difficult regimes and seeking to exercise influence away from the limelight.[24]

The issue, of course, did not go away. Alex Todd told the Foreign Office minister David Owen in October 1976 that the Society was still under continual pressure from Fellows and others to break off relations with the Soviet Academy. In his Anniversary Address that year, Todd defended his policy of keeping the door open, arguing that the Society, as distinct from individual Fellows, had no particular mandate to intervene in the case of scientists suffering retribution for opposing

---

[21] C/28(74).

[22] For example: John Ziman, 'The problem of Soviet scientists', *Nature*, 246 (7 December 1973), 322–3, and Eric Burhop, 'The problem of Soviet scientists: a reply to John Ziman', *Nature*, 248 (12 April 1974), 542. Also OM/129(73) and OM/130(73).

[23] CM 11 October 1973, minute 18, and OM/109(73).

[24] C/234(73); CM 17 January 1974, minute 8; CM 7 March 1974, minute 7; CM 4 April 1974, minute 10; CM 23 May 1974, minute 38(v); RMA459. See also Jennifer Goodare, *Representing science*, chapter 5. One incidental scientific benefit was that the Soviet Academy gave the Society a sample of moon rock in 1974, and Soviet scientists were invited to a major discussion meeting on lunar science the following year: OM/75(74).

their governments. When engaged in political activities, scientists had no privileged position distinct from that of other citizens. Owen endorsed Todd's approach.[25]

FCO ministers were comfortable with the way that the Society's publicly funded exchange schemes with the Soviet Union during the Cold War served both scientific and political objectives.[26] The FCO recognised that the exchange schemes had a general political value and helped meet the need to broaden contacts between East and West, especially following the 1975 Helsinki Conference on Security and Cooperation in Europe. It therefore added £3,000 to the £40,000 provided to the Society through the British Council and the £40,000 in the PGA to support exchanges with the Soviet bloc. The total of £83,000 covered the UK share of a nominal forty-nine person/months per year exchange in each direction, with the sending side covering travel costs and the receiving side all in-country costs.[27]

However, increasing restrictions on Soviet scientists wishing to travel outside the Soviet Union threatened relations between the Society and the Soviet Academy. Only one out of twenty-five scientists specifically invited by the Society actually reached Britain during 1978 and 1979. The arrest of Sakharov on 22 January 1980 and his subsequent exile to Gorky was for many western observers the final straw.[28] The Society's Foreign Secretary, Michael Stoker, immediately expressed the Society's concerns to the Soviet Academy. Todd made a speech[29] criticising the treatment of dissidents in the USSR, and rejected a request from the Soviet Academy to expand the exchange scheme. He also wrote to all Fellows on 21 February 1980 warning that scientific links were likely to suffer as a result of Soviet action against dissident scientists. Even then, however, the Society concluded that for it actively to sever links altogether would not be in the best interests of science or scientists in the UK or the USSR. If the exchange programme collapsed, it would be because individual UK scientists decided not to participate, not because the Society closed it in protest.[30]

---

[25] Meeting with Owen: OM/153(76); Alexander Todd, 'Anniversary Address, 1976', 462. See also leader in *The Times*, 18 February 1977, and subsequent letters, 3 March and 14 March.

[26] See meetings with Frank Judd in January 1978 (OM/18(78)) and Nick Ridley in November 1979 (OM/127(79)).

[27] CM 13 October 1977, minute 8, and CM 30 November 1977, minute 7.

[28] This occurred a month after the Soviet invasion of Afghanistan, and thus at a time when East/West tensions were again running high.

[29] As leader of the British scientific delegation to the Helsinki Agreement Scientific Forum rather than as President of the Royal Society, but he later explained in a letter to *The Times* (8 November 1980) that the speech also represented the view of the Royal Society Council. TODD Acc 1021, Box 32. See also *Nature*, 283 (21 February 1980), 709.

[30] OM 14 February 1980, minute 2(a); OM/15(80); Michael Kenward, 'Anglo-Soviet scientific exchanges frozen', *New scientist* (28 February 1980), 637.

In the event the door was kept open, with considerable difficulty. Two hundred and sixty-five Fellows, with some Swedish colleagues, constituted themselves the International Committee for Victimised Soviet Scientists and placed an advertisement in *The Times* in November 1980 calling for pressure on the Soviet Government to honour its Helsinki commitments concerning human rights and freedom of the individual.[31] The Society added its voice to the mix, though mainly through official diplomatic channels. Todd's successor as President, Andrew Huxley, told the Soviet Ambassador bluntly how problematic it was to keep scientific exchanges going in light of the mistreatment of dissident scientists.[32] The Society continued to highlight individual dissident cases to the Soviet Academy,[33] and to protest at the Academy's chronic maladministration, which threatened to make the scheme unworkable. 'Straight talking' with Soviet officialdom was the order of the day. The exchange schemes continued to run, though Todd's decision to reject the Soviet proposal for expansion was confirmed, and inflation was allowed to erode the sums available to Soviet visitors for in-country expenses. In his 1982 Anniversary Address, Huxley observed: 'Even without restrictions on exchange schemes, the oppressive treatment of scientists regarded as dissidents in the USSR is harming scientific interchange between that country and Britain.' Relations between the Soviet Academy and the NAS were at least equally cool.[34]

NATO partners were not exempt from the Society's opposition to restrictions on communication between scientists. The Society alerted the UK Government and raised with the NAS cases of British scientists being refused visas to enter the United States for scientific conferences in areas of research deemed by the American authorities to be of defence interest, such as materials, artificial intelligence, microelectronics, optical engineering, superconductivity, computing, cryptology and biotechnology.[35] Huxley's successor as President, George Porter, used

---

[31] *The Times*, 4 November 1980, p. 5. The 265 Fellows included four current members of the Royal Society Council.

[32] OM 6 November 1980, minute 2(e); OM 14 May 1981, minute 2(c).

[33] John J.P. Deverill, 'Scientific exchange with the USSR under the agreement between the Royal Society and the USSR Academy of Sciences', in Craig Sinclair, ed., *The status of Soviet civil science* (Martinus Nijhoff Publishers, 1987), 255–77.

[34] C/102(82); David Dickson, 'US cuts back on official exchanges with USSR', *Nature*, 283 (7 February 1980), 513; OM 23 March 1982, minute 6; OM 30 November 1982, minute 2(b).

[35] Peter Warren to David Phillips, 28 October 1983: David Phillips papers, MS Eng. C.5501, O.55; RMA622; OM 12 January 1984, minute 2(a). Also records of meetings with the FCO minister Malcolm Rifkind, 6 February 1985: OM/20(85), and with the NAS in June 1985: OM/78(85). And further exchanges in 1987: RMA622 and RMA1379.

his first Anniversary Address to defend the right of all scientists to travel for scientific discussions and international scientific meetings.

Like his predecessors, Porter combined spirited criticism of Soviet policies[36] with continued personal engagement with his opposite numbers. He led a Royal Society delegation to the USSR in May 1986 and received a return visit from the Soviet Academy in November 1987, on both occasions having detailed discussions about individual refuseniks. He was undeterred by the experience: 'Attitudes to freedom in science, and consequently beyond science, can only be changed by dialogue of this kind, painfully slow and depressing as it may be.'[37] The 1987 visit by the Soviet Academy was its first at presidential level for twenty years, a product perhaps of Gorbachev's new policies of glasnost and perestroika, and it was recognised in the UK as a move towards greater warmth between the two bodies. The FCO responded by providing funds for a significant expansion of the exchange schemes from 1988.[38] Porter assured an anxious Foreign Office that the exchanges benefited the UK as well as the USSR: notwithstanding American concern about militarily useful technologies leaking to the USSR, the West could learn from Russian expertise in, for example, lasers and space technology.[39]

The context for this science diplomacy was transformed in 1988 and 1989 as Gorbachev lost control of the forces he had unleashed and reform turned to revolution across the USSR and its satellites. Already in June 1989, Royal Society Officers were discussing how best to open up the European Science Exchange Programme (see Chapter 9) to the newly emerging democracies of Eastern Europe.[40] Hungary was invited to join the Programme in October that year, having taken down its border fence with Austria, overhauled its constitution and agreed to democratic elections. Other countries followed. When George Porter led his second Royal Society delegation to Russia in December 1989, a month after

---

[36] In addition to many other initiatives, Porter was one of 272 Fellows and Foreign Members of the Royal Society (including six past and future Presidents) who by March 1986 had lent their names to an international campaign to free the Soviet dissident scientists Yuri Orlov and Anatoly Shcharansky: RMA198.

[37] George Porter, 'Address at the Anniversary Meeting on 1 December 1986', *Science and public affairs*, 2 (1987), 4; also C/177(86). During his 1986 trip Porter tried, unsuccessfully, to visit Sakharov, and the Soviet Academy sought to justify his exile; three years later, Sakharov and his wife Yelena Bonner visited the Royal Society.

[38] See briefings for meetings with the FCO minister Lord Glenarthur in March 1988 and March 1989: RMA622; and minutes of the latter meeting: OM/55(89). Glenarthur met the Soviet Academy President, Gury Marchuk, during the latter's 1987 visit to the Royal Society.

[39] Meeting with the Foreign Office minister Tim Eggar, February 1986: OM/33(86) and RMA622; Anniversary Addresses 1986 and 1988.

[40] OM 6 June 1989, minute 2(d).

the fall of the Berlin Wall, the context had changed utterly, and so had the mood of the Russian Academy. As Porter later commented:

The relations between members of the Royal Society and the Soviet Academy were most amicable and the bear-hugs became quite strenuous. We agreed to double our science exchange programme ... and there was even a proposal from our Physical Secretary [Francis Graham-Smith] that we might exchange, for a trial period of one year, Mr Gorbachev and Mrs Thatcher, which was greeted with wild enthusiasm ... Rather to our surprise the Soviet Academy raised no objection to our forging independent agreements with other Soviets.[41]

The Royal Society faced a considerable challenge in discerning how best to promote the interests of science, and of international scientific colla-boration, during the turbulence and uncertainty that accompanied the break-up of the Soviet Union and its satellites.[42] There was urgency in securing the success of the multiple revolutions across the region, and there was competition in establishing good relations with the new regimes. But the collapse of communism led, in the short term, to eco-nomic chaos and loss of the infrastructure that, however inefficiently, had underpinned the research effort. Building on the relationships it had established over decades of keeping the door open, and on its politically neutral reputation, the Society sent delegations to many of the countries affected to see at first hand what was needed – much as it had done in Europe at the end of the Second World War (see Chapter 9). In addition to finding extra money to enlarge and tailor the exchange programmes, it arranged for joint projects to be initiated and for equipment and journals to be made available. It also provided advice and help for the restructuring of science in individual countries and for national academies to adapt themselves to new roles outside the parameters of the old centrally planned economy.

The Society was therefore in a good position to respond to the Prime Minister's request in September 1991 for an analysis of the state of the science academies, and of the scientific strengths and weaknesses, in the various components of the former Soviet Union.[43] In fact, it was in a very good position, because the previous week it had recruited Terry Garrett to head its international activities – Garrett having served no fewer than three terms as Science Attaché or Science Counsellor in Moscow, the most recent immediately before joining the Society's staff. The request had, of course, been prompted by the Society itself,[44] as a device for

[41] George Porter, 'Address at the Anniversary Meeting on 30 November 1990', *Science and public affairs*, 6 (1991), 5.
[42] Interview with Tony Epstein.
[43] John Major to Michael Atiyah, 11 September 1991: RMA111.
[44] Interviews with Michael Atiyah and Terry Garrett.

giving higher profile to the scientific dimension of the political upheavals then under way. The Society's new President, Michael Atiyah, had for long favoured the Society engaging strongly with scientists working in adverse circumstances. He was alarmed by stories of Russian research scientists leaving the country in large numbers or switching to unskilled jobs in response to repeated salary and budget cuts. The report was ready for the Prime Minister by the end of the year.[45] It led directly to major increases in funding from the FCO and the Office of Science and Technology (totalling £2 M over three years) in support of the Society's efforts to keep Russian science connected to the best work under way in the West.[46]

IIASA

The International Institute for Applied Systems Analysis was conceived as a multilateral effort at using science to keep the door open during the Cold War. The initiative began with Lyndon Johnson accepting a proposal along such lines in June 1966 and asking the President of the Ford Foundation, McGeorge Bundy, to take it forward.[47] The Russian Premier, Alexei Kosygin, soon signalled his support, and Bundy did the rounds of potential partners during 1967. The Institute would be funded by governments but had formally to be non-governmental,[48] so the members were national science academies or specially invented entities such as national committees for applied systems analysis. In the UK,

[45] OM 14 November 1991, minute 2(c); OM 29 November 1991, minute 2(a); OM/100 (91); CM 29 November 1991, minute 25; C/21(92); OM/25(92); Royal Society, *Academies of sciences in the constituent republics of the Former Soviet Union: a current appraisal* (January 1992).

[46] Meetings with the FCO minister Mark Lennox-Boyd, 20 February 1992 [OM/40(92)] and 25 February 1993 [OM/23(93)]; OM 14 May 1992, minute 2(f); OM 18 June 1992, minute 2(e); CM 16 July 1992, minute 35; Michael Atiyah, 'Address at the Anniversary Meeting, 29 November 1991', *Notes Rec R Soc*, 46 (1992), 162–3; Michael Atiyah, 'Address at the Anniversary Meeting, 30 November 1992', *Notes and records of the Royal Society of London*, 47 (1993), 114.

[47] Alan McDonald, 'Scientific cooperation as a bridge across the Cold War divide: the case of the International Institute for Applied Systems Analysis', *Annals of the New York Academy of Sciences*, 866 (1998), 55–83; Solly Zuckerman, *History of the International Institute for Applied Systems Analysis*, paper ASA(P)(72)6, September 1972: copy at RMA425; Howard Raiffa, *History of IIASA*, talk given at IIASA 23 September 1992 (transcript on IIASA website); J. Rennie Whitehead, *Memoirs of a boffin* (published online, 1995), chapter 14. Interviews with Peter Cooper, Terry Garrett and Peter Warren.

[48] For example, Russia wanted East Germany to be a member, but other participants such as the Federal Republic of Germany and the United States did not recognise East Germany and therefore could not work with it. Setting IIASA up as non-governmental circumvented such issues.

Bundy therefore sounded out the Royal Society, and Blackett discussed the matter with his opposite numbers at the US National Academy of Sciences. At that stage they had too little information to form a view about the potential value of the initiative, but expressed interest.[49]

The Social Science Research Council was initially expected to be the UK member of IIASA, in part because the Royal Society was thought not to be interested – although the government had never actually asked the Society. The Department of the Environment (DoE) was to be the source of the UK's financial contribution to IIASA. Following discussion with the DoE, and urged on by both the UK Chief Scientific Adviser Solly Zuckerman and his successor Alan Cottrell, the Society accepted the government's eventual invitation to take on this role provided it could conduct it on its own terms. The Society was sympathetic to the political context of IIASA, but it was a touch sceptical about IIASA's scientific agenda, which had a strong social science flavour beyond the Society's normal disciplinary range. So the Society reserved the right to stop representing the UK in IIASA at a later date, and it declined to provide an interim secretariat for IIASA. It did, though, agree to host the meeting that formally launched IIASA. This took place in October 1972, five years after the idea had first been mooted, and IIASA then established itself at Schloss Laxenburg near Vienna.[50]

The UK Cabinet backed the diplomatic objectives of IIASA, and the Foreign Office was a strong supporter. But by 1977 the Treasury was asking questions about value for money and direct benefit to the UK, and the DoE had to struggle to secure an increased subvention to IIASA.[51] A new IIASA five-year research plan attracted criticism from friend and foe alike, some of it stemming from wariness of systems analysis as lacking intellectual rigour.[52] The Royal Society and the NAS discussed IIASA's overall performance, the NAS being slightly the more complimentary but both bodies highlighting the uneven standards achieved by the various research programmes. A review drafted largely by the Royal Society Treasurer, John Mason, in 1981 was fairly even-handed but included some strong criticisms of both management style and research performance.[53]

---

[49] Blackett to Seitz, 23 May 1967; Gene Sunderlin to Ronald Keay, 30 June 1967; Keay to Sunderlin, 16 July 1967: all at RMA1344.

[50] OM/43(71); OM/109(71); OM 11 November 1971, minute 3(b); OM/115–126(71); OM 30 November 1971, minute 2(d); CM 13 January 1972, minute 13; C/34(72); CM 10 February 1972, minute 15.

[51] Martin Holdgate to Ronald Keay, 26 January 1977: RMA1344.

[52] See, for example, John Mason's comments ahead of a meeting with the NAS in May 1977: RMA1002.

[53] Copies at RMA1002.

Just as that review was being considered within the Society, the Reagan Administration, not known as an advocate of détente, announced that it was withdrawing funding from IIASA for 1982. The NAS lobbied against the decision, though allegations of Soviet spying activities around IIASA did not help. Eventually the NAS secured funding for one more year, but no more. At that point the Royal Society sought assurances from DoE that its own funding for IIASA would continue, commenting that the case for this rested more on the political benefits of adherence and the maintenance of scientific contacts between East and West than on the scientific merits of IIASA's output – whose social science character it disliked. But IIASA was no longer a DoE priority, and it became a matter of finding one or more Whitehall departments that cared sufficiently to come up with the budget beyond 1982. The Foreign Office decided not to step into this particular breach, arguing that other bodies like UNESCO and the Economic Commission for Europe provided sufficient opportunities for East/West interactions.[54] There ensued a period of brinkmanship on both sides of the Atlantic.

In the United States, the National Science Foundation stuck to the government's decision not to fund American membership of IIASA. However, an IIASA support group rallied to the cause and raised the necessary money from private sources, and the American Academy of Arts and Sciences, an independent organisation based in Cambridge, Massachusetts, agreed to take over from the NAS as the adhering body. When the UK Government finally took the decision to pull out, there were thoughts that something similar might happen, but it was not to be. The Royal Society remained unimpressed by IIASA's forward strategy;[55] no other body was able to take the Society's place;[56] and not enough alternative sources of funds were identified. So, at the end of 1982, the UK controversially dropped out of IIASA,[57] while the United States, under new arrangements, remained a member. For some years after that there was intermittent pressure on the UK, and on the Royal Society, to reconsider, but the decision stood. The political context, of

---

[54] CM 5 March 1981, minute 28(11); 'Vienna institute hit by espionage charges', *Nature*, 290 (30 April 1981), 725–6; Ronald Keay to Martin Holdgate, 30 November 1981, and Holdgate to Keay, 29 December 1981: RMA1002; CM 14 January 1982, minute 7.

[55] A previous Foreign Secretary, Kingsley Dunham, was one of several Fellows who felt the Society could have done more to encourage British experts to participate actively in IIASA's research and drive up the standards, rather than sit back and criticise from a distance. Dunham to Keay, 2 October 1982: RMA1002.

[56] The Fellowship of Engineering offered to be the adhering body, and was accepted as such by the IIASA Council, but was unable to extract financial support from the Department of Industry, the Foreign Office or any other department. Robin Caldecote to Patrick Jenkin, 8 November 1982, and Jenkin to Caldecote, 25 January 1983: RMA1002.

[57] Interviews with John Ashworth and Peter Cooper; numerous papers at RMA1002, 1308.

course, changed and with it the rationale for IIASA's existence, but IIASA continues its scientific work to the present day.

The multilateral approach to keeping the door open through scientific collaboration, on this example, would seem to be even more challenging than the bilateral approach, given the greater number of variables that had constantly to be juggled. But one criterion for success, which the Royal Society understood very well, was that diplomatic objectives by themselves were not enough. The science itself had to be worthwhile. If it would not work as science, it would not work as diplomacy.

China

The UK was the first non-communist country to recognise the People's Republic of China, as early as 1950, though for twenty years diplomatic relations were conducted only at the chargé d'affaires level. The United States, however, at the outset imposed an embargo on trade with communist China and severe restrictions on its own citizens from travelling to China. These were finally lifted in connection with Nixon's visit to China in 1972, and full diplomatic recognition followed in 1979. In the late 1950s, the period of Mao's ill-fated 'Great Leap Forward', while the West was seeking to weaken links between China and the USSR, China sought to drive a wedge between the UK and the United States.[58] So the UK's dealings with China, even on scientific collaboration, demanded the utmost care.

Against that background, the Royal Society in late 1958 rebuffed a roundabout approach from the Ford Foundation to mount a scientific mission to 'Red China' with Foundation funding. It was not willing to front a covert American initiative. The fact that the Society was not American was an asset in its dealings with China at a time when Sino-American relations were at a low ebb, and it did not want to jeopardise that asset.[59] The Society was, however, willing to work directly with the Chinese Academy of Sciences (CAS)[60] on its own account, provided the

---

[58] Victor S. Kaufman, *Confronting communism: US and British policies towards China* (University of Missouri Press, 2001), chapter 5. At this time the West seems to have regard communist China with greater wariness than it regarded the USSR: Rosemary Foot, *Trading with the enemy: the USA and the China trade embargo* (Oxford University Press, 1997).

[59] OM 1 December 1958, minute 3(a); OM/95(58). Five years later, the Society refused to act as go-between with the Chinese Academy of Sciences to facilitate an invitation for Harrison Brown, the NAS Foreign Secretary, to visit China. OM/43(63); OM 20 June 1963, minute 2(b). And in 1968 it absented itself from another NAS attempt to engage with CAS: OM 10 October 1968, minute 3(c).

[60] CAS was often referred to by its Latin name Academia Sinica until about 1978. Not to be confused with the Taiwan Academy of the same Latin name.

Foreign Office did not object. So Cyril Hinshelwood wrote to the Foreign Secretary, Selwyn Lloyd, in January 1959 about inviting CAS to send a delegation to the UK. The Foreign Office took the view that Britain's self-interest lay in prioritising good relations with the United States. A holding reply indicated that the government would favour making overtures to CAS

provided that anything we do is done with the knowledge and approval of the United States Government, since we do not want to imperil our atomic energy exchanges with the United States or to lay ourselves open to Congressional criticism on the grounds that we are allegedly exchanging scientific information with Communist China.[61]

After consulting the British Ambassador to Washington, Lloyd asked the Society to hold off inviting CAS until Congress had approved the nuclear materials agreement later in the year. The Society acquiesced in this.[62] The Lhasa uprising against Chinese occupation in Tibet further added to the complexity of judging when it was politically appropriate to issue an invitation. Eventually, CAS accepted an invitation to send a four-man delegation to the Society's tercentenary celebrations in July 1960, and a larger group came over the following year for a more extensive visit. The American State Department was briefed in each instance.

Meanwhile, Cyril Hinshelwood, expert collector of Chinese pottery as well as expert chemist, spent a month in China in the summer of 1959 at the personal invitation of the Chinese Academy of Sciences (Figure 8.2). He was one of the early western scientists to visit post-revolutionary China. Describing himself as 'an enthusiastic sinophile', he wrote an upbeat account of his tour in *New scientist* and looked forward to a 'resumption of cordial interchange' between Britain and China.[63] This visit came to be seen as the launch pad for the Society's relations with CAS.

The success of the Chinese visit to the UK in 1961 prompted a return invitation for a Royal Society delegation for the following year. The Foreign Office saw this as the first step in establishing a programme of scientific exchanges. It regarded science as the most feasible medium for engaging with China at that stage, and experience with the USSR

---

[61] Patrick Dean to Cyril Hinshelwood, 24 January 1959, and subsequent correspondence: RMA222. The government refused to defray the costs of the proposed visit from CAS.
[62] The Royal Society Treasurer, William Penney, Deputy Chairman of the UK Atomic Energy Authority, was heavily involved in negotiating the nuclear materials agreement with the Americans.
[63] Cyril Hinshelwood, 'A visit to China', *New scientist*, 6 (5 November 1959), 858–60; CH/15/1–3. Neither the Royal Society Council nor the Officers formally discussed the visit, and to that extent it was not an 'official' RS visit. Hinshelwood travelled on his own.

Figure 8.2 Cyril Hinshelwood joking with Guo Moruo, President of the Chinese Academy of Sciences, during his visit to China in 1959. © The Royal Society

suggested that the formidable practical hurdles were manageable with patience and goodwill. With an eye to long-term prospects of trade with China, industry might be persuaded to help with the costs. The 1962 Royal Society visit (Figure 8.3) led to the start of an exchange scheme with CAS, with a strong emphasis on early-career Chinese scientists visiting the UK, many with interests in applied research. In contrast to the formality of the USSR scheme, CAS insisted on its scheme being informal and direct with the Royal Society rather than with or through the British Government. The Foreign Office (via the British Council) nevertheless agreed to cover the Society's costs. The first Chinese scientists under the scheme arrived in the UK in autumn 1963.[64]

It was not plain sailing. Like the formal USSR scheme, the informal Chinese scheme was in practice marked by administrative miscommunications and cultural confusions that took time to resolve. Once the

[64] CM 9 November 1961, minute 14; CM 13 December 1962, minute 11; RMA222; GLB/66/13–15, notably H.W. Thompson, *Visit to China, 22 September–8 October 1962*; House of Commons Question on China (cultural relations): Hansard, 692 (no 79), question 20, 23 March 1964.

Figure 8.3 Royal Society delegation to China, September–October 1962 (l to r): George Lindor Brown, C.H. Waddington, Tommy Thompson, Herbert Powell, Gordon Sutherland. © The Royal Society

teething issues were sorted out, heads of the departments in which Chinese postdoctoral visitors were placed reported mostly in glowing terms on progress achieved, but sometimes despairingly about attitude or aptitude.[65] It was a learning experience for both sides, and mutual patience was key. But sharing knowledge was not universally recognised as a good thing. In 1964 the scheme was lambasted by a South Carolina Democrat senator named Olin Johnston, who saw the whole idea of exchanges as 'traitorous to the free world'. 'Such foolishness', he declared, 'is costing American lives on the battlefield, and I should hope the United States would find a way to put an end to it.' The Society stressed in response that visiting scientists did not have access to classified information, and observed drily that many senior scientists in China had been trained in the United States.[66]

An official Royal Society visit to China by two members of Council, Patrick Blackett and Roy Clapham, in September 1964 strengthened relations with CAS and boosted the exchanges. The Foreign Office briefed the delegation that, as a result of its dispute with the USSR and the withdrawal of Soviet technological aid, China was wanting to build up its trade and cultural links with the capitalist world, and that science was relevant to this.[67] The following summer, Paul Gore-Booth, now back as Permanent Secretary at the Foreign Office after his stint in India, restarted the internal debate about non-political dialogue with China. Not all his colleagues were sympathetic. Benjamin Strachan, for example, a career soldier before he became a diplomat, disparaged the Society's championing of scientific diplomacy:

I fear that the importance which the Royal Society attach to their link with the Academia Sinica stems principally from their pride in it as a demonstration of how science transcends politics (a well worn myth whose naivety seems impervious to experience . . .). The Chinese are of course desperately anxious to acquire Western technological know-how, and the bulk of their effort to achieve this is done through 'private' visits – presumably because these are less subject to control or scrutiny by HMG.[68]

Some of Strachan's colleagues took a similar line, and Gore-Booth had to moderate his enthusiasm a little.

But the debate was, anyway, overtaken by the Cultural Revolution, with its pervasive currents of anti-intellectualism, which took off in May 1966. The President of CAS, Kuo Mo-jo, was among many attacked or

[65] See, for example, paper CH/3(66): copy at RMA222.
[66] Undated cutting from *The Daily Telegraph* kept by Patrick Blackett: PB/7/2/4/10.
[67] RMA222 and C/201(64). The delegation resisted pressure to take a BBC camera with them.
[68] TNA FO 924/1529. ·

purged from their positions. Scientific contacts with foreign countries were severely curtailed. Relations between CAS and the Royal Society cooled rapidly. CAS abruptly and unceremoniously ordered all postdoctoral students and visiting researchers to return immediately to China, to the dismay of their British hosts. Tommy Thompson, who as a member of the Society's 1962 delegation to China had been enthralled by what he experienced and who had strongly favoured building stronger relations with CAS, was left apologising to numerous heads of department facing disrupted research programmes.[69] The Society was sufficiently irritated by the whole process that for some years it refused to make the first move towards reopening contacts with CAS. The door was left open in this instance, but it was up to the Chinese to knock on it.

Hints began to emerge in 1969 that Chinese hostility to contacts with the West might gradually be lessening, and that scientific exchanges might be one medium through which international connections could be re-established. But slowly: the Foreign Office minister Fred Mulley told the Royal Society that prospects of renewed relations with China were 'not at all good'.[70] It needed careful handling, not least because of the Taiwan problem. The FCO warned the Royal Society that any systematic engagement with the scientific authorities in Taiwan would sabotage the chance of progress with China. The Society heeded the FCO's advice.[71] A later Foreign Office minister, Lord Lothian, briefed the Society's leadership in February 1971 that the thaw in China's relations with the West 'still had a long way to go'. He suggested, nevertheless, that prospects of eventual trade benefits implied reasonable efforts should be made to reinstate scientific contacts, which he thought would be 'valuable in dispelling mistrust of the West'.[72] The Society had already been cultivating the new Chinese Chargé d'affaires in London, inviting him to its annual reception in June 1970 and having periodic meetings. It had also drawn up a list of twenty-three Fellows who were keen to make scientific visits to China,[73] and raised funding from the Leverhulme Trust to support a new scheme of visits for senior scientists. Exchange arrangements prior to the Cultural Revolution had been markedly

---

[69] HWT B.405–408; RMA222; Geoffrey Oldham, 'Chinese science and the Cultural Revolution', *Technology review*, 71 (October 1968), 23–9.
[70] Meeting 25 April 1969: OM/46(69).
[71] C.J. Audland to David Martin, 31 October 1969: TNA FCO 55/234.
[72] RS and FCO reports on meeting with FCO. 12 February 1971: OM/19(71), HWT B.185.
[73] On the Society's role as gatekeeper for scientists seeking to visit China, see Jon Agar, '"It's springtime for science": renewing China-UK scientific relations in the 1970s', *Notes and records of the Royal Society of London*, 67 (2013), 7–24. Also HWT B.410–413, and RMA222.

one-sided, with many more Chinese visitors to the UK than vice versa, and the Society wanted to redress the balance. Friendly overtures continued through 1971, and visits started up again on a tentative scale.

An official Royal Society delegation led by the President, Alan Hodgkin, visited China in May 1972 and had talks with the CAS leadership. Mao's China had been recognised at the UN in October 1971 and had been accepted in that capacity by the UK in March 1972, so the visit took place in an atmosphere of relative goodwill towards the UK. It also took place three months after Richard Nixon's much-publicised visit, which had marked a warming in China's relations with the United States. Hodgkin's delegation reported back in glowing terms about its experience of Chinese hospitality, but in sombre terms about the massive disruption to scientific activity caused by the Cultural Revolution. CAS wanted Britain's help in re-establishing its ability to promote fundamental research, which had been disparaged as irrelevant by the Cultural Revolution. David Martin concluded that it would take 'some time and much patient understanding' before a full-scale exchange programme could be operational.[74] A return visit by CAS, hosted by the Royal Society, quickly followed, the first multi-professional scientific delegation to leave China since the Cultural Revolution. The FCO set considerable store on the Society's ability to engage in this way with its Chinese counterpart.[75]

Hodgkin observed that the Cultural Revolution was 'still very much in force' in May 1972. It continued, with declining intensity, until Mao's death in September 1976 and the subsequent fall of the Gang of Four. Exchanges between CAS and the Royal Society, reinvigorated by the 1972 discussions, continued to develop steadily during this period.[76] Tommy Thompson, having completed his stint as the Society's Foreign Secretary, kept a close eye on progress as Chairman of the Great Britain/China Committee and the Great Britain/China Centre, and led another Royal Society team to China in 1974. Taking

[74] C/122(72); Alan Hodgkin, *Chance and design: reminiscences of science in peace and war* (Cambridge University Press, 1992), 380–3, and John Gardner, 'The Gang of Four and Chinese science', *Bulletin of the atomic scientists*, 33 (September 1977), 24–30. A copy of the delegation's report was sent to the FCO.

[75] FCO telegram 601, 23 August 1972: TNA FCO 34/153. The CAS delegation also visited Sweden, Canada and the USA: FCO briefing, October 1972 TNA FCO 55/917, and *The Times*, 7 October 1972. One member of the CAS delegation, Chang Wen-yu, had studied with Rutherford in Cambridge in the 1930s. For FCO reliance on the Royal Society in the exchange context, see telegram 1388 to the Ambassador in Peking, 19 December 1972: TNA FCO 34/153, and John Addis (British Ambassador in Peking) to Ronald Keay, 18 December 1973: HWT B.416.

[76] OM/88(75), and interview with Geoffrey Allen.

his cue from the invention of ping-pong diplomacy in 1971, Thompson also exploited his position as Vice-Chairman (later Chairman) of the Football Association to arrange for the newly appointed Chinese Ambassador in 1972 to watch Arsenal draw 1–1 with Chelsea at Highbury stadium in north London. But he was thwarted in later efforts to arrange for Arsenal to play a Chinese team, on the pedantic grounds that China was not a member of the FA.

The post-Mao succession struggles in China were resolved with the consolidation of Deng Xiaoping's power during 1977. Science was given pride of place in the new dispensation as the linchpin of the 'four modernisations', and the Chinese leadership acknowledged that progress in science demanded commitment to international cooperation. John Ashworth in the Central Policy Review Staff (part of the Cabinet Office) rapidly developed plans for expanding the UK's scientific engagement with China, in active competition with France and Germany in particular on both scientific and trade objectives. The FCO minister Frank Judd briefed the Royal Society in January 1978 on the political niceties of dealing with the new China, and various contacts in Whitehall kept the Society up to date as official thinking developed during the year. The Society accepted the emerging consensus that it was time for an intergovernmental agreement on scientific and technological cooperation to complement its own informal arrangements with CAS: British firms seeking commercial links with China strongly favoured such an approach.[77]

The Society's Foreign Secretary Michael Stoker led a delegation to China in July 1978. The West was busy making up for lost time. Stoker's visit was a week before the Secretary of State for Education and Science, Shirley Williams, went there, a month before the Trade Secretary, Edmund Dell, and about the same time as the American President's science adviser, Frank Press. The Chinese Academy undertook a return journey to the UK in November. The upshot was a new and more detailed agreement between the Society and CAS for a considerably enlarged exchange scheme that built on the relations developed between the two bodies over the previous two decades. It was still an independent, non-governmental agreement, but was set in the framework of the intergovernmental agreement signed at the same time – the first treaty between the UK and China since 1949.[78]

John Ashworth commented in January 1979 that Britain had been slow off the mark in developing scientific and technical relations with the

---

[77] Jon Agar, 'Springtime for science'; interview with John Ashworth; HWT B.422–424; OM/18(78); RMA221, 222.

[78] Interview with Michael Stoker; OM/106–9(78); Robert Walgate, 'China wants 800,000 scientists by 1985', *Nature*, 274 (10 August 1978), 525.

USSR. He was keen to do better with China.[79] One response was an outbreak of coordination initiatives, with the Department of Trade, the DES and the Royal Society all getting in on the act. The Society established both a short-lived coordinating committee and a China Information Service in an attempt to systematise and make available the knowledge it was gleaning about Chinese science from various sources, including reports written by UK scientists returning from exchange visits.[80] It had also convened a meeting in December 1978 with counterparts in France, Germany, Sweden and the United States to compare notes on their arrangements for exchanges with China.[81] The standard pattern was found to comprise approximately balanced exchanges at fairly senior levels, coupled with provision for much larger numbers of Chinese research students to spend significant periods working in the host countries at the expense of CAS or equivalent bodies. The US arrangements were the least formalised: its diplomatic relations with China were not fully restored until the beginning of 1979. But the United States then became by a large margin the destination of choice for Chinese research students working abroad.

The American Government saw exchanges – across the fields of science, technology, culture, sports and journalism – as key to combating mutual prejudice and signalling the new determination to normalise relations with China in the years after Nixon's 1972 visit. Academics were happy to go along with this.[82] Similar attitudes prevailed in the UK. In addition to diplomatic objectives, trade – and therefore international competition – was an explicit focus of government policy in this context. Openings for UK firms to export high-technology goods to China were keenly sought as the Chinese economy became more accessible. The Royal Society argued frequently that such openings were likely to follow personal contacts formed through scientific

---

[79] Coordinating meeting convened by the Board of Trade, 11 January 1979: TNA AT 82/139.

[80] OM/16(83).

[81] Minutes of meeting, 11 December 1978: RMA1377; OM 14 December 1978, minute 2 (i), and C/55(79).

[82] Zuoyue Wang, 'US-China scientific exchange: a case study of state-sponsored scientific internationalism during the Cold War and beyond', *Historical studies in the physical and biological sciences*, 30 (1999), 249–77; Zuoyue Wang, 'Transnational science during the Cold War: the case of Chinese/American scientists', *Isis*, 101 (2010), 367–77; Kathlin Smith, 'The role of scientists in normalising US-China relations: 1965–1979', *Annals of the New York Academy of Sciences*, 866 (1998), 114–36; John Richardson, 'Exchanges in the process of "normalisation": US government perspective', and A. Doak Barnett, 'Exchanges in the process of "normalisation": an academic view', both in Anne Keatley, ed., *Reflections on scholarly exchanges with the People's Republic of China, 1972–1976* (Committee on Scholarly Communication with the People's Republic of China, 1978).

exchanges.[83] In the short term, the purely scientific benefit of exchanges was seen as accruing more to China than to the UK in most, though by no means all, fields of research. Apart from some in the Treasury, few people in Britain regarded that as a real problem in the great scheme of things.

But the UK seemed not to be fully exploiting its natural advantages of language (English was then widely taught in China), of relative political acceptability and of scientific reputation. A British Council report in early 1982 concluded that, in comparison with the opportunities for commercial and political gain, its investment in scientific relations with China was 'under-funded, inadequately staffed and uncoordinated'. Percy Cradock, Ambassador in Peking, strongly endorsed the report, and the FCO accepted its key points. Its author, Julian Schweitzer, briefed the Royal Society in detail.[84] The Society discussed with Robin Nicholson, Government Chief Scientific Adviser, how it could help, but accepted that the coordinating role that Schweitzer had called for should be undertaken at government level rather than by itself.[85] A Royal Society delegation to China in April 1982 found that the introduction of full-cost fees for foreign students in British universities, and the high bench fees charged to researchers in some instances, had created serious problems and were impeding Britain's position in the competition for good students. The full-cost fees issue was mitigated in 1983 when the government made £46 M available through FCO to offset its negative impact on relations with selected countries; but by 1985 the Society was badgering the government over the fact that France was attracting three times as many Chinese students and researchers as the UK, and Germany six times as many.[86]

The Society established additional exchange agreements with the Chinese Association for Science and Technology (CAST) and the Chinese Academy of Medical Sciences in 1982, and secured significant funding for several years from the oil giant BP to use alongside government funding for its exchanges with CAS. It also organised a major joint project with CAS, in the form of a geotraverse of Tibet, which was seen as something of a coup.[87] A further coup was associated with the

---

[83] Letter from the RS Foreign Secretary Arnold Burgen to Lord Belstead, minister, 26 July 1982: RMA221.

[84] Percy Cradock to Alan Donald (FCO), 1 February 1982, and related notes: RMA172. The same message characterised another British Council report four years later, and a report by the science attaché in Peking, Simon Featherstone, also in 1986: RMA172 and RMA1300.

[85] OM 13 May 1982, minute 2(a); OM/62(82).

[86] Arnold Burgen to Geoffrey Howe, 17 July 1985: RMA172. See also correspondence with Olek Zienkiewicz, January–March 1986: RMA172.

[87] The Society's Deputy Executive Secretary, Peter Warren, was a geologist and a natural enthusiast for the project. CAS turned down approaches from the French and others in

Queen's visit to China in October 1986. The industrialist Lord Rhodes, then over ninety years old, and the Society dreamt up a new scheme to bring young Chinese scientists to Britain; Rhodes rapidly extracted most of the money needed from his business contacts, and the Society extracted the remainder from the Office of Overseas Development; Rhodes persuaded the Queen to present the scheme – the China Royal Fellowships – to the people of China as her gift during her State visit, and the Society agreed to administer it. The scheme was for three years in the first instance and funded 30 one-year visits per year; it was then extended through successive reviews, and was eventually coalesced into a wider Royal Society scheme in 2008.[88] The industrial sponsorship was significant given the clear view in official circles that government S&T strategy towards China should be focused on securing economic benefit for the UK. The first four China Royal Fellows to arrive in Britain, and the industrial sponsors of the scheme, were introduced to the Queen, the Duke of Edinburgh and the presidents of CAS and CAST at a Royal Society reception in March 1987.

The FCO minister, Lord Glenarthur, told the Society in March 1989 that scientific relations between the UK and China were developing rapidly.[89] But progress was severely jolted when the Chinese pro-democracy protest movement, initiated in the spring of 1989, was crushed in the vicious government crackdown on and after 4 June centred on Tiananmen Square. Scientists were divided on how best to express their revulsion at Tiananmen Square, much as they had been over the 1968 invasion of Czechoslovakia. The NAS, for example, suspended all its Chinese exchange programmes.[90] The Royal Society, in contrast, sent a message of 'concern and goodwill' to the CAS President, Zhou Guangzhao. The British Academy wrote analogously to the Chinese Ambassador. Learning from the experience of the Cultural Revolution in 1966 when the (many fewer) Chinese students working abroad were all abruptly recalled to China, the Society rapidly took steps to enable visiting Chinese researchers such as the China Royal Fellows to extend their stays in the UK. The Society also agreed to continue 'scientist-to-scientist' exchanges with bodies like CAS and CAST that it could technically categorise as non-

order to keep this a purely Sino-British initiative. See notes on Andrew Huxley's March 1984 visit to China: RMA1559, and interview with Peter Warren.

[88] George Porter to Robert Fellowes, 20 December 1985: RMA622; OM/37(86). For the haphazard origins of the scheme, see John Deverill interview.

[89] George Porter, 'Address at the Anniversary Meeting on 30 November 1987', *Science and public affairs*, 3 (1988), 12; OM/55(89).

[90] Richard P. Suttmeier, 'Scientific cooperation and conflict management in US-China relations from 1978 to the present', *Annals of the New York Academy of Sciences*, 866 (1998), 137–64.

governmental. But it cut off contacts with the Ministry of Geology and other explicitly governmental bodies and, following the UK Government's lead, suspended high-level visits that might be construed as support for the Chinese Government. The Society explained its approach publicly in terms of its 'long tradition of helping individual British and overseas scientists to develop and maintain contact for their mutual benefit, irrespective of political considerations'. Individual scientists, of course, could and did cancel planned trips to China.[91]

Zhou Guangzhao was in a vulnerable position. CAS was a prominent organisation with de facto close connections to government. But it also ran two universities, and many students and others working for CAS were sympathetic to the protesters. These included Fang Lizhi, the dissident physicist and cosmologist who was one of the leaders of the pro-democracy movement. Fang and his wife Li Shuxian took refuge in the American embassy on 5 June, and were to stay there over a year. He had been elected to CAS membership in 1980 at the early age of 44; CAS eventually revoked that status because of his political activities. He became an icon for the times, a focus of heated attention for all sides of the political debate. On 19 June 1989 the Canadian Committee of Scientists and Scholars issued a statement likening him to Mahatma Gandhi, Martin Luther King and Andrei Sakharov. The Chinese press attacked him with corresponding vigour. In the end the Royal Society helped to break the impasse, in delicate consultation with the NAS, by offering him a Royal Society Research Professorship at the Institute of Astronomy in Cambridge 'in conformity with the Society's tradition of freedom for scientists'. Fang accepted the offer. It took another five months of careful diplomacy to secure safe passage from the American embassy, and Fang Lizhi and his wife arrived in Cambridge at the end of June 1990. At the end of the year they moved on to the United States, first Princeton and then Tucson, as had always been their intention.[92]

Fang Lizhi was able to give the Society uniquely informed commentary on its post-Tiananmen policies towards China.[93] Since scientists in China

---

[91] OM 6 June 1989, minute 1; RMA1559; Hansard, 6 June 1989, col 30; OM 13 July 1989, minute 2(a); RS press release, 7 August 1989.

[92] Cong Cao, 'The Chinese Academy of Sciences: the election of scientists into the elite group', *Minerva*, 36 (1998), 325; RMA1379 and RMA1559; George Porter, 'Anniversary Address, 1990', 5–6; Orville Schell, 'China's Andrei Sakharov', *The Atlantic monthly*, May 1988, 35–52; James H. Williams, 'Fang Lizhi's big bang: a physicist and the state in China', *Historical studies in the physical and biological sciences*, 30 (1999), 49–87; William Bown, 'A dissident view on life, the universe and democracy', *New scientist* (21 July 1990), 19; Fang's obituary in *The Economist*, 14 April 2012.

[93] Fang Lizhi to Michael Atiyah, 25 December 1990, and further correspondence, 9 January, 31 March and 18 April 1991: RMA172.

needed contact with their peers in the rest of the world, he endorsed the Society's instinct to keep exchanges going. But he worried that the Chinese authorities would 'use this normalisation to do propaganda for a thorough forgetfulness of their violation of human rights', the more so once the world's attention moved on to other conflicts such as Iraq's invasion of Kuwait. He therefore urged the Society also to publicise the cases of scientists still being persecuted for their political activities and to intercede for them with the Chinese Government whenever possible.

The European Union rescinded most of its post-Tiananmen embargoes on dealings with China in October 1990, and the process of normalisation gradually gathered momentum. The FCO agreed in January 1991 that the status quo could be restored in all areas except defence. It recognised the academic benefits of the Royal Society's exchanges as legitimate in their own right, while stressing that where possible S&T relations with China should also have both commercial and diplomatic objectives. The Society's President, Michael Atiyah, received a delegation from CAS in April 1991 and, having alerted the FCO minister Mark Lennox-Boyd to his intentions, took the opportunity to raise the cases of a number of scientists still being held by the Chinese Government.[94] Zhou Guangzhao invited Michael Atiyah to make a return visit in April 1992, and Atiyah again ensured that dissident scientists were included on the agenda. Neither that nor the Society's 1991 signing of an exchange agreement with the National Science Council of Taiwan seemed to impede the Society's good relations with its Chinese counterparts. In his 1992 Anniversary Address, Atiyah commented: 'Despite the continuing tight political control, which we naturally deplore, scientists in China are keen to strengthen their links with the international scientific community.'[95] At his next meeting with Lennox-Boyd, in February 1993, the talk was of the phenomenal growth in China's GNP, foreign trade and S&T expenditure, and of the need for the UK to redouble its efforts in science diplomacy to keep abreast of developments.

Scientific traffic between the UK and China escalated in subsequent years, through an increasing diversity of mechanisms. The Royal Society's contribution to the flow of scientists on study visits, research fellowships and joint projects to and from China ran to over 100 per year in the 1990s, funded by a mix of private and public sources. Atiyah's

[94] OM/23(91); OM 14 March 1991, minute 2(h); OM/39(91) (meeting with Lennox-Boyd, 27 March 1991); *New scientist*, 4 May 1991. Atiyah later told Lennox-Boyd that the response from CAS 'had not been very positive': OM/40(92).
[95] Michael Atiyah, 'Anniversary Address, 1992', 114.

Figure 8.4 Paul Nurse presenting the King Charles Medal to the
Chinese Premier, Wen Jiabao, June 2011. © The Royal Society

successor, Aaron Klug, led an extensive visit in 1997 to reinforce the
Society's visibility in China. The Foreign Secretary Julia Higgins found,
on a 2003 visit, that the Society was well known among the Chinese
scientific community, with its scientific journals and its policy reports
receiving wide attention and some being translated into Chinese. The
UK Research Councils opened a collective office in Beijing in 2007, and
the Royal Society seconded a member of staff to it. At the instigation of
Brian Heap, Julia Higgins' predecessor as Foreign Secretary who had
been particularly active in cultivating links with China, the Society's
journal *Philosophical Transactions* invited Chen Zhu, CAS Vice-
President, to guest-edit a special issue in 2007 on biological science in
China. In September 2013, coinciding with a visit to CAS by the Royal
Society President Paul Nurse, a further special issue, edited by the CAS
President Bai Chunli, focused on nanotechnology in China. The first
scientist living in mainland China (as distinct from Hong Kong or
Taiwan, or scientists of Chinese origin based in the United States) to
be elected to the foreign membership of the Royal Society was Zhou
Guangzhao, in 2012; Chen Zhu followed in 2013, and Bai Chunli in
2014.

During his final visit to the UK as Chinese Premier, in June 2011,
Wen Jiabao made a point of including the Royal Society in his itinerary.

The Society gave him its King Charles II medal (Figure 8.4).[96] The President, Paul Nurse, praised him for overseeing 'one of the most ambitious programmes of national research investment the world has ever seen', and extolled the role of scientific cooperation based on openness, free exchange of ideas and mutual respect in increasing enlightenment and improving the quality of life. In a remarkable acceptance speech, Wen Jiabao not only alluded to Deng Xiaoping's modernisation programme and the role of science and technology in generating prosperity, but also promised that the new prosperity would be shared by 'both urban and rural residents' and by 'the Chinese people of all ethnic groups'. 'Tomorrow's China', he continued, would be:

a country that fully achieves democracy, the rule of law, fairness and justice ... a more open, inclusive, culturally advanced and harmonious country ... drawing on the strengths of fine foreign cultures ... The Chinese Government encourages large Chinese companies, research-oriented universities and research institutions to increase cooperation with their British counterparts. It also encourages more exchange of top-level talents and joint research between our two countries.

Even as rhetoric, it was a powerful endorsement of the capacity of science to shape, and be shaped by, international relations.

South Africa

South Africa declared itself a republic and left the Commonwealth on 31 May 1961. A year later, the British Government determined that South Africans were neither British subjects nor Commonwealth citizens. At that point the Royal Society Council concluded that South Africans could no longer be elected as Fellows, a category then reserved for 'British subjects or citizens of Eire' and, from 1965, for 'British subjects or Commonwealth citizens or citizens of the Irish Republic'.[97] South Africans were also excluded from the Society's Commonwealth Bursaries, and similar funding schemes. They did, though, became eligible for the Society's initiatives aimed at non-Commonwealth countries, such as Leverhulme Visiting Professorships (of which three were awarded

---

[96] The King Charles II medal is reserved for foreign heads of state or government who have made outstanding contributions to furthering scientific research in their own countries. Wen Jiabao was only the fourth person to receive it, preceded by Emperor Akihito of Japan (1998), Abdul Kalam of India (2007) and Angela Merkel of Germany (2010).

[97] CM 12 April 1962, minute 3. Between 1945 and 1962, five scientists resident in South Africa were elected to the Fellowship – fewer than Canada (17), Australia (14) or the USA (9), but more than India (4) or New Zealand (2). In 1962, five scientists based in South Africa were candidates for the Fellowship; the three of South African citizenship were withdrawn, while the two of British citizenship remained on the list. CM 8 November 1962, minute 5.

during 1970–5 to scientists wanting to work in South Africa). And South Africans could in principle be elected as Foreign Members, though in fact none were during the thirty-three years that South Africa remained outside the Commonwealth.

On 6 November 1962 the UN General Assembly passed Resolution 1761 condemning apartheid and calling for sanctions. An arms embargo followed, but western nations generally refused to back wider economic sanctions at that stage. The Royal Society's response was to keep scientific links as open as possible. Fellows were well aware that many of the South African intelligentsia were strongly and publicly opposed to the apartheid policies of their government. So President Patrick Blackett received the Secretary of the Royal Society of South Africa in 1968, and the Society was formally represented at the RSSAf centenary celebrations in 1977. When UNESCO was being pressurised in 1971 to break off relations not only with South Africa but also with NGOs like ICSU that continued to maintain links with South Africa, the Society vigorously voiced its concern. The UK Government warmly supported the Society: cutting off relations was not in its view the most effective way of combating apartheid.[98]

South Africa was represented on international bodies like ICSU through its Council for Scientific and Industrial Research (CSIR). A proposal by CSIR in 1982 for an exchange agreement with the Royal Society exposed the delicacies of dealing with official government bodies in the apartheid context.[99] The proposal itself was modest enough – about five research visits in each direction per year, of two to three months duration. But the Officers were worried about 'a likely adverse reaction by some Fellows and by others to any formal arrangements', so they pushed hard to secure the benefit of the visits without the potential awkwardness of a formal agreement. CSIR went along with that. The Society gave minimal publicity to it in the UK: Fellows were simply asked to alert possible applicants, and the visits were handled under a generic category of exchanges with countries having no formal agreement with the Society. After the first year (1983–4), the South African 'quota' was, predictably, under-subscribed. In the same vein, the Officers agreed in June 1984 to a meeting with the CSIR President, Chris Garbers, but would not give him a platform to lecture on the merits of South African science.

[98] Meeting between the Society's Officers and the FCO minister Lord Lothian, 12 February 1971: OM/19(71) and HWT B.185.

[99] OM 15 July 1982, minute 2(f); OM 19 July 1982, minute 1(a); OM 4 November 1982, minute 2(a); CM 4 November 1982, minute 12; OM/85(82); OM/113(82); RMA459 and 714.

A congress of the International Union for Prehistoric and Protohistoric Sciences (IUPPS) scheduled to be held in Southampton in 1986, under ICSU rules banning exclusion of bona fide scientists, further exposed the challenges of keeping the door open. The IUPPS secretary, Peter Ucko, Southampton City Council, Southampton Students Union and the Southampton branch of the Association of University Teachers all voted for scientists resident in South Africa to be barred from attending the IUPPS congress – in keeping with the prevailing zeitgeist, but in direct contravention of the ICSU rules. Such action by the local organisers of a congress was unprecedented. IUPPS responded by moving its event to Mainz, where it was assured of being able to implement its principles of openness to all. Ucko and his supporters thereupon established a breakaway body, the World Archaeological Congress, which staged its inaugural event in Southampton in September 1986, the date originally scheduled for IUPPS.

The Royal Society President, Andrew Huxley, was unequivocal in condemning this episode. Together with Randolph Quirk, President of the British Academy, he wrote to *The Times* on 27 November 1985 to defend the principle of free circulation and to sympathise with 'South African scholars who have striven over the years to resist apartheid at home, only to find themselves confronted by apartheid in Britain'. And he used his final Anniversary Address three days later to stress the importance of free circulation, reminding his audience that he had been equally critical of the exclusion of two Soviet scientists from entry to Australia for the International Congress of Biochemistry at Perth in 1982.[100] He was particularly vexed by the thought that the victims of the IUPPS controversy were much more intimately engaged in the daily struggle against apartheid than the comfortable citizens of Southampton. *Nature* endorsed the Royal Society's line, and organisers of future international congresses in the UK were left in no doubt about what was expected of them if they wanted Royal Society support.[101]

The UK Government behaved ambivalently. On the one hand, it told the Royal Society in 1987 that it had no intention of interfering with the rights of individuals to travel between South Africa and the UK, and declared itself 'neutral' on South Africans taking part in international scientific meetings in the UK. On the other hand, it advised the Society against organising

---

[100] Following Huxley's intervention on that occasion, the Australian Government guaranteed that no scientist would be excluded from the International Congress of Physiological Sciences in Sydney in 1983.

[101] Andrew Huxley, 'Address at the Anniversary Meeting, 30 November 1985', *Supplement to Royal Society News*, (December 1985), vi; OM/10(86); C/14(86); Joseph Palca, 'South African exclusion causes academic schism', *Nature*, 319 (13 February 1986), 524; *Nature*, 319 (20 February 1986), 608; Steve Connor, 'Science and sanctions', *Nature*, 320 (21 August 1986), 19–20.

bilateral scientific meetings with South Africa, and briefed the National Physical Laboratory to prevent a CSIR researcher attending a scientific conference that it was organising. The Society protested in vain at this latter incident.[102] Such were the political niceties of the time.

During the later 1980s, the South African Government under P.W. Botha relaxed elements of the apartheid system in moves that were seen by some as the prelude to more far-reaching reform. But the South African Science Counsellor in London, Niels Hauffe, advised the Royal Society in May 1988 that it would be premature for George Porter as the Society's President to 'place his reputation, and that of the Royal Society, on the line' by making a formal visit to South Africa. Porter briefed the Fellows accordingly at a meeting in November that year: official contacts that might appear to give comfort to the South African Government were still to be avoided.[103] The eventual abolition of apartheid required the replacement of Botha by F.W. de Klerk in 1989, the release of Nelson Mandela from prison on 11 February 1990, multiracial democratic elections in 1994, and much else. The FCO minister William Waldegrave told the Society in March 1990 that the UK Government regarded de Klerk as a genuine reformer, and that both the UK and the European Community as a whole were now supportive of using scientific, cultural and educational contacts as positive means of rolling back apartheid. The Society modestly increased its indicative funding for individual exchanges with South Africa. Waldegrave's successor, Mark Lennox-Boyd, told the Society of further political progress a year later, and again in 1992,[104] but the immense difficulties of creating genuinely multiracial establishments and undoing the cumulative educational deficits of previous decades were quickly apparent.

The Society moved cautiously, anxious not to jump the gun by, for example, an inappropriate choice of institutional partner. A British-born FRS working in South Africa, Frank Nabarro, who was closely involved in moves to create a single South African academy and many similar initiatives, provided the Society with a stream of briefings on the developing situation and expressed his frustration that the Society was not moving further or faster in normalising its relations with South Africa. He was particularly vexed that the Russian and Dutch academies had already paid official visits.[105] By February 1993, the Society's President and Foreign

---

[102] Meeting with the FCO minister Tim Eggar, 1987 (OM/14(87)) and correspondence at RMA622.

[103] Niels Hauffe to Stephen Cox, 19 May 1988: RMA459; C/237(88).

[104] OM/50(90), OM/39(91), OM/40(92).

[105] CM 18 April 1991, minute 10; OM 19 June 1991, minute 2(a); Frank Nabarro to John Enderby, 8 November 1991; Jackie Gordon to Frank Nabarro, 13 December 1991; Anne McLaren to Frank Nabarro, 1 April 1992: RMA459 and RMA1944.

Secretary, Michael Atiyah (Figure 8.5) and Anne McLaren, were briefing Lennox-Boyd on their plans to visit South Africa the following year.[106]

South Africa rejoined the Commonwealth on 1 June 1994. The Prime Minister John Major made a formal visit towards the end of September 1994 – the first such visit since Harold Macmillan in 1960.[107] His entourage included groups representing several areas of public life – business, finance, sport, science – where the government wanted to reanimate relations with South Africa. The idea was that each group would participate in certain plenary events (including meetings with Mandela and de Klerk) and would then carry out its own programme. The science group comprised Michael Atiyah, the Chief Scientific Adviser, Bill Stewart, and a senior official, Robin Ritzema. They found that the UK's recent experience in producing the 1993 White Paper *Realising our potential* was of particular interest to the new South African science minister Ben Ngubane, who was faced with undertaking a thorough review of his own country's science policy. Michael Atiyah concluded from the visit that, from a Royal Society perspective, 'the main issue is how to preserve the high standards of research while at the same time broadening the whole educational base'.[108]

The pace of official interactions between the UK and South Africa accelerated.[109] Ngubane was in London in November 1994 to continue his talks with the Royal Society and others. The science minister David Hunt was one of numerous UK ministers to visit South Africa, going with Bill Stewart in February 1995 to take forward the agreement on scientific cooperation that Major and Mandela had signed the previous September. He confirmed with Ngubane a three-year scheme with an annual budget of £200,000 to encourage collaboration between British and South African research scientists. The UK end would be administered by the British Council. During the same trip, the South African Foundation for Research and Development (FRD) approached Bill Stewart with a proposal for an additional five-year, £1 M p.a., scheme focused on expanding university-level exchange programmes and jointly funded by the UK and South Africa. Stewart responded positively, and indicated that the Royal Society would be the most appropriate body to implement the UK end of the programme because of its experience in running research exchanges. When he wrote to Michael Atiyah to tell him of this, he added that the UK share of the costs might be met equally by the government and the Society.

---

[106] OM 11 February 1993, minute 2(d); OM/23(93).
[107] RMA236 and RMA1944; OM 6 October 1994, minute 2(a) and CM 6 October 1994, minute 40; also interview with Bill Stewart.
[108] Michael Atiyah to John Major, 26 September 1994: RMA236.
[109] RMA459, 967, 1718, 1810, 1944.

Figure 8.5 Michael Atiyah. © Anne Purkiss

The Society was perturbed to be told it had to find £250,000 p.a., but only mildly so: it was committed to the cause, and the China Royal Fellowships provided an encouraging precedent on fundraising possibilities.[110] Michael Atiyah responded accordingly to Hunt and Stewart at a meeting on 15 March 1995. However, he stressed that he first wanted to clarify the issues that a new scheme would need to address. A week later he and Anne McLaren led a Royal Society team on a long-planned visit to South Africa.[111] The upshot of that visit was not simple exchanges but a decision to work with the FRD to build up centres of excellence in particular research areas (chosen by FRD) in five historically disadvantaged universities,[112] via linkages with selected UK universities, visits in each direction, collaborative research projects, and staff and postgraduate student development. The scheme ran for ten years, with the UK science minister David Sainsbury praising it in 1999 as 'a model of international bilateral cooperation'. Keeping the door open in this instance was the prelude to close and long-term engagement with the partner country.[113]

Argentina

The Royal Society started an exchange programme with CONICET, the Argentine Science Council, in 1969, and a more formal arrangement was put in place in 1977. Scientific collaboration with Argentina in this period also took the form of joint expeditions, with both botanical and geological transects of Patagonia being initiated in 1975. These activities took place against the background of the dire human rights situation prevailing in Argentina before and, particularly, after the 1976 military coup. Individual scientists visiting Argentina reacted according to their personal experiences. The engineer Adam Neville, for example, commented in 1976 on military dominance of university administration, but argued for a stronger British presence so that Argentinians, still favourably disposed towards the UK, should seek their scientific education in the UK rather than the United States. The Society's Foreign Secretary, Michael Stoker,

---

[110] The Society eventually extracted additional funding from the government, and found the rest by raiding some of its existing publicly funded activities.

[111] This visit coincided with one by the Queen and the Duke of Edinburgh, at the beginning of which the Queen appointed Nelson Mandela a member of the Order of Merit. That provided an opening for Michael Atiyah (OM, 1992) to send Mandela a letter of congratulation and put in a plug for the importance of science and technology for the future of South Africa.

[112] Interview with Brian Heap.

[113] Julia Higgins, 'The Royal Society in Africa', unpublished paper given at the conference on 'The Royal Society and science in the 20th century', 22–23 April 2010.

who devoted a substantial part of his 1978 Argentine visit to human rights issues, recorded bluntly: 'Coming from the UK there is a lot that makes one sick.' He did not propose a halt to the exchanges, but he did reject an Argentine request to increase their scale.[114]

Argentina invaded the Falkland Islands on 2 April 1982. At the time of the invasion the Society was in discussion with the Falkland Islands Foundation over a proposed ecological survey of the Falklands, and was contemplating a detailed proposal by Tony Fogg and others for a programme of biological research in the Islands.[115] However, in terms of science and international politics the focus was on the geographer Edward Shackleton, son of the explorer. In 1976 he had written a detailed analysis of the economic potential of the Falklands for the government, only for most of his recommendations to be rejected as too expensive. In 1982, with the Argentine forces nearly defeated, Margaret Thatcher asked him to update his report. He repeated and augmented his original recommendations, and this time, in a radically altered political environment, they were accepted and implemented. The Society offered to help him with the update, for example on the scientific aspects of offshore resources, but he had been given just a few weeks to complete the task and the Society's offer was politely declined. Seven years later Shackleton was elected to the Society as an honorary (Statute 12) Fellow.[116]

When the invasion occurred, the Royal Society promptly cancelled the handful of exchange visits in the pipeline under its scheme with CONICET. On this occasion there was no door that could sensibly be kept open. The British Council closed its Buenos Aires office and recalled its staff. On 23 June 1982, the week after cessation of hostilities, the Overseas Development minister Neil Marten advised the Society to make haste slowly in reactivating its exchange programme. The Honorary President of CONICET, Luis Leloir, was a Foreign Member of the Royal Society, and the following month Michael Stoker's successor as Foreign Secretary, Arnold Burgen, was corresponding with Leloir about the great importance of international collaboration for the practice of science. The two bodies started helping individual scientists secure entry visas and supporting a few ad hoc visits. The military dictatorship in Argentina came to an end with the October 1983 general election, and

---

[114] RMA534; OM/134(78); interview with Michael Stoker. Beyond the politics, most visitors commented warmly on the friendliness of their hosts.

[115] C/110(79); RMA555.

[116] George Jellicoe, 'Lord Edward Arthur Alexander Shackleton', *Biographical memoirs of Fellows of the Royal Society*, 45 (1961), 486–505; OM 17 June 1982, minute 2(g); George Brock, 'Freedom, but what else can we do for the Falklands?', *The Times*, 16 June 1982, 12.

with FCO encouragement the Royal Society/CONICET exchange programme then gradually revived, supported by an updated agreement.[117]

The Society used its access to government to resolve one particular problem. Exchanges of scientific publications and samples had been impeded since the Falklands invasion by a general embargo on the import of goods from Argentina. Andrew Huxley lobbied Keith Joseph, the DES Secretary of State, in September 1983, pointing out that the UK stood to be accused of blocking the free flow of scientific information. Joseph secured the requisite exemptions, though the general embargo remained in force a while longer.

Individual visitors to Argentina and a formal Royal Society delegation in 1985 found themselves warmly welcomed and were struck by the keenness of Argentinian scientists to establish or re-establish connections with their British counterparts. They were also impressed by the quality of some of the research they saw, particularly in biology and medicine, despite the difficult conditions. The key to collaboration was to concentrate on the science, and to acknowledge human rights issues, but strictly to leave issues like the sovereignty of the Falkland Islands to the politicians. As the respective governments wanted to normalise relations so far as possible, it was useful that the scientists at least could work together. Diplomatic relations were restored in 1990.

[117] RMA1372, 1488; OM/80(83); OM 12 January 1984, minute 2(c).

9   Europe: competition and collaboration

Is it that we are unwilling to become good Europeans?[1]

The French President, General de Gaulle, twice blocked the UK's entry to the European Economic Community, in January 1963 and again in November 1967. Entry was finally effected on 1 January 1973. But where Britain's politicians were being thwarted, the scientists were making headway in building a European dimension to their work. Perhaps the stakes were lower and compromise therefore easier to secure. Perhaps the growing sense of European-ness found more ready expression among scientists than among other groups. Collaboration among scientists at European level was, nonetheless, still beset with political considerations of various kinds, centred for the most part on competition for control.

This chapter explores how the Royal Society responded to Howard Florey's challenge, in his valedictory Anniversary Address on 30 November 1965, that it should immerse itself more wholeheartedly in European science. In 1945, the Society had shown strong European instincts. Within weeks of the end of the war in Europe, it had sent envoys to Belgium, Czechoslovakia, Denmark, France, the Netherlands and Norway to explore how it could help them to get science going again on a peacetime footing. These visits were greatly appreciated, and the Society was gratified to learn of the esteem in which Britain, and British science, were held. The envoys' overarching finding was that 'the scientific world is looking to the West and particularly to this country [Britain] to take the lead in re-establishing and further developing scientific relationships'.[2] However, by the time the Officers were reviewing their post-Trend strategy

---

[1] Howard Florey, 'Anniversary Address, 1965', 430. Patrick Blackett acknowledged Florey's influence in highlighting the need to engage more actively with West Europe in a speech to the NAS in April 1966: PB/7/2/4/10.

[2] CM 17 May 1945, minute 7; 'Visits to liberated countries by representatives of the Royal Society', *Notes and records of the Royal Society of London*, 4 (1946), 82–99. These visits were stimulated by a report from Patrick Blackett to the Royal Society Council: CM 22 February 1945, minute 14 and Appendix B. That report also called for scientific attachés to be appointed to some British embassies.

in October 1964, the Society's international work was focused mostly at the global level. ICSU and all its associated international scientific unions and special committees, to which the Royal Society adhered on behalf of the UK, tended to dominate its international agenda, not least because of the many staff it then employed to deal with the British National Committees associated with each body.[3] The International Geophysical Year had been a further prominent aspect of the Society's international work during the 1950s, and other scientific expeditions featured on the agenda subsequently. And, as described in Chapter 8, the Society was busy during the late 1950s and early 1960s trying to develop relationships with the Soviet Union and China. So Howard Florey's charge that, among these international activities, the Society was neglecting its European neighbours outside the Soviet bloc was well founded.

In the postwar decades, European science was manifested in formal intergovernmental collaboration in major institutions such as CERN and the various space agencies. In addition to their scientific objectives, such institutions provided context for developing relations between science and foreign policy and for negotiating the American postwar influence on European science and European politics.[4] But postwar European science was also manifested in growing collaboration between individual scientists in different parts of Europe. It was there that the Royal Society concentrated much of its response to Howard Florey's challenge. It also invested significant effort in the development of European groupings around particular disciplines and around science as a whole. And it worked with national science academies in other European countries to engage with the policy-making process at European level. The Society was a good enough European that it allowed a film crew from German television to film five minutes of a Council meeting in 1972 in connection with the UK's entry into the European Economic Community on 1 January 1973,[5] but it needed occasional prodding to keep the enthusiasm going.

European Science Exchange Programme

The idea of promoting scientific collaboration in Europe through exchange schemes was very much in the air from the late 1950s onwards.

---

[3] See Chapter 10.

[4] John Krige, 'The politics of European scientific collaboration', in John Krige and Dominique Pestre, eds., *Science in the twentieth century* (Harwood Academic Publishers, 1997), 897–918; John Krige, *American hegemony and the postwar reconstruction of science in Europe* (MIT Press, 2006). At one stage the NAS toyed with the idea of setting up a European office, possibly based at the Royal Society, but it came to nothing: OM 9 May 1963, minute 2(a).

[5] CM 2 March 1972, minute 18.

For example, NATO appointed a Science Committee in 1957 to foster collaboration among the scientists of NATO countries, in the interests of strengthening long-term bonds between those countries.[6] The Committee's first action was to establish a programme of fellowships for researchers, mostly under thirty-five years old, to work for various periods in laboratories in NATO countries other than their own. This was run initially with start-up funding from the Ford Foundation, whose Director of International Affairs, Shepard Stone, was keenly alert to the cultural value of science in international relations. Between 1959 and 1969, 7,000 NATO fellowships were awarded, including 1,400 to British scientists. It was the largest scheme of its kind. Shepard Stone was instrumental in channelling Ford money also to CERN and to Niels Bohr's Institute for Theoretical Physics in Copenhagen to support fellowship schemes, with the same cultural agenda in mind.

From 1958 the Royal Society had a small scheme of visiting professorships to bring 'distinguished Commonwealth or foreign' scientists to work in the UK for 6–12 months, and since the end of the Second World War it had had a rather larger scheme to support the costs of British scientists wishing to travel overseas to attend conferences or visit laboratories.[7] Both schemes could, of course, be used to engage with West European countries, but in the mid 1960s the Society had no scheme aimed specifically at West Europe. It did, though, feel quite territorial about its role as purveyor of travel grants to academic scientists. This was about to be tested.

The Council for Scientific Policy had a full discussion on international scientific relations at its very first meeting, in January 1965. CSP was chaired by Harrie Massey, later to become the Society's Physical Secretary. Howard Florey was a member of CSP, and took the opportunity to set out the Royal Society's stall on promoting international collaboration.[8] He was particularly anxious to head off a suggestion by some of his CSP colleagues that extra money should be given direct to universities for travel grants, fearing that that might dampen the growing demand for the Society's grants. Massey was sympathetic to the Society's

[6] Andreas Rannestad, *NATO and science: an account of the NATO Science Committee 1958–1972* (NATO Scientific Affairs Division, 1973); John Krige, *American hegemony*, chapter 7; John Krige, 'The Ford Foundation, European physics and the Cold War', *Historical studies in the physical and biological sciences*, 29 (1999), 333–61; CSP(ISR)(66)17. SRC ran the UK participation in the programme. See Brian Heap interview for the later evolution of the programme.

[7] OM/91(58); John S. Rowlinson and Norman H. Robinson, *The record*, 7, 63, 135. The travel grant scheme had a budget of £47,000 for FY 1966–7.

[8] OM 29 January 1965, minute 2; OM 4 March 1965, minute 2(b); OM 1 April 1965, minute 2(a); OM/31(65) (copy of CSP minute); OM/32(65).

position and its claim to be better able to select the most meritorious applicants, and invited it to submit costed proposals for its international ambitions. The Society's response included a bid for a doubling of its travel grant scheme – but, at this stage, still nothing with a specifically West European flavour.

Later in 1965 David Martin had a long talk with Harrie Massey about international relations.[9] This led to a joint RS/CSP review group led by Massey which, as mentioned in Chapter 7, soon morphed into a standing committee of the CSP.[10] Patrick Blackett and Tommy Thompson, newly elected President and Foreign Secretary respectively, took leading roles in the review, as did David Martin. For much of 1966, the review provided a forum for high-level debate on how to promote international collaboration, particularly at European level.[11]

One idea on the table in 1966, from a group of OECD science ministers, was to build up a series of explicitly European 'centres of excellence' in selected fields. Another idea, pushed by Blackett, was for the Royal Society to attract leading scientists to Britain by running select residential conferences on the Gordon conference model.[12] But a good deal of the review group's time was spent on a proposal from the Department of Education and Science for an 'International Convention on European fellowships for growing points in science'.[13] This, naturally, would be an intergovernmental initiative, designed both to attract financial backing from other governments and to ensure a reasonable return to the UK from the Treasury's own contribution. The fellowships would be aimed at postdoctoral researchers and be tenable for periods of between three months and two years in institutions in countries other than the fellow's own. They would be concentrated in about five favoured areas of research, selected by an advisory board (possibly constituted via existing OECD machinery) and reviewed every few years. This element of dirigisme was counterbalanced by the hope that, contrary to the OECD proposal, centres of excellence would emerge in a bottom-up way through the choices of the appointed fellows. The DES calculated that the UK contribution to its proposed scheme might be about £150,000 p.a., or 20 per cent of the total cost, with eleven other countries providing the rest.

[9] OM/86(65); OM 30 November 1965, minute 2(a).
[10] At the same time the Royal Society set up a new International Relations Committee under Tommy Thompson to animate its own thinking on international policy.
[11] Papers relating to the CSP review group are at HWT C.68–72, and at TNA ED 214/32–49.
[12] Tommy Thompson had just been instrumental in establishing the EUCHEM (European Association for Chemical and Molecular Sciences) conferences on Gordon conference lines, with participants drawn mainly from European countries.
[13] CSP(ISR)(66)10, 10 May 1966. See also TNA FO 371/189400.

The DES proposal was intended to provide Europe's leading postdoctoral researchers with an attractive alternative to emigration to the United States, which the government saw as a real threat. The proposal was also intended, explicitly, to puncture what officials saw as the growing demand for European or international centres in an increasing number of research areas – the problem being that suggestions for such centres seemed to be proliferating without thought as to how their budgets would be found or controlled. This was a particular issue in areas of research that did not inherently require multinational levels of investment in equipment, and where both Treasury and DES tended to regard new international proposals simply as ruses by the scientists to extract more money from the public purse. A pressing example of this was what DES described as the 'EMBO-inspired movement for the Europeanisation of basic biology' – of which more in a moment.

The Royal Society, and the CSP review group as a whole, wanted greater flexibility and, above all, insisted that any scheme had to be run by scientists rather than by civil servants or politicians. Member countries should manage their own funds earmarked for the scheme, rather than put them into a common pool to be managed centrally. The emphasis on preselected 'growing points' was strongly resisted. There was concern that potential partner countries might see the scheme as a means of strengthening British science at the expense of European science by attracting talented researchers from the Continent to the UK, and the CSP therefore stressed the need to ensure a full quota of British fellows going from the UK to institutions elsewhere.[14]

Meanwhile, goaded by Florey's challenge and urged on in private meetings with Harrie Massey, the Royal Society did not wait patiently for the CSP review to run its course. At the beginning of 1966, just after he became Foreign Secretary, Tommy Thompson sent a questionnaire to a cross-section of the Fellowship to establish the nature and extent of informal international collaboration mediated simply through personal contact. The 200 replies confirmed the demand for extra exchanges and other forms of collaboration, and provided detailed information about needs and opportunities in individual fields of research.[15] A quick review of formal schemes run by the British Council, NATO, SRC and the Ciba Foundation among others found nearly 1,000 visitors per year from West Europe to the UK and about a quarter that number in the opposite direction. It also discerned substantial unmet demand. So the Society

[14] OM/59(66); OM 16 June 1966, minute 3(a); CSP(66)37; CSP(ISR)(66)20; HWT C.71.
[15] RMA303.

proposed a package of initiatives for itself to undertake: about 50 one-year fellowships per year at postgraduate and postdoctoral level in either direction between the UK and Continental Europe, twenty bursaries for shorter stays by mid-career scientists, funds for forty visits by senior scientists and residential conferences complementing the newly established EUCHEM scheme – with a total indicative budget of £750,000 over five years.[16]

Blackett, Thompson and Martin fed the Society's developing thinking into the CSP review group in April 1966. When the DES proposal surfaced a month later, there was no suggestion that the Society should ease up on its own plans. Instead, and with the strong support of Frank Turnbull, then DES Deputy Secretary, and other members of the CSP review group, the Society decided to treat the programmes as complementary. A meeting of all concerned with Jack Embling, Frank Turnbull's successor at DES, in November 1966 agreed that the Society should run the DES scheme (by then called the European Science Fellowship Programme, ESFP) in parallel with its own, and that SRC should be represented on the relevant Royal Society committee and cooperate at staff level. The cooperation extended to joint application forms and assessment procedures in those areas where the Royal Society and SRC/NATO schemes overlapped.[17]

For the Society, refining its objectives for European collaboration went hand in hand with raising money for them. Its intention was to fund at least the start-up phase of its programme from private sources.[18] Already in December 1965, his first month as President, Blackett was speculating whether he could persuade the Ford Foundation to provide start-up funds. He had an exploratory meeting with Joe Slater, deputy to Shepard Stone at the Foundation, in February 1966, at which he presented the Society as the natural home for an exchange scheme, on the basis of its links with academies and leading individual scientists throughout West Europe and its inherent ability to attract 'better British scientists for exchange visits than the British Council, NATO, SRC etc.'. The formal application was submitted to the Ford Foundation in April 1966, and formal notification that Ford had agreed a grant of $200,000 (£71,000) over three years came through in January 1967, with informal notification coming through a couple of months earlier. That same grant

[16] OM/14(66); C/48(66); RMA282.
[17] OM 16 June 1966, minute 3(a); correspondence between David Martin, Tommy Thompson, Nicholas Kurti (RS assessor on SRC) and William Francis (SRC Secretary): HWT C.71-C.72; OM/115(66); C/106(67) and correspondence at RMA899.
[18] HWT C.71.

round, incidentally, produced $150,000 for the Niels Bohr Institute to strengthen international scientific exchange, and $280,000 for Queen's College, Oxford (where Howard Florey was Provost) to assist a fellowship programme for scholars from Continental Europe.[19] European exchanges were high on Ford's agenda.

The Society also secured other grants for its European plans. One of the first actions that the Czechoslovakian émigré and founder of the Pergamon Press, Robert Maxwell, had taken on being elected a Labour MP in 1964 was to badger the Foreign Office to do more to stimulate relations between British and foreign scientists. Thompson visited him in June 1966 to sound him out for a possible donation. He was surprised to find Maxwell well informed about the Society's bid to Ford. Maxwell proved interested, and a donation of £10,000 duly arrived a few months later.[20] A third donation came from the Wates brothers (of the construction family), who followed an initial grant of £3,000 in April 1966 with an agreement, in October 1966, to give the Society £30,000 over three years for the European programme (and substantial support for other projects in later years).[21]

During 1966 Thompson travelled extensively in West Europe talking to his opposite numbers, and found considerable enthusiasm for the Society's proposals. So he convened a meeting at the Society on 1 December 1966 of academy presidents to build collective support.[22] Belgium, Denmark, France, Germany, Italy, the Netherlands, Norway, Spain, Sweden and Switzerland were represented, and Howard Florey, Harrie Massey and Jack Embling all took part. By then, crucially, Thompson could be pretty confident of having sufficient private funding to get the scheme off the ground on his own terms. Reinforcing the momentum, Blackett used his Anniversary Address on the eve of the meeting to stress the importance of a distinct European dimension to the practice of science.[23] The British could be good Europeans even if they were excluded from the EEC.

A follow-up meeting was hosted by the Deutsche Forschungsgemeinschaft, the largest independent German research funding organisation, at Bad Godesberg on 28 April 1967 to sort out the practicalities of implementation. Again, the Society did not dally. The private funds it had secured allowed it to make an immediate start.[24] In

[19] OM 16 December 1965, minute 2(a); OM/22(66); HWT B.209-B.210; RMA282.
[20] TNA FO 371/178054; HWT B.521; CM 13 October 1966, minute 31. The £10,000 was paid through a personal covenant over seven years.
[21] OM 5 May 1966, minute 2(d); RMA282.   [22] C/167(66).
[23] P.M.S. Blackett, 'Anniversary Address, 1966', xii–xiii.
[24] Council was impressed how quickly Thompson and the staff moved to implement the exchange programme: CM 9 February 1967, minute 25(iii).

mid January, Thompson wrote to Fellows and Foreign Members with details of the Society's new European schemes, encouraging them to stimulate applications from their colleagues. He put simple and flexible administrative arrangements in place. Seeing how the Society's plans were surging ahead, Jack Embling abandoned his efforts to promote the DES's intergovernmental ESFP initiative at the beginning of February. Instead, he gave the money earmarked for that initiative (£50,000 in the start-up year, £200,000 the following year) to the Royal Society to use essentially for its own programme, provided comparable contributions could be obtained from other countries.[25]

At the opening of the Bad Godesberg meeting (Figure 9.1), Thompson was therefore able to announce that the new scheme was already under way, with fourteen postgraduate fellowships and forty-five short visits to and from the UK having been awarded in the first round from private funds, and with the promise of much more to come as public funds came into play. Participants were reported to be very impressed, and keen for the Society to continue its leadership of the initiative. The important issue now was the balancing funds: all participating organisations (national academies or equivalent bodies) were urged to secure funding to cover, or help cover, in-country costs for elements of the scheme being met from public sources. Sharing costs in this way had been a key element of the DES/OECD plans. Most Bad Godesberg participants signalled their intention to do so, and within a week of the meeting Thompson told the Ford Foundation how its grant was successfully leveraging large sums from the public purse. Embling was delighted with this progress.[26]

The overall pattern of the exchange scheme had more or less settled down by the time of the third meeting of partner bodies, in Amsterdam in November 1967. It was, in essence, a series of parallel bilateral agreements between the Royal Society and analogous bodies in each of sixteen West European countries.[27] Postgraduate and postdoctoral fellowships of six months to two years duration were funded mainly by government, with balancing contributions from each partner country, while shorter study visits (one week to six months) and specialised research conferences

[25] OM/4(67), OM 12 January 1967, minute 3(a), CM 12 January 1967, minute 8; C/21 (67); RMA225, 562, 914.

[26] HWT C.74, RMA225, 771. See also David Martin's report on the meeting to the CSP Committee on International Scientific Relations, 19 May 1967: RMA562.

[27] Israel had wanted to take part, but was deemed to fall outside the West European focus of the ESEP. Instead, a parallel bilateral programme was established along similar lines in 1967, with help from Marcus Sieff and Victor Rothschild and, later, with public funding. Correspondence at RMA197 and HWT B.229, B.231, B.245; OM 2 March 1967, minute 3(d); C/52(67) and CM 13 April 1967, minute 24; OM/69(67); Report of Council for the year ended 31 August 1970.

Figure 9.1 The Bad Godesberg meeting, 28 April 1967, to prepare the launch of the ESEP (l to r): P. Gaillard, B. Segre, H. Curien, C. Müller Daeker, Tommy Thompson, Avon Meualt, J. Speer, David Martin, Nadia Slow, A. Tiselius
Photographer stamp on reverse: Georg Munker, Koln, 53 Bonn, Germanesnstrasse 127

were funded mainly by the private sources that the Society had secured. Each partner managed the spending of its financial input and the selection of its own participating researchers. It remained the Society's hope that a fully multilateral programme as originally envisaged by the OECD – exchanges between any two countries in the group – would eventually come into being,[28] but the quickest and most pragmatic route to that was to get a programme centred on the UK up and running, and then to hope that other countries would establish analogous programmes that would coalesce in due course. However, there turned out to be consistently no appetite for this,[29] and the programme remained focused on the UK.

[28] The French, in particular, were dismayed to realise, late in the day, that the scheme was not fully multilateral from the outset. See note on Thompson's visit to CNRS, 25 October 1967: HWT B.231. On awkward negotiations with French partners throughout the development of the ESEP, see Jennifer Goodare, *Representing science*, chapter 3.

[29] O/110(69); C/157(72) [minutes of IRC meeting, 20 July 1972, especially minute 8]; *Report of the Working Group on scientific interchange* Cmnd 4843 (January 1972), 50; OM/100(79). In contrast, the Commonwealth bursaries scheme (see Chapter 10) was multilateral.

Figure 9.2 (l to r) Patrick Blackett, HM Queen Elizabeth II, HRH Duke of Edinburgh and Howard Florey at the opening of the Society's new headquarters at Carlton House Terrace, 21 November 1967. © The Royal Society

What to call the scheme was, as ever, a matter of disproportionately intense debate, started in Bad Godesberg and concluded in Amsterdam. A wish to avoid national associations led to a strong bid for a Latin name such as Stipendium Europaeum or Scientia Europa. But the final choice, 'for the time being', was European Science Exchange Programme (ESEP). The name stuck, as temporary expedients often do.

Two days after the November 1967 Amsterdam meeting, at the formal opening of the Royal Society's new premises at Carlton House Terrace (Figure 9.2), the 'international relations of science, particularly in relation to Europe' topped the list of activities that Patrick Blackett predicted would benefit from the improved facilities. And in his Anniversary Address at the end of that month, he rehearsed the origins of the ESEP, which he described as a 'sign of spontaneous movement towards a greater cohesion of European science'. Brushing aside occasional vexations, he added: 'One very welcome discovery during all these negotiations has been the warmth of the welcome among European scientists for the initiative in these matters taken by the Royal Society.'[30] Personal contacts fostered over the years, and its tacit pre-eminence among European academies – as well as healthy relationships with the DES and with charitable foundations – had enabled the Society to get its programme off the ground. These attributes would also prove valuable as the Society turned to other aspects of building European science.

Molecular biology

Molecular biology was something of a preoccupation for those thinking about the European dimension to doing science. The European Molecular Biology Organisation was established in 1963 as a private, self-governing body of individual researchers.[31] The Royal Society as such was not directly involved. EMBO was supported for the first three proof-of-concept years mainly by the Volkswagen Foundation. Its Europe-wide programme of short-term and long-term fellowships established during this period allowed fellows to work at European institutes other than their own and was warmly appreciated. It was a practical and fairly uncontroversial contribution to stemming the movement of talent to the United States, where the subject was blossoming.[32]

What was controversial, however, was EMBO's proposal for a European Molecular Biology Laboratory (EMBL). A large central laboratory was not obviously justified by the need to share the costs of major equipment (unlike CERN), and some feared it would draw talent away from national centres that were still trying to establish themselves. Moreover, unlike EMBO, EMBL would have to be a publicly funded intergovernmental body and, unless new money was forthcoming, it would draw money away from existing publicly funded research programmes and from other areas of research. Both the Royal Society and the CSP initially hedged their bets on the EMBL proposal, and then hardened their opposition to it. From a national perspective it was at least partly a timing issue: molecular biology had to be more thoroughly

[30] 'The formal opening of the Society's new home at 6 Carlton House Terrace', *Notes and records of the Royal Society of London*, 23 (1968), 5; P.M.S. Blackett, 'Anniversary Address, 1966', x–xii.

[31] John Tooze, 'A brief history of EMBO', in *European Molecular Biology Organisation 1964–1989* (EMBO, 1989), 8–19; Michel Morange, 'EMBO and EMBL', in John Krige and Luca Guzzetti, eds., *History of European scientific and technological cooperation* (European Communities, 1997), 77–92; Georgina Ferry, 'Fifty years of EMBO', *Nature*, 511 (10 July 2014), 150–1.

[32] John Kendrew told Florey in 1961 that on a recent visit he had been 'quite astonished by the boom in these subjects all over the States'. Kendrew to Florey, 5 April 1961: HF/1/17/13/1.

embedded in the UK before the UK could benefit from the establishment of a European centre such as EMBL.[33]

By 1968, however, the mood among most European countries was strongly in favour of an eventual EMBL, and continued UK opposition carried the risk of the UK being tarred as bad Europeans. As Albert Neuberger put it in a letter to Patrick Blackett:

Almost all countries on the Continent of Europe were [earlier] looking to the United Kingdom for leadership, and what we provided was aloofness and an apparently uncooperative attitude which our friends on the Continent found difficult to understand ... If we stand aloof now we shall not be able to have any influence on future developments in Europe.[34]

What was on the table at that stage was a two-step process: first, the creation of an intergovernmental European Molecular Biology Conference (EMBC), mainly to provide financial support for the EMBO fellowship programme when the Volkswagen money ran out at the end of 1968, and then EMBL.[35] The Royal Society convened a meeting of Fellows in March 1968 to take stock. After the meeting, both the Society and CSP conceded that the UK had failed to persuade its European partners of the virtues of a loose, non-governmental federation, and they advised the government to acquiesce in the establishment of the intergovernmental EMBC. But they hoped that EMBC would be part of 'a more general international arrangement for scientific research' rather than isolated support for one area of research, and they kept their options open on the EMBL. The UK Government accepted their advice, and, as if atoning for its previous niggardliness, was one of the first to make a financial contribution to keep the fellowship programme going.[36] EMBC was formally ratified in April 1970.

That did not mean that EMBL was a done deal. The debate about having a European laboratory at all, and about various forms that it might take, raged on. The Royal Society convened another meeting in October

[33] John Krige, 'The birth of EMBO and the difficult road to EMBL', *Studies in history and philosophy of biological and biomedical sciences*, 33 (2002), 547–64; OM 6 May 1965, minute 2(d), OM/38(65), OM/45(65), OM/60(65), RMA1364; CM 14 July 1966, minute 34(ii).

[34] CSP(68)10; Neuberger to Blackett, 23 February 1968: RMA1364.

[35] For a rehearsal of the UK perspective on some of the issues at stake, see C/45(68) (minutes of IRC meeting, 1 March 1968).

[36] CSP meetings, 9 February 1968 and 22 March 1968; CM 7 March 1968, minute 22(iv); RMA1364; 'EMBO – the year of decision', *New scientist*, 37 (29 February 1968), 458–60; 'Britain warms to EMBO', *New scientist*, (27 February 1969), 437. With the Society's agreement, part of the UK's contribution to EMBC for exchanges was drawn from the government allocation to the Royal Society for ESEP: Thompson to Appleyard (DES), 16 May 1967: RMA197.

1969 to provide a forum for the protagonists to put their cases. John Kendrew, by then Secretary General of both EMBO and EMBC, was one of several who submitted carefully drafted papers to the meeting as well as participating in the discussion. The meeting revealed strong support for European research collaboration in general (evidenced by numerous personal statements and by the rapid success of the ESEP), but an 'overwhelming weight of argument' against the EMBL model of collaboration in the context of molecular biology. Despite requests to respect the confidentiality of the occasion, it was sufficiently controversial that someone thought it worth leaking to *Nature*, which promptly published an account of the meeting.[37]

A meeting convened by EMBO in Konstanz the following month to take forward the EMBL proposal therefore began in a pessimistic mood. A more modest proposal gradually emerged from that meeting, more clearly differentiated from existing national laboratories.[38] The initiative was then developed on that basis, and began to build support. In March 1970 the MRC put its weight behind EMBL.[39] The Royal Society refrained from further intervention on the matter. Formal intergovernmental agreement on EMBL was reached in May 1973, and ratification was completed the following year. John Kendrew became the first Director General of EMBL.

European Physical Society

While the Royal Society was a key part of the conversation around European molecular biology, it was just a very modest part of the finance when it came to organising European physics. The European Physical Society was launched in September 1968 after two years of discussion between national physical societies. The UK input was led by the Institute of Physics, much the largest national physical society in Europe following its amalgamation with the Physical Society. The IoP approached the DES for a financial contribution to get the European body launched. DES asked the Royal Society for advice. Patrick Blackett and Tommy Thompson discussed the initiative with James Taylor, the IoP President, and concluded that it was a good idea, that it would not duplicate the ESEP and that it merited a degree of financial support. Jack Embling worried about setting a precedent for other groups of scientists

---

[37] CM 9 October 1969, minute 36; CM 6 November 1969, minute 18; *Nature*, 224 (1 November 1969), 406–7; RMA914. The Biological Secretary Bernard Katz complained strongly that the *Nature* article misrepresented his views.

[38] John Tooze, 'A brief history of EMBO', 14.

[39] John Gray (MRC Secretary) to Jack Embling, 3 March 1970: RMA914.

who might also wish to establish European societies and pester DES for money, but he agreed that the Royal Society could use some of its existing public funds to help. The Society duly gave IoP £250, and Patrick Blackett was one of the star turns at the EPS inaugural congress.[40]

A couple a years later, the IoP came back for more, the EPS in the meantime having grown in ambition and scope. Thompson was broadly sympathetic to making a grant, but cautious since he was aware of at least a dozen analogous European initiatives in different disciplines and was keen to treat them equitably. Embling was even more worried than before about the danger of setting an unsustainable precedent. But eventually he relented, on condition that other major countries also contributed to the EPS budget, that the grant tapered rapidly to zero and that it did not set a precedent for any future bids. The upshot was a grant from the Royal Society to the IoP of £4,500 over three years while the EPS achieved financial self-sufficiency.[41]

European Research Council

The proliferation of European initiatives in particular areas of science was, for some, a problem, especially where initiatives ignored national needs or distorted the overall balance of effort between disciplines. This prompted suggestions for overarching mechanisms to look at European science as a whole, against concern over how best to build a European scientific community and promote fundamental research at European level in the face of competition from, especially, the United States. One such suggestion, in 1968 from a group chaired by Olivier Giscard d'Estaing (brother of the future French President), was for a European Institute of Science and Technology to carry out intensive research programmes and provide high-level training. This fared no better than the attempt by James Killian, President Eisenhower's Science Adviser, to establish an International Institute for Science and Technology in Europe a few years previously.[42] A second suggestion, also in early 1968, was put to the group of OECD science ministers by the DES Secretary of State, Patrick Gordon Walker, and was for a European Research Council. The CSP set up a small working group to develop

---

[40] Henk Kubbinga, 'European Physical Society (1968–2008): the early years', *Europhysics news*, 39 (2008), 16–8; HWT B.179; Embling to Martin, 21 August 1967: HWT C.74; CM 18 January 1968, minute 18.

[41] OM/30(70); Thompson to Sam Edwards, 4 May 1970, and Edwards to Thompson, 12 May 1970: HWT B.179; CM 11 June 1970, minute 9.

[42] HWT C.75; John Krige, *American hegemony*, 208–25. The European Institute proposal resurfaced, briefly, in 2006.

the concept; the Royal Society was represented on it by Tommy Thompson.[43]

Gordon Walker's ERC proposal originated in a discussion between Jack Embling, John Gray (Secretary of the MRC) and a Swedish scientist, Professor Engström.[44] It was essentially a response to the concern expressed by the Royal Society and others about lack of 'a more general international arrangement for scientific research'. Brian Flowers, Chairman of the SRC, was supportive. He hoped an ERC would gradually acquire a budget and executive functions, enabling it to implement a balanced approach to European collaboration. It would also show the British as good Europeans at a time when they were still knocking on the door of the EEC.

The ERC proposal had several subtexts. One was the hope on the part of some DES officials that it would render the EMBC redundant. Another, added by officials during the CSP discussions, was that it might take over the ESEP as well. There was much intricate debate about how to attach the ERC to an existing intergovernmental body of the right shape, so as to avoid the time-consuming process of creating a new one. There was also concern to keep it distinct from a proposal then being championed by Harold Wilson for a European Centre for Technology. But in the end it all ran into the sands.[45] A very different European Research Council came into being in 2007, as a grant-giving agency for basic research under the auspices of the EU Framework Programme. At the insistence of the Royal Society and many other commentators, that particular ERC was controlled by scientists rather than administrators and its funding decisions were based on the scientific quality of the proposals rather than political or other criteria.

European Science Foundation

The European Science Foundation emerged out of the same broad context that gave rise to the European Research Council proposal: a wish from the late 1960s onwards to strengthen scientific research across Europe and to create a European scientific identity, and a wish also to treat science as a coherent whole rather than as disciplinary fragments.[46]

---

[43] CSP(ISR)(67)12 (copy at HWT C.74); CSP(IRF)(68)1 (copy at HWT C.75).

[44] CSP(IRF)(68)2nd meeting, 16 May 1968 (HWT C.76).

[45] CSP(68)44: TNA FCO 55/58; CSP(IRF)(68)5 and CSP(ISR)(68)2nd meeting, 30 May 1968 (HWT C.76).

[46] For general background on the European Science Foundation, see Brian Flowers, *The ESF – the first twenty years!* (ESF, 20 November 1984); Gérard Darmon, 'European Science Foundation: towards a history', in John Krige and Luca Guzzetti, eds., *History of European scientific and technological cooperation* (European Communities, 1997), 381–

The wishing was being done with varying intensity at both national and multinational levels, by scientific leaders, national and international science administrators, and politicians. Inevitably, one of the main themes of the debate was the balance between science-led and policy-led approaches, bottom-up and top-down. Many of the issues were close to the Society's heart, and the Society stayed close to the action.

There were two strands. The first was growing recognition of shared interests, particularly among the national research councils of Western Europe, and a growing sense of collegiality across borders. This crystallised at a conference in Aarhus in February 1972, when the research councils of 13 European countries agreed to collaborate in establishing a European Scientific Research Council dealing with basic research. Brian Flowers, then Chairman of the SRC, told the conference: 'We can build a little bit of Europe through science ... Collaboration should be seen as a way of life in Europe.' European collaboration was not just about large projects.

The second strand was more directly political. Altiero Spinelli, a leading European federalist, was the European Commissioner responsible for industry, research and technology. In 1972 he launched an attempt to develop a common European policy on scientific and technological research.[47] He wanted to tie European Community research activities more closely to socio-economic needs, and in that context envisaged a gradual harmonisation of national R&D policies. At the same time, he recognised that fundamental research needed 'support rather than organisation, stimulus rather than planning'. To help deliver that, his proposal included the creation of a (nominally) independent European Science Foundation which would support and coordinate existing research centres and generally stimulate research cooperation at EC level, for example by facilitating individual mobility, supporting research meetings, promoting centres of excellence and reviewing proposals for major capital projects. Spinelli's ESF would also advise the Commission on R&D policy, and would be funded by the Commission.

When consulted by the DES, the Society's immediate reaction to Spinelli's proposal was a guarded welcome for some aspects (support for rather than takeover of existing organisations, recognition of need for gradual evolution of structures, apparent acceptance of a bottom-up approach to the management of fundamental research), and distinct

402; *Voices of European science* (ESF, 1994); interviews with Tony Epstein, Brian Flowers, Brian Heap, Julia Higgins and Peter Warren.

[47] Commission of the European Communities, *Objectives and instruments of a common policy for scientific research and technological development* (CAB/VII/74/72-E; CEC, 31 May 1972). See also IR/7(72) (Thompson's annotated copy at HWT B.24) and IR/10(72).

wariness of others (unnecessary structures, excessive consultation, expansion of the Commission's Joint Research Centre). The Society characteristically stressed the need for ESF to be led by scientists rather than by administrators or politicians. It hoped that the ESF might encourage further exchange schemes of the ESEP type, but was emphatic that ESF should not take over ESEP itself. Overall, Spinelli's proposal was something that, in Alan Hodgkin's words, 'gave us and our colleagues in the Research Councils much concern'.[48]

Under Kingsley Dunham, Thompson's successor as Foreign Secretary, the Royal Society reviewed the overall pattern of its European activities in the summer of 1972 and identified opportunity to extend its own leadership role – the more so because of the UK's imminent entry to the European Community.[49] To that end, it invited its analogues from a dozen European countries to a meeting on 1 December. It also invited the most senior scientific policy-makers in the UK. In order to advertise its credentials as a thoroughbred European, it put on an exhibition to remind participants of its long historical links with the scientists and academies of Europe.

The 1 December meeting was dominated by Spinelli's ESF proposal.[50] From experience of other scientific collaborations, particularly in fundamental research, it was evident to participants that success depended on schemes being driven by the needs of the science rather than by political agendas, and on the management being largely in the hands of scientists. Moreover, the natural unit for cooperation in fundamental research was West Europe rather than the EC (then too small) or Europe as a whole (too difficult so long as the Iron Curtain remained in place). The real problem about the Spinelli ESF was that it might be set up by EC governments without proper involvement of the scientific community and in disregard of what was going on already. Participants strongly favoured a different approach altogether, based on the interactions that had multiplied in recent years among national academies and research councils, both informally and, for example, in connection with running the ESEP. Fruitful European collaboration, they believed, was more likely to grow from such experiences than from a structure imposed from the top and dogged by the political requirements of *juste retour*. Participants went home to brief their governments accordingly.

[48] David Martin to Edwin Appleyard, 9 September 1972: HWT B.25; Alan Hodgkin, 'Address at the Anniversary Meeting, 30 November 1973', *Proceedings of the Royal Society of London. Series B, Biological sciences*, 185 (1974), x.

[49] C/157(72) [minutes of IRC meeting, 20 July 1972, especially minute 8], IR/15(72).

[50] IR/17(72)a (summary of the meeting).

From that point, momentum built behind this alternative approach to fostering European research collaboration. The Royal Society decided, after further discussion with the UK Research Councils, that it should aim to divert the European Commission's coordinating zeal away from fundamental research and towards the science that affected legislation in such matters as standards, pollution and agriculture.[51] The Advisory Board for the Research Councils, which replaced the CSP at the beginning of 1973, set up a group to consider European scientific collaboration in general, and the Spinelli proposal in particular. Dunham represented the Royal Society on this group.[52] By spring 1973 the Spinelli proposal was stalling, because of resistance from the scientific community and because a summit meeting of EC Member States had called for a wider approach to a common policy on science and technology, focused on coordination of national policies and joint implementation of projects of interest to the Community, especially in the more applied areas. On fundamental research, the ABRC came down in favour of the looser approach advocated by the Royal Society, emphasising the needs of the science and making full use of existing structures.

The follow-up meeting to the Royal Society's 1 December 1972 event was hosted by the German research organisation, the Max Planck Gesellschaft, at Munich on 13–14 April 1973. Spinelli's successor at the European Commission, Ralf Dahrendorf, was present. At this and a further meeting in Paris on 24–25 September 1973, the initiative passed decisively to the scientists. Academies and research councils agreed to establish what they briefly called the European Fundamental Science Foundation as an independent organisation with a Europe-wide membership to promote transnational collaboration in fundamental research. The EFSF would be funded by its members, possibly with some input from the Commission, and it would seek a close relationship with the Commission, including offering policy advice. It would concentrate on fundamental rather than applied research, to the extent that such a categorisation had meaning in practice. It would be concerned with promoting all fields of research, including social science and humanities, and with catalysing international and interdisciplinary collaboration, mobility, networking and coordination. It would have a relatively modest budget and therefore would not be drawn into the hazardous business of financing major collaborative projects. The Royal Society was sufficiently

---

[51] OM/27(73), OM/30(73), OM 1 March 1973, minute 2(c).
[52] ABRC(E)(73)1, ABRC(73)21, draft minute of 6 April 1973 ABRC meeting: copies of these and related papers at RMA1029.

enthusiastic to propose that the new body be headquartered in London.[53] With a British President (Brian Flowers) and German Secretary General (Friedrich Schneider), however, the Foundation opted for a French site (Strasbourg) for its headquarters, and held its founding conference there on 18 November 1974. Its inauguration was attended by forty-two national academies and research councils from fifteen countries.

Brian Flowers had initially wanted membership restricted to research councils, to the exclusion of national academies.[54] But the Royal Society worked with the leaders of the main scientific bodies throughout Europe to ensure that academies could play a full part in the debate about how to promote European science, both before and after the 1 December 1972 meeting. The Foundation therefore included academies, though Kingsley Dunham complained at the founding conference that academy interests were insufficiently accommodated.[55] Moreover, against the Royal Society's express advice,[56] Brian Flowers secured the backing of the UK delegation for his own nomination as President, and academies were very poorly represented in the initial membership of the ESF Executive Council. It was not an equal partnership, and the Society thought this boded ill for a body that was supposed to be independent and science-led.[57] At UK level, the Society acquired a slightly enhanced stake in ESF activities by agreeing for a while to coordinate the input from the seven UK members (five research councils, Royal Society and British Academy).

Flowers later described his wish to develop ESF without national academies as 'an error of judgement'. That erroneous judgement would be repeated. In 1993, in a move reminiscent of the Aarhus initiative, the heads of the national research councils formed their own group (EuroHoRCs). This provided an alternative mechanism for one of ESF's sociological functions. At its 2003 General Assembly, ESF expressed the view that research councils, as the major funding agencies,

---

[53] C/178(73) (minutes of IRC meeting 10 September 1973); CM 11 October 1973, minute 19; IR/13(74).

[54] For Tommy Thompson's private thoughts about the research councils over-reaching themselves, see his 'Comments on recent proposals about a European Science Foundation', 12 June 1973: HWT B.30. He was disparaging of Kingsley Dunham's efforts to defend the Society's interests against the encroachments of the research councils and government departments, and scathing of the proposal to nominate Flowers as the inaugural ESF President: Thompson to his fellow chemist Guy Ourisson, 14 May and 27 May 1974: HWT B.298.

[55] CM 30 November 1974, minute 12.

[56] The Officers would have preferred Harrie Massey as President: OM 11 July 1974, minute 2(a). The SRC felt obliged to support Flowers: interview with Geoffrey Allen.

[57] See correspondence at OM/125(74) and OM/129(74); OM 7 November 1974, minute 2 (c); and OM 30 November 1974, minute 2(b).

were its primary members and that it was sufficient for academies to be represented on the ESF Governing Council through ALLEA (see below), which had only non-voting observer status. Coupled with increasing dissatisfaction with ESF's apparent mission drift and excessive budget increases, this provoked the Royal Society (and the Royal Irish Academy) to give notice of withdrawal from ESF membership at the end of 2006. ESF hastily reviewed its attitude to academy members, but the withdrawals went ahead.

ESEP (continued)

In 1970 the Council for Scientific Policy decided to review the ESEP, which it had earlier helped into existence. The working group carrying out the review for CSP was, oddly, chaired by the ESEP's instigator Tommy Thompson[58] and included David Martin among its members. The Royal Society's own evidence to the review, signed off by Thompson in his Foreign Secretary guise, documented the success of the scheme to date and made a case for continued expansion. As the review progressed, Thompson found himself having to fend off suggestions from Edwin Appleyard of DES that the EMBO fellowship scheme should be expanded to cover all areas of science (thus competing head on with the ESEP), or that the ESEP should exclude biology from its remit because EMBO was already covering it sufficiently.[59] The working group's report, submitted to the Secretary of State in July 1971, of course recommended that the ESEP be expanded.[60] The ground had already been prepared a year earlier at a private dinner hosted by the Prime Minister, Harold Wilson, for Blackett and other Fellows, when Thompson in particular had made a play for a doubling of the ESEP budget over the next three years.[61] The government was sympathetic, and the ESEP budget grew from £255,000 in 1970 (of which half from the DES and half in balancing contributions from the partner countries) to £509,000 in 1973 and over £1 M by 1979.[62] The 2,000th award under the ESEP was made in 1974.

[58] Papers relating to the review may, therefore, be found at HWT C.82–91. Thompson served as the Society's Foreign Secretary until 30 November 1971. For some touchy exchanges when Thompson thought the ESEP was being unjustly attacked, see HWT C.87.

[59] OM 5 November 1970, minute 3(a). Jack Embling repudiated Appleyard's suggestions in a meeting with the Officers the following month: OM/114(70).

[60] The report was later published as Cmnd 4843.

[61] TNA PREM 13/3422, HWT B.528. Wilson had been elected to the Society in 1969 under Statute 12 (see Annex).

[62] The Society was to some degree a beneficiary of the government's decision, a year before the UK's entry to the European Community, to spend £6 M on improving cultural

The ESEP was then much the largest scheme available to UK scientists for postdoctoral exchanges with West Europe, and at the end of the decade it accounted for half the total number of fellowships for Western European exchanges at postdoctoral level.[63]

The subsequent development of the ESEP was to some degree shaped by competition for the niche it occupied and by changing external circumstances. During the mid 1980s, the French partners put out feelers about the possibility of ESF taking over the ESEP in its entirety and expanding it to all countries represented in ESF, but this was turned down.[64] Eastern European countries were invited into ESEP as they moved towards democratic government from 1989 onwards, and the high quality of some of their research, especially in the physical sciences, surprised their western colleagues.[65] Though confident that ESEP was successful within its own terms, the Society was worried that Europe was still fragmented scientifically and therefore not competing as strongly as it could with the United States or Japan. But it was also worried by the spectre of 'fortress Europe'. It therefore stressed the importance of much stronger connections not only among the scientists of the European Community but also between them and scientists beyond the EC. By way of practical encouragement, it published a guide to the various funding schemes then available (at a time when such information was relatively difficult for individual researchers to obtain), and analysed the factors making for successful collaboration in research.[66]

From the late 1980s, the European Community set out to stimulate collaboration among the scientists of its Member States through a series of programmes. The Royal Society thought that EC provision at postgraduate and postdoctoral levels particularly was inadequate.[67] So its 1991 corporate plan included a commitment to the continued expansion of the ESEP where the demand existed. However, successive EC Framework Programmes then put increasing resources into promoting

relations with Europe, as announced to the House of Commons by Geoffrey Rippon on 6 March 1972.

[63] C/215(74) and OM/100(79). Note also the speech by Thomas Brimelow (FCO Permanent Secretary) at the Society's 1974 annual dinner: Thomas Brimelow, 'Anniversary dinner 1974', 9.

[64] Memo Nadia Slow to Peter Warren, 3 August 1984: RMA1218.

[65] Interview with Brian Heap.

[66] SEPSU, *Guide to European collaboration in science and technology* (December 1987; second edition, December 1990); SEPSU, *European collaboration in science and technology: pointers to the future for policy makers* (SEPSU Policy Study No 3, February 1989). Both documents were drafted by Malcolm McOnie, on secondment to SEPSU from the chemical company ICI.

[67] Royal Society evidence to Commons Select Committee on Education, Science and the Arts on science policy and the European dimension, 2 November 1989.

Figure 9.3 Bob May admitting Tim Berners-Lee to the Fellowship, 2001. © The Royal Society

personal mobility and collaboration, under such headings as the Stimulation Programme, Human Capital and Mobility, Training and Mobility of Researchers and, especially, the Marie Curie programmes of grants and fellowships. Scientists in some of the ESEP countries turned increasingly to such sources, and the ESEP budget was trimmed accordingly.[68]

The year 2000 saw the 10,000th ESEP award.[69] The ESEP at that time accounted for 20 per cent of the Society's annual spend on international exchanges and projects of £5 M from public sources. It was valued for its light bureaucracy and its willingness to fund joint projects as well as exchanges. But, as Bob May (Figure 9.3) pointed out in his 2004 Anniversary Address, its functions had by then been largely taken over by a package of European Community schemes. The Society was deliberately taking a more low-key role in the administration of the ESEP as a formal, unified scheme, concentrating instead on other modes of collaboration and other geographical priorities. None of the other partners wanted to fill the

[68] OM 14 October 1993, minute 2(d), OM 6 October 1994, minute 2(f).
[69] Aaron Klug, 'Anniversary Address 2000', 174.

gap. Given the ready availability of alternative sources of funding, the ESEP as a distinct initiative had reached its natural end.[70]

National academies on the European scale

There have been several attempts to establish Europe-wide academies with individual memberships.[71] Much the most successful, and the only one with which the Royal Society was associated, is the Academia Europaea. The initial drive here came from the UK science minister, Peter Brooke. He was due to attend a meeting of science ministers convened by the Council of Europe in September 1984. In discussion beforehand with the Chief Scientific Adviser, Robin Nicholson, and also with the Royal Society (Andrew Huxley and Arnold Burgen, President and Foreign Secretary respectively), Brooke saw a stronger and more explicit European identity for the scientific community as the problem to be solved, and a 'European Royal Society' as a key element in the solution. As well as serving Brooke's wish to make a personal impact at the Council of Europe, such an initiative could, thought Nicholson, meet a basic political need:

I am very concerned about the French bid for scientific and technological leadership in Europe which is being pushed so hard by the Mitterrand administration. In some ways it is laughable compared with our own and Germany's claims but if France goes on behaving as though she were leader and continues to fuel the claim with cash, people will eventually believe it and then it will happen. In my judgement this would have very adverse consequences on the UK's economic competitiveness in a future world dominated by science and technology led industries. Thus, if the idea of a European Academy of Sciences has any merit, there are advantages in the UK taking a firm initiative at the start.

It was a good example of playing international collaboration for national advantage.[72]

Brooke duly had his political success at the Council of Europe meeting, where his academy idea was greeted with warmth by most participants,

[70] Interview with Julia Higgins.

[71] For example, the Académie Européenne des Sciences, des Arts et des Lettres (Paris, 1980); the Academia Scientiarum et Artium Europaea (Salzburg, 1990); and the European Academy of Sciences (Belgium, 2002).

[72] Robin Nicholson to Peter Brooke, 10 September 1984: D.C. Phillips papers, MS Eng. c.5526, O.222; interviews with Arnold Burgen, Robert Hinde and John Meurig Thomas. Nicholson's comment reflects a very different situation from that experienced by Ian Heilbron, the Royal Society's envoy to France in October 1945, who reported: 'The French are very favourably disposed towards us and there is no doubt that they will follow any lead given by the Royal Society.' 'Visits to liberated countries by representatives of the Royal Society', *Notes and records of the Royal Society of London*, 4 (1946), 99.

including the French, and with confusion by the Germans. His proposal to call it the Leonardo Academy went down well. *Nature* welcomed the initiative. Brian Flowers, President of ESF and thus likely to be affected by any such initiative, initially damned it with faint praise. The Royal Society, too, responded with muted enthusiasm, especially since one task that the proposed academy might acquire was that of running a European exchange programme. Nonetheless, Brooke, recognising that success would depend on scientists having ownership of the new academy, invited the Society to take the initiative forward.[73]

Further discussions with Nicholson followed. The DES did not share its minister's enthusiasm, and the Society too made haste slowly. In 1985 Peter Brooke was succeeded as science minister by George Walden, and any thought of the Leonardo Academy running an exchange programme dropped off the agenda. But Walden retained the broad notion of a European academy, and the Society agreed to help gauge the levels of enthusiasm within the wider scientific community. It first tested the waters through a meeting in June 1986 with a small group of Fellows including Brian Flowers and John Kendrew, and then in October 1986 held a larger meeting with scientific representatives from some of the main Council of Europe countries. This found strong support for a 'European Society of Scientists', with a membership comprising the most eminent active individual scientists in Europe, tasked with developing a European identity in science. It was recognised late on that the new body would have to cover a far wider disciplinary range than the Royal Society – humanities and social science as well as natural science. ESF endorsed these moves, and took a hand in bringing the new body to fruition as a complement to itself. The linguistically neutral name Academia Europaea was agreed in December 1987. Arnold Burgen secured initial funding and ensured that its headquarters were located in the UK, thus assuaging Robin Nicholson's earlier concerns. The inaugural meeting was held in Cambridge in September 1988. Arnold Burgen, having finished his term as the Royal Society's Foreign Secretary, became the Academia's first President.[74]

---

[73] Copies of Brooke's 17 September 1984 speech and the accompanying press release at RMA579 and RMA1218; *Nature*, 311 (27 September 1984), 286; David Dickson, 'The Leonardo project', *Times higher education supplement* (28 September 1984); Brian Flowers, *The ESF – the first twenty years!*, 10–11; Brooke to Huxley, 30 September 1984: OM/111(84); OM 11 October 1984, minute 2(c); Peter Warren to John Goormaghtigh (ESF), 18 October 1984: RMA1218.

[74] OM 13–14 July 1985, minute12; Peter Brooke to George Porter, 30 October 1985: RMA579; OM/144(85) (meeting with DES officials); OM/11(86) (meeting with Walden); OM/66(86); OM/107(86); OM 6 November 1986, minute 2(a); OM/126 (87); Anne Buttimer, 'Academia Europaea: founders and founding visions', *European*

In March 1990 Royal Society Officers told the Foreign Office minister William Waldegrave that the Academia Europaea was up and running as an independent body (despite some residual ambivalence from the Académie des Sciences in Paris).[75] The Society continued to support the Academia from public funds, both for its accommodation (by custom, it fell to the host country to cover this) and for some of its activities, while the Academia took steps to diversify its funding base and establish its place in the matrix of European organisations.

The Academia Europaea brought together some leading individual researchers of geographical Europe. ESF brought together the national academies and major public research funding agencies of geographical Europe. That did not exhaust the possibilities for creating European groupings. An initiative by the Dutch, French and Swedish national science academies led eventually, in March 1992, to the decision to set up a standing conference of academies under the name of ALLEA (ALL European Academies).[76] ALLEA established a pattern of regular information exchange, biennial meetings, and periodic reports on such topics as intellectual property and the social relevance of basic research, and by 2013 had a membership of 55 academies drawn from 43 countries.

However, this did not give the national academies an effective presence on the European policy stage, and by the end of the 1990s the Royal Society was keen to have that presence. Specifically, it was concerned that the European Commission was experimenting with various mechanisms for obtaining scientific advice, but that neither the expertise nor the independence of that advice were guaranteed.[77] And the Society wanted a mechanism that could in principle reach all the main EU policy-making bodies, not just one of the Commission Directorates.

Following a review of ALLEA in 1999, a small group of academies, convened by the Royal Society, decided that a new and separate mechanism was needed to deliver this advisory function.[78] A period of robust debate followed among the key players about the details. All were united,

---

review, 19 (2011), 153–253; Arnold Burgen, 'Academia Europaea: origin and early days', *European review*, 17 (2009), 469–75.

[75] OM/50(90). The Académie had earlier sought to delay the launch meeting to allow for additional consultation: Arnold Burgen pointed out that consultations had been going on for four years, and stuck to the planned timetable. OM/101(88).

[76] OM 15 January 1987, minute 2(f); OM/61(90); Arnold Burgen to Paul Germain, 24 July 1990: OM/97(90); OM/41(91); OM 12 March 1992, minute 2(c). Interview with Julia Higgins.

[77] The most extreme example was the Science Commissioner, Édith Cresson, who appointed Philippe Berthelot, her dentist and close friend, to advise her about science in 1995; but there were also significant shortcomings in more properly constituted advisory mechanisms.

[78] See interviews with Brian Heap, Julia Higgins and Bob May.

though, in condemning an opportunistic proposal by the French science minister Roger-Gérard Schwartzenberg that the Commission should establish yet another European Academy of Science and Technology during the French presidency of the European Council (July–December 2000) as its source of advice. His ministerial predecessor, Claude Allègre, had tried the same idea: neither came to anything.

It was eventually agreed to establish what became known as the European Academies Science Advisory Council (EASAC) as a distinct body. EASAC would focus on expert advice in areas of public policy (environment, energy, health, etc.) with a scientific component. The Council comprised the president or another senior figure of the national academy (or equivalent) of each EU Member State. This inclusive model was key to policy influence: it was politically easier for an EU-wide body such as the Parliament or Commission to receive advice from another EU-wide body than from a single country. Both ALLEA and the Academia Europaea were also invited to be members, thus giving them a stake in the success of EASAC. Individual EASAC members were empowered to make decisions without having to refer back to their academies, so that EASAC could act relatively quickly. The Council itself decided what projects should be undertaken; the work was done by expert groups. It was then reviewed and signed off by the Council. Costs were shared among the members.

EASAC was launched at a high-level meeting of its member academies in Stockholm in June 2001. The inaugural Chairman was the Foreign Secretary of the Swedish Academy, Uno Lindberg. The Royal Society, which had more experience than most academies in systematic policy advice work, provided the secretariat for the first eight years, after which it passed to the Leopoldina, by then formally identified as the national academy of Germany. Ahead of the launch there had been extensive market research among the potential recipients of EASAC's advice, and good working relationships were established with, particularly, the European Parliament and Commission. EASAC's work programme was a mix of commissioned and self-initiated (and thus self-funded) projects in European science policy generally and in policy-relevant aspects of the biosciences, energy and environmental sciences. In addition to generating substantive and often influential reports, EASAC organised workshops and lectures for the European Parliament and prepared commissioned briefings on scientific aspects of draft legislation. The creation and early success of EASAC owed much to Royal Society leadership, and the Society remained a strong supporter of it. In this it showed itself a good, indeed a leading, European.

10   Doing science globally

It sometimes happens that a major advance . . . takes place abroad . . .[1]

The globalisation of the practice of science through the twentieth century is one of its most striking features.[2] By the end of the century more scientific research was being done in more countries, and being done more collaboratively,[3] than could possibly have been imagined at the start of the century. The geographical balance of the total effort was transformed in the process. The Royal Society's participation in and response to these trends is the subject of this chapter.

The Royal Society has always been keen to interact with scientists all over the world – especially those capable of producing 'major advances'. It is fundamental to its existence that this should be so. Achievement in science at the highest level is determined by reference to global benchmarks, not simply national ones, and a body seeking to recognise and harness such achievement has to be attuned to the best in the world. Individual Fellows, almost axiomatically, engage closely and personally with the (other) world leaders in their fields. Correspondingly, the corporate Royal Society has to engage with the best of world science.

That engagement has taken various forms. In the first couple of centuries of the Society's existence, it was manifested most obviously through the far-flung authorship and readership of the *Philosophical Transactions*,

---

[1] *The encouragement of scientific research in the UK* (Royal Society, 1961). For context, see Chapter 2.

[2] Useful guides to this complex topic include Derek de Solla Price, *Little science, big science* (Columbia University Press, 1963); Susan E. Cozzens, 'The discovery of growth: statistical glimpses of twentieth-century science', in John Krige and Dominique Pestre, eds., *Science in the twentieth century* (Harwood Academic Publishers, 1997), 127–42; and Royal Society, *Knowledge, networks and nations: global scientific collaboration in the 21st century* (March 2011).

[3] To take one measure: in just twenty years, between 1988 and 2008, the proportion of all papers in science and engineering journals that were co-authored rose from 40 per cent to 64 per cent, and the proportion that were internationally co-authored rose from 8 per cent to 22 per cent. Michael Jubb, 'The scholarly ecosystem', in Robert Campbell et al., eds., *Academic and professional publishing* (Chandos Publishing, 2012), 72–3.

through the welcome extended to foreign scientists visiting London, and through election to the Fellowship of scientists living outside Britain. In the twentieth century, the Society's sense of itself as a body with global reach found additional forms of expression. For example, the Society played a very active part in many of the international groupings that sprang up at non-governmental level to promote particular scientific initiatives. It took a series of steps to help individual UK-based scientists meet and work with their overseas colleagues. It set out to help developing countries strengthen their science. And the core Fellowship became markedly less centred on the UK during this period.

The Society's response to the globalisation of science was, in a sense, to globalise its accolade function, both through the election process and through its choice of international partners and initiatives. The Society had leverage beyond the UK because its explicit or implicit endorsement was sought by scientific communities beyond the UK. At the same time, it recognised a duty to help individual British scientists access the major advances taking place abroad. In the 1960s and 1970s, the Society was devoting 35–40% of its resources to international activities – much its largest single area of expenditure until the growth of the University Research Fellowships scheme (Chapter 3) started to dominate the budget. This level of commitment came naturally. The Society was conditioned to operate globally for good scientific reasons, over and above the political drivers discussed in previous chapters. It did not aspire hubristically to be the world's academy, but it did aspire to a leadership role that was recognised globally and that benefited science in the UK.

The Society was present in various ways in many Commonwealth countries, including in sub-Saharan Africa; it was present in Europe; it was present, so far as it could be, in politically sensitive countries like China and the USSR; its Fellows were increasingly to be found working in the United States, and since the 1960s there were regular meetings at Officer level with the National Academy of Sciences. But 'major advances' in science could occur in any part of the world. Although the Society might choose at particular times to concentrate its attention on particular regions, strict geographical prioritisation was not part of its approach. Thus the identification of high-priority countries or regions did not go hand in hand with a list of countries in which, conversely, it had little or no interest. The balance of the Society's commitment to various countries in terms of budgets or exchanges was as much a response to local demand and myriad local decisions as a reflection of strategic intent. This provided a useful degree of flexibility and was, arguably, a response to the growing global competition for talent. Officers did, though, keep an eye on the resulting patterns, and sometimes sought to secure extra funding for areas where they wanted to stimulate or accommodate extra demand. In that sense the Society saw itself as a global body open to opportunity without geographical limit, constrained only by resources and (occasionally) by the operation of international sanctions. A corollary was that the Society was recognised globally as the non-governmental voice of British science.

An international Fellowship

The criteria for election to the Fellowship include not only scientific achievement but also nationality.[4] The Royal Society of London is the UK's national academy of science, and Fellowship is restricted to British or Commonwealth nationals. The only other category of membership is Foreign Membership. It took until 1978 to evolve the simple definition that Foreign Membership is for scientists of exceptional achievement who do not meet the nationality requirements for Fellowship.

The founding Charters did not have a specific category for them, but people 'residing in forraigne parts' were elected to the Society from the outset. By 1700, 18 per cent of the Fellowship was of foreign nationality. When foreigners constituted almost one third of the Fellowship, around 1765, the Society took steps to constrain their numbers, mainly because foreign candidates were often not well known in Britain and their merits could therefore be harder to assess, and also because they were not required to pay subscriptions.[5] By 1787 the number of foreigners in the Fellowship had dropped below 100, out of a total of more than 550. The distinct category of Foreign Member was then introduced to designate those who were 'neither natives nor inhabitants of his Majesty's dominions', and their number was limited to 100. By the end of the Napoleonic Wars (1815), there were fewer than fifty Foreign Members, and the limit on their number was halved. There it remained until 1945, when at the urging of A.V. Hill it was replaced by an upper limit on the number elected annually. That limit was set at four; it was subsequently raised to six in 1982, to eight in 2006 and to ten in 2013. Foreign Membership remained roughly constant at 8–11 per cent of the total throughout the twentieth century.

Of the forty-seven Foreign Members in 1900, virtually all were European (mostly French or German) and just 10 per cent were American. Nine countries were represented in the total. By 1950, Americans constituted

---

[4] For more information about the election process, see Annex.
[5] Henry Lyons, *The Royal Society 1660–1940. A history of its administration under its charters* (Cambridge University Press, 1944), 343.

Table 10.1 *Geographical distribution of the Fellowship by country of residence*

	Total Fellows	Total Foreign Members	Fellows in UK (% total FRS + ForMem)	Fellows elsewhere in Commonwealth (% total FRS)	Fellows in USA (% total FRS)	Fellows elsewhere (% total FRS)
1900	459	47	437 (86.4%)	16 (3.5%)	3 (0.7%)	3 (0.7%)
1925	443	36	405 (84.6%)	31 (7.0%)	3 (0.7%)	4 (0.9%)
1950	517	54	461 (81.7%)	41 (7.9%)	8 (1.5%)	7 (1.4%)
1975	790	78	662 (76.3%)	79 (10.0%)	39 (4.9%)	10 (1.3%)
1990	1,070	100	852 (72.8%)	109 (10.2%)	91 (8.5%)	18 (1.7%)
1995	1,137	106	849 (68.3%)	131 (11.5%)	134 (11.8%)	23 (2.0%)
2000	1,203	112	898 (68.3%)	128 (10.6%)	151 (12.6%)	26 (2.2%)
2005	1,285	131	946 (66.8%)	143 (11.1%)	170 (13.2%)	26 (2.0%)
2010	1,356	145	995 (66.3%)	159 (11.7%)	166 (12.2%)	36 (2.7%)

over 40 per cent of the Foreign Membership, and in 2010 they exceeded 65 per cent. By then, scientists of European nationality were down to 31 per cent of the Foreign Membership, and sixteen countries were represented in total. There is a degree of capriciousness in the nomination of candidates for Foreign Membership, and a significant time lag in the process (Fellows are now generally elected in their early/mid fifties, whereas Foreign Members tend to be elected about ten years older) even so, these figures suggest at best a gradual and partial response to global trends.

It is not only Foreign Members who live outside the UK. Trends in the geographical distribution of Fellows are shown in Table 10.1. The proportion of Fellows + Foreign Members living outside the UK grew from less than one seventh to more than one third of the total between 1900 and 2010. Within that trend, the proportion of Fellows living in (non-UK) Commonwealth countries grew steadily up to 1975, and slowly thereafter.[6] Growth switched instead to the United States, with the proportion of (expatriate) Fellows living there increasing rapidly from 1975 and overtaking numbers in Commonwealth countries in the mid 1990s. In part, this reflected the brain drain during and after the 1960s, and to a smaller extent it reflected Fellows moving to the United States after normal retirement from full-time posts in the UK. Fellows living outside the Commonwealth and the United States were often based at international centres such as CERN, EMBL or various astronomical facilities.

[6] In the nineteenth century, Fellows living in such countries were generally British expatriates. During the twentieth century, they were increasingly local citizens.

The Commonwealth

The British Commonwealth is not obviously a natural geographical unit for the conduct of science. The existence of a Commonwealth dimension to the Society's life is, rather, an accident of history and politics, embedded at the outset and part of the package ever since.

Citizens of the Empire/Dominions/Commonwealth have always been eligible for election as ordinary Fellows: they have never had a separate membership category. This has been a source of periodic anxiety to Fellows based in the UK. For example, in 1961 Carl Pantin commented to the Biological Secretary Lindor Brown: 'Quite evidently, we are going to see a very great increase in Commonwealth people potentially of Royal Society calibre.'[7] Pantin's worry was the eighteenth-century one about foreign candidates generally, the difficulty for a London-based organisation of comparing the merits of Commonwealth candidates against British candidates. By 1967 Officers were worrying, rather, that Commonwealth candidates would in future come to outnumber British ones.[8] But so far debates on whether to have a separate category of Commonwealth Fellow, or to treat Commonwealth citizens as candidates for Foreign Membership, or to constrain their competition with UK-based candidates in some other way, have invariably affirmed the status quo. That would appear to reflect the majority wishes both of Commonwealth scientists themselves and of scientists based in the UK. There may be elements of cultural imperialism in the ultimate professional accolade for a Commonwealth scientist being in the gift of a London-based body,[9] but refusals to accept election or to remain a Fellow on those grounds are rare.[10]

The Commonwealth is not constant. For example, Eire left the Commonwealth in 1948, and India became a republic in January 1950. In April 1951 the Royal Society Council determined to restrict Fellowship to those of British nationality, but sought advice from the

---

[7] Carl Pantin to Lindor Brown, 20 September 1961: HF/1/17/1/17.

[8] OM 6 April 1967, minute 2(b).

[9] R.W. Home, 'The Royal Society and the Empire: the colonial and Commonwealth Fellowship. Part I: 1731–1847', *Notes and records of the Royal Society of London*, 56 (2002), 307–32; R.W. Home, 'The Royal Society and the Empire: the colonial and Commonwealth Fellowship. Part II: after 1847', *Notes and records of the Royal Society of London*, 57 (2003), 47–84; Roy MacLeod, 'The Royal Society and the Commonwealth: old friendships, new frontiers', *Notes and records of the Royal Society of London*, 64 (2010), S137–49.

[10] One example is Arthur Lee of Massachusetts, elected to the Fellowship in 1766, who resigned his Fellowship after America gained its independence since he no longer regarded himself a British national: R.W. Home, 'The Royal Society and the Empire: Part I', 311.

Commonwealth Relations Office about what the term meant in relation to the citizens of Eire and India. Following that advice, Council decided on a 2:1 vote to treat the citizens of Eire, equally with British subjects, as eligible for election as Fellows rather than Foreign Members. This was made explicit in the Statutes, which from 1951 restricted eligibility for the Fellowship to 'British subjects or citizens of Eire'. The Society also continued to elect Indian citizens as Fellows, despite dispute over whether they were, or were not, 'British subjects': A.V. Hill and, for example, C.V. Raman insisted they were not, while the Home Office insisted they were.[11] When South Africa left the Commonwealth in 1962, by contrast, its citizens were regarded as ineligible for election as Fellows. This was all tidied up in 1965, when after further long debate the rubric defining eligibility for election as a Fellow was amended to 'British subject or Commonwealth citizen or citizen of the Irish Republic', and in 1981 simply to 'citizens of British Commonwealth countries and citizens of the Irish Republic'.[12]

Such debates are really about how UK-centric the Society should be. Apart from the election process, the Society's engagement with the Commonwealth as a whole (as distinct from its engagement with individual Commonwealth countries) has waxed and waned since the Second World War, somewhat in line with the fluctuating profile of the Commonwealth in British public life. In recent decades, the Commonwealth has been mostly in the background rather than the foreground of the Society's debates about strategy.[13] During and after the War, however, it was a different matter.

A prominent highlight was the 1946 Empire Scientific Conference, hosted by the Society in London with the help of a substantial Treasury grant.[14] The planning had started five years earlier. Opened by the King and running for three weeks with extensive publicity, it almost rivalled the 1960 tercentenary for show. Its impact was as much symbolic as scientific. It held up science in the Empire as an example of what could be achieved by friendly international collaboration, and it held up the Society as the continuing keystone in such an undertaking.

The – post-imperial – sequel to the 1946 conference was the British Commonwealth Scientific Conference hosted by Australia in 1952. One

---

[11] See, for example, correspondence in OM/33(57), OM/5(61) and OM/44(65). Raman eventually resigned his Fellowship on the issue, shortly before he died: CM 4 April 1968, minute 20.

[12] Variations on the nationality debate arise over questions of exactly how residency qualifications or rules on dual citizenship are to be interpreted.

[13] Interviews with Arnold Burgen, Brian Heap, Bob May and Peter Warren.

[14] Roy MacLeod, 'The Royal Society and the Commonwealth', S140–2; John S. Rowlinson and Norman H. Robinson, *The record*, 5–6; A.V. Hill, *Memories and reflections*, chapter 12.

outcome from that was the creation of the Australian Academy of Science, closely modelled on the Royal Society. A second outcome was the Commonwealth Bursaries Scheme, stimulated in part by the example of the Fulbright Foundation.[15] The scheme was run by the Royal Society with funding in the first instance from its own Parliamentary Grant and from the Nuffield Foundation, and subsequently also from various participating Commonwealth countries and from the Commonwealth Foundation. Bursaries allowed Commonwealth scientists of proven ability to work in other Commonwealth countries for up to one year with a view to improving their general capacity to 'extend the bounds of knowledge', and provided travel costs and sufficient living allowance 'to avoid frustration'. A review after 25 years concluded that the scheme had made valuable contributions both to science and to the cohesion of the Commonwealth. By then 600 bursaries had been awarded, increasingly to applicants from the developing parts of the Commonwealth. In 1984 Commonwealth bursaries morphed into the Developing Countries Fellowships Scheme, still with substantial support from the Nuffield Foundation.[16] The Society also established several bilateral schemes with individual Commonwealth countries, such as India (Leverhulme Visiting Professorships, introduced 1962), Canada (1983), New Zealand (1984), Australia (1985) and Australia and New Zealand (Endeavour fellowships, 1990). Where there were no specific exchange schemes, it still took care to nurture relations with the scientifically active members of the Commonwealth, for example through frequent visits in each direction at Presidential and Foreign Secretary level and through a variety of joint projects and networking initiatives.

Howard Florey, a keen supporter of the Commonwealth who remained 'instantly recognisable as an Australian' throughout his long career in Oxford,[17] used his first Anniversary Address in 1961 to press the case for increasing 'as much as possible all activities which impinge even remotely on maintaining contact with colleagues both from the Commonwealth and from other countries'. Two years later he commented: 'With the blurring of the conception of the British Commonwealth one of the functions of the Society, which may become more important with the passage of time, is to see that our close associations with actively growing and

---

[15] CM 12 February 1953, minute 4 and Appendix A; CM 15 October 1953, minute 34(vi) and Appendix C; Royal Society, The Royal Society Commonwealth bursaries scheme, 1954 to 1978 (1979); and, for related schemes, 'The travelling scholar and international understanding', Nature, 173 (2 January 1954), 1–4.

[16] OM 19 May 1983, minute 2(e).

[17] Henry Harris, 'Howard Florey and the development of penicillin', Notes and records of the Royal Society of London, 53 (1999), 249–50.

sympathetic countries are fostered in every possible way.'[18] On the eve of a further Commonwealth science conference hosted by the Society in Oxford in 1967, the Officers reflected on the nature of these close associations: 'The Royal Society should not be regarded as a "mother" Society with "branches" in overseas countries.'[19] Like many British institutions in this period, the Society was quietly negotiating the transition from matriarch to *primus inter pares*.

The Oxford conference was instigated by Patrick Blackett, Florey's successor as President. It focused on broad issues of science policy, particularly the role of science and technology in the economic progress of developing countries. Blackett devoted most of his 1968 Anniversary Address to rehearsing its achievements and generally talking up the value of Commonwealth science. The Prime Minister, Harold Wilson, told his friend Blackett that he was keen to promote science at Commonwealth level,[20] and the Society soon found itself trying to organise a further conference. This time it wanted to concentrate on science policy for more advanced countries, and accordingly restricted participation to Australia, Canada, New Zealand, India and Pakistan. In the event, issues relevant to India and to Pakistan were addressed separately, and just the first three countries participated in a smaller session at the Society in January 1971. The CSP took note of their respective experiences in dealing with policy challenges common to them all.[21] Plans for a much more inclusive conference of Commonwealth science academies and similar bodies to be held in 1980 proved impracticable and foundered for lack of a determined product champion.[22]

Commonwealth issues stayed in the background during the 1980s, though individual Commonwealth countries enjoyed favoured status and personal links were assiduously maintained. The Society's first two corporate plans (1986 and 1987) mentioned the Commonwealth only in passing, in the context of elections to the Fellowship.[23] The third, more polished, plan in 1990 highlighted the need to enthuse the next generation of scientists about the value of intra-Commonwealth links. The new

---

[18] Howard Florey, 'Address at the Anniversary Meeting, 30 November 1961', *Proceedings of the Royal Society of London. Series B, Biological sciences*, 155 (1962), 314; Howard Florey, 'Address at the Anniversary Meeting, 30 November 1963', *Proceedings of the Royal Society of London. Series B, Biological sciences*, 159 (1964), 404.

[19] OM 6 April 1967, minute 2(b). On the 1967 conference, see interview with John Deverill.

[20] TNA PREM 13/3422 – minute of a dinner in February 1970 with Patrick Blackett and other Royal Society representatives.

[21] OM 14 May 1970, minute 3(b); OM/52(70); OM 11 June 1970, minute 2(d); OM/63 (70); CSP(ISR)(71)2, copy at HWT C.80.

[22] OM 19 July 1979, minute 2(b).

[23] Roy MacLeod, 'The Royal Society and the Commonwealth', S145.

Endeavour fellowships with Australia and New Zealand were seen as part of that. They proved short-lived. Faced with sharp cuts in its Parliamentary Grant a year later, the Society reassigned the Endeavour fellowships budget, sacrificing the old Commonwealth to the more urgent needs of countries of the former Soviet Union. And so it continued: no wish to alter the Commonwealth basis of the Society's constitution, but no great appetite either to exploit that basis in any formal way other than through the election process. The 2012–7 strategic plan, however, broke the pattern with an explicit aspiration that the Society should strengthen its activities with colleagues across Commonwealth countries. The first outcome of that was a proposal to hold a further Commonwealth science conference, in India at the end of 2014.

Capacity building

Where the Commonwealth dimension found its most concrete expression was in the Society's work on capacity building. This materialised into an explicit strand of the Society's portfolio around the early 1960s, as Britain's former colonies moved towards independent nationhood and membership of the Commonwealth, and their need grew for indigenous scientific capability in support of educational and economic goals. One of many challenges was to deal with the loss of established channels of science funding from Britain, which was an unintended consequence of independence. In response, the British Government established the Overseas Research Council in 1959 to advise the Privy Council on science in relation to development, and established the Department of Technical Cooperation in 1961 to work directly with developing countries. In the same period both ICSU and UNESCO stepped up their efforts to help developing countries build their scientific capacity, and the Royal Society participated in some of the initiatives associated with this. The Society also established its own administrative machinery for development issues.[24]

Given the strong personal and professional links of many Fellows, it was natural that the Society should get involved to some extent in the scientific aspects of development, and that in doing so it should concentrate its efforts on Commonwealth countries.[25] But what could it do in

---

[24] The Society's Subcommittee on Developing Countries, created in 1964, was absorbed into the International Relations Committee in 1970.

[25] Not only Fellows. Ronald Keay, appointed to the Society as Deputy Executive Secretary in 1962, had previously worked twenty years for the Colonial Forestry Service, including a long spell in Nigeria, and was one of several key shapers of the Society's capacity-building strategy.

practice to make a real difference? The answer was a variant on its accolade function: it worked not just to develop capacity but also, implicitly, to provide indigenous scientists through association with the Society with a sense of belonging on equal terms to the global scientific community. The key was personal contact. So the Society arranged for senior scientists from the UK to visit and work with scientists in the new Commonwealth countries and for the latter, reciprocally, to visit the UK; it subsidised their attendance at international meetings; it ensured official Royal Society representation at milestone events such as the inaugural meeting of the East African Academy of Science in 1962; and it supported the creation of institutions such as research centres and new national academies.[26] The Society recognised the need in such initiatives to collaborate with other bodies, both to secure funding and to ensure overall quality. It worked closely with the Department of Technical Cooperation in building contacts with individual countries and in guiding the Department's selection processes for its own schemes;[27] it worked with private foundations such as the Leverhulme Trust; and it liaised closely with the NAS, which was promoting its own capacity-building initiatives.

A 1965 paper for the Society's International Relations Committee characterised capacity building as 'somewhat peripheral' to the Society's core business, but justified a modest level of effort on the grounds that it could produce considerable benefit.[28] Twenty years later, in discussion with the British Council, there was greater commitment: the Society talked of supporting good science in developing countries both as part of its generic wish to promote science per se and as part of what it recognised as its 'wider responsibility to assist economic development by supporting and strengthening the scientific capability of developing countries'.[29] By the early 1990s, about 7% of the Society's £4.5 M spend on international activities was focused on developing countries.

As the mainstream Science Budget in the UK became increasingly directed towards wealth creation and UK benefit, the Society's ability to invest in capacity building depended more and more on contributions from private sources and from the Overseas Development Administration (later the Department for International Development, successors to the Department of Technical Cooperation). A review of the Society's strategy

---

[26] Paper I/6(64): TNA OD 11/193; David Martin to Howard Florey, 30 October 1964, and an undated draft (c. December 1964) by Howard Florey: HF/1/17/2/40; paper I/1(65), 27 January 1965 (copy at HWT B.1).

[27] CM 7 November 1963, minute 31; CM 5 March 1964, minute 7; and Patrick Linstead to Andrew Cohen (Permanent Secretary at the DTC), 27 August 1964: TNA OD 11/193.

[28] Paper I/2(65): TNA OD 11/193.   [29] OM/135(85).

in 2001 struggled with the issue of balancing capacity building against engagement with countries that were already scientifically strong. But this was to some extent a false dichotomy, and the Society found itself vigorously defending the need for developing countries to undertake their own high-quality basic research. This was opposed by the Development Secretary, Clare Short, who wanted to focus her Department on supporting primary education in developing countries to the exclusion of university-level teaching and research. The Society won that particular argument, and Hilary Benn succeeded Clare Short at DfID.[30] He quickly created a new post of Departmental Chief Scientific Adviser, to the Society's delight, and recruited Gordon Conway, President of the Rockefeller Foundation and newly elected to the Fellowship, as its first occupant.[31]

The UK hosted the July 2005 meeting of the heads of the G8 governments, at the Gleneagles Hotel in Perthshire. The Prime Minister, Tony Blair, let it be known well in advance that he wanted the meeting to include major sessions on climate change and on capacity building. He appointed a 'Commission for Africa' to prepare the ground for the latter. The Society lobbied hard to ensure that the Commission paid attention to the role of science and engineering in development. It also initiated the production of a joint statement by the science academies of the G8 nations and the Network of African Science Academies that highlighted the 'fundamental importance of science, technology and innovation in tackling a wide range of problems facing Africa and other developing regions' and reminded developed countries of their responsibilities to help build science in Africa. The statement was widely used by participants in their preparations for Gleneagles, though the official communiqué at the end of the meeting focused more on governance and trade issues.

These experiences re-energised the Society's commitment to capacity building, which had become focused mainly on post-apartheid South Africa (see Chapter 8). Following extensive consultation with individual countries and with other organisations active in capacity building in sub-Saharan Africa, the Society launched a series of initiatives.[32] One focused on two Commonwealth countries, Ghana and Tanzania, and set out to foster linkages between selected UK research centres and their counterparts in Ghana or Tanzania, centred on joint research projects in priority

[30] Interview with Bob May, and Bob May, 'Anniversary Address, 2004'.
[31] At one stage this CSA post was destined to be set at too low a level in the DfID hierarchy to be effective, but pressure from the Society ensured that it was established at a level equivalent to Permanent Secretary.
[32] Interviews with Lorna Casselton and Julia Higgins; Julia Higgins, 'The Royal Society in Africa'; Martin Rees, 'Anniversary Address, 2006', 81; and associated press releases.

Figure 10.1 Martin Rees presenting the first African Pfizer award, in 2006, to Alexis Nzila of Kenya for his work on the mechanism of anti-malaria drug action. © The Royal Society

areas. The aim was to transfer skills to the African centres, with maximum operational flexibility suited to the needs of each specific collaboration. There was also provision for an initial networking phase to help potential collaborators identify each other. The Leverhulme Trust, which had supported capacity-building activities by the Society over the previous fifty years, provided £3.3 M in 2008 to launch what became known as the Leverhulme/Royal Society Africa Awards, and extended the grant for a further five years in 2011.

A second strand of capacity-building work was launched in 2008 with funding from Pfizer. This concentrated on developing national academies of science in Ghana (the oldest academy in sub-Saharan Africa, founded in 1959), in Tanzania (the youngest academy, founded in 2004), and in Ethiopia (which was thereby enabled to launch its academy in 2010). In partnership with the Network of African Science Academies, the Royal Society/Pfizer programme provided training, mentoring and project support. Crucially, it helped the academies to leverage influence with their national governments and to play a more effective role in promoting science. The Society's own experience in, for example, providing policy

Figure 10.2 Members of the Network of African Science Academies visiting the Society, October 2006 (l to r): Stephen Agong, Gabriel Ogunmola, Wieland Gevers, Mohammed Hassan. © The Royal Society

advice and promoting public engagement, adapted to local circumstances, was a fertile source of ideas for the other academies. From 2006 Pfizer also sponsored the Royal Society/Pfizer award in biological science for research contributing to capacity building (Figure 10.1). The winner, selected by the Society, gained a prize of £5,000 and very considerable personal prestige, plus a grant of £60,000 towards future research.

A third strand was targeted across the sub-Saharan region as a whole. Drawing on the Leverhulme experience and funded by DfID at £15 M, the Royal Society/DfID Africa capacity-building initiative aimed at strengthening the research capacity of universities and research centres in sub-Saharan Africa by supporting the development of sustainable research networks. Each network included one centre in the UK and three centres across two or three African countries (Figure 10.2). The scheme was launched in 2012. The Society's 2012–7 strategic plan pledged to maintain its engagement with capacity building, with a focus on Africa and on early-career scientists, on national academies of science, and on collaborative research projects.

Global exchanges

A good deal of the Royal Society's international work was focused on helping scientists in the UK to access research opportunities in other parts of the world. Its 1960 submission to Lord Hailsham, quoted at the beginning of this chapter, included a strong bid for additional resources for its travel grants scheme. Both the Foreign Office and the Council for Scientific Policy backed the Society to broaden its role in facilitating the international movement of scientists: its political independence and its reputation for discerning individual excellence fitted it for this. The ESEP and bilateral arrangements with the Soviet Union and China entrenched the practice of managing reciprocal exchanges in the Society's repertoire.

At the end of his presidential term in 1970, Patrick Blackett boasted: 'We now have exchange agreements which cover nearly every country in the world where worthwhile scientific research is carried out.'[33] One obvious country missing from the list was the United States, because of the plethora of exchange schemes already run by other organisations; however, about one third of the Society's travel grants were awarded to people wishing to visit the United States for short periods.[34] Continued growth of the exchange programme prompted Alex Todd to add in 1978: 'These exchange agreements, which extend worldwide, are of the greatest importance and provide an effective method for the free exchange of information and ideas which is the life-blood of science ... It may well be that maintenance and development of overseas activities is one of our most important functions.'[35]

By 1982 the Society had formal exchange agreements with some forty countries outside the Commonwealth and informal arrangements with many others, underpinning around 1,000 research visits of between two weeks and 12 months duration annually. Concern arose both about complexity of administration and about the difficulty of ensuring consistent standards of scientific merit across all those schemes. Assessment procedures were therefore rationalised and an overarching International Exchanges Committee established to address these issues.[36] The 1986 corporate plan, summarising these moves, asserted that the Society was able to support any 'worthwhile application for a visit to any part of the

---

[33] P.M.S. Blackett, 'Address at the Anniversary Meeting, 30 November 1970', *Proceedings of the Royal Society of London. Series A, Mathematical and physical sciences*, 321 (1971), 5. For a list, see OM/37(70).

[34] OM/100(79), OM/3(81).

[35] Alexander Todd, 'Address at the Anniversary Meeting, 30 November 1978', *Proceedings of the Royal Society of London. Series B, Biological sciences*, 204 (1979), 8. For an updated list of the Society's exchange agreements, see OM/6(77) and OM/84(78).

[36] OM/106(82), OM 4 November 1982, minute 2(f).

world', and a 1992 review of strategy confirmed that the Society 'was particularly well placed to implement international exchange schemes and should continue to give that high priority ... Benefits accrued not only to science, but for UK industry and in the cultural and political sphere.'[37] In 1992 the Society supported a total of 1,372 exchanges to and from the UK, of which 30 per cent were with 'greater Europe and Israel', 28 per cent with the former Soviet Union, 14 per cent with Australasia, Canada and India, 13 per cent with China and Hong Kong, 6 per cent with Japan, 5 per cent with Latin America and 4 per cent with other parts of the world.[38]

By the turn of the century, the established pattern of schemes for international collaboration was due for revision. The schemes, underpinned by detailed and individualised agreements with numerous partner organisations and funded by diverse sources, then included travel grants (primarily for UK scientists going to overseas conferences, up to two weeks), study visits or short visits (for travel in either direction to initiate or develop research collaborations, 2–26 weeks), a limited number of exchange fellowships with particular countries (in one or other direction, to spend time undertaking research in laboratories outside the home country, six months up to three years), and joint projects (to support short periods of travel in either direction over a two-year period within the context of an established collaboration between two research centres). The complexity of this provision was no longer sustainable. One of the drivers for revision was a decision to move these schemes from the group of staff handling international relations to the group handling grants and research fellowships, which led to a markedly greater emphasis on procedural efficiency. The introduction of newly commissioned software called e-GAP for computerised processing of applications reinforced that trend.

The upshot of extended discussions was consolidation into essentially two types of provision: incoming and outgoing. The web of bilateral arrangements was swept up into this consolidated approach, which reflected the increasingly fierce global competition to attract the best available talent.[39] Incoming fellowships were mostly coalesced into a single scheme of Newton International Fellowships, for postdoctoral researchers from any part of the world to work in the UK for up to two years, with generous subsistence and research allowances. Borrowing from the example of the Humboldt Foundation, the scheme allowed Newton fellows to be supported for up to ten years after their tenure to

[37] Royal Society, *Corporate plan 1986–1996*, 19; OM 17–18 October 1992, minute 1(3).
[38] Royal Society, *Annual report 1993* (1993).
[39] Lord Sainsbury of Turville, *The race to the top: a review of Government's science and innovation policies* (HMSO, 2007).

continue networking with their British colleagues. At the behest of the UK Government, which provided the bulk of the funding (£13.4 M over an initial three years from 2008, enough for a steady state of 100 active fellowships), the scheme covered engineering, humanities and social science as well as natural science, and was run jointly with the British Academy and the Royal Academy of Engineering.[40] The first round of the Newton scheme, in 2008, elicited 700 applications from 78 countries (dominated by China and India) and resulted in fifty appointments from twenty-two countries.

Outgoing grants were also merged into a single scheme, the International Exchanges Scheme, allowing for one or more visits by UK-based applicants to an overseas research partner over a period of up to two years. In its first year, 2011–12, this scheme had a budget of £2.3 M and attracted 1,216 applications for an eventual total of 186 awards.

Global scientific bodies

In the postwar period, ICSU[41] was the largest global scientific body to which the Royal Society adhered, in terms of cost and staff effort: in 1977, for example, the British National Committee for ICSU estimated that the Society spent 10 per cent of its Parliamentary Grant-in-Aid and 12 per cent of its staff resources on its ICSU operations. ICSU was and is a non-governmental body, though funded largely by public money. The Society adhered on behalf of the UK. It was a role that even DSIR, when trying in 1963 to constrain the Society's responsibilities in international science policy, did not begrudge it.

The Royal Society had been pivotal in the creation of ICSU, in 1931, and of the bodies that preceded it. It remained a keen supporter of ICSU in the 1960s[42] and proud of its involvement, regularly boasting of the

---

[40] The Royal Academy of Engineering dropped out after a year, but the discipline coverage of the Newton fellowships remained unchanged.

[41] The International Council of Scientific Unions, rebranded in 1998 as the International Council for Science. Founded in 1931, with origins going back to 1899, its members are national bodies like the Royal Society and international unions covering specific fields of science. Peter Alter, 'The Royal Society and the International Association of Academies, 1897–1919', *Notes and records of the Royal Society of London*, 34 (1980), 241–64,; and Frank Greenaway, *Science international: a history of the International Council of Scientific Unions* (CUP/ICSU, 1996).

[42] In 1965, the ICSU secretariat, based in Rome, was reported to be looking for a larger venue. The Society already housed the secretariats for two ICSU initiatives (the International Quiet Sun Year and the International Biological Programme), and offered to accommodate the ICSU secretariat as well. In the event, however, ICSU stayed on in Rome for another seven years and then moved to Paris where its main collaborator, UNESCO, was located. OM 30 November 1965, minute 3(a); David Martin to Frank Turnbull, 6 December 1965: HWT C.65; OM 16 December 1965, minute 2(e); OM 13 January 1966, minute 2(d).

elaborate committee structure it had in place to deal with ICSU and all its member bodies. Tommy Thompson was President of ICSU for a three-year term just before becoming the Society's Foreign Secretary, and President of the chemistry union IUPAC afterwards, and other Officers were similarly involved in their respective disciplinary unions. The Society's active engagement with ICSU operations gave British scientists ready access to a large number of international initiatives. It also gave the Society a useful forum in which to exercise leadership among its peer group of national academies. The margins of ICSU meetings provided many valuable opportunities for the Society's representatives to conduct other business and gossip with their colleagues from other countries.

However, international bodies such as ICSU attracted a degree of controversy in Treasury circles. A senior OECD official estimated that in 1963 there were at least 250 non-governmental and 50 intergovernmental international organisations engaged in scientific activities, many of recent origin.[43] Such profligacy attracted attention. Treasury officials of only moderately cynical bent would minute each other about the way devious non-government scientists argued the case for new international bodies or dreamt up International Years – often sponsored by ICSU – in order to bypass normal budgetary processes and commit the government to additional spending on science.[44] Such cunning behaviour by scientists was to be found across the Atlantic as well. President John F. Kennedy told the NAS at its centenary meeting: 'Every time you scientists make a major invention, we politicians have to invent a new institution to cope with it and almost invariably, these days, it must be an international institution.'[45]

This produced much anxious fretting about 'the problem of international scientific organisations' in the mid 1960s. The ACSP and its successor, the CSP, each conducted investigations,[46] and DES, the Foreign Office, the Royal Society and others submitted evidence in defence of their particular viewpoints. It was not just a matter of avoiding

---

[43] Jean-Jacques Salomon, *International scientific organisations* (OECD, 1965), 20.

[44] For example, W.W. Morton to Frank Turnbull, 1 October 1964: 'We began with the International Geophysical Year and then had IQSY. Now we have an International Biological Programme and the prospect of an International Hydrological Decade. The fashion is on ...' [TNA CAB 124/2983]; exchanges between the Society and DES: OM/19(65) and OM/20(65); DES paper 'The problem of international scientific organisations', CSP(66)1 (copies at HF/1/17/2/45 and TNA FO 371/189399); and Jean-Jacques Salomon, 'International scientific policy', 425. On IQSY, see interview with Chris Argent.

[45] Jean-Jacques Salomon, 'International scientific policy', 411. The 1964–74 International Biological Programme was a topical example: see interview with George Hemmen.

[46] *Annual report of the ACSP 1961/62*, Cmnd 1920 (January 1963), 1–5, and minutes of the CSP working party on international scientific relations during the spring of 1966.

wasteful duplication. The Royal Society worried about governmental bodies like OECD getting too closely involved with non-governmental scientific activities. The Foreign Office was concerned that British officialdom's instinctive hostility to international scientific initiatives would gain the country an 'essentially negative and laggard' reputation abroad and damage its wider diplomatic interests. DES insisted, to the contrary, that initiatives lacking sufficient scientific merit should not be undertaken purely on political grounds, and it was uncomfortable about entrusting large initiatives to the care of a non-governmental body like the Royal Society to manage via the relatively hands-off mechanism of a grant-in-aid.[47] International science could not be a private matter for scientists.

Anxiety about the budgetary impact of these international initiatives was not the preserve only of Treasury officials. In the UK, money for international scientific ventures tended to come from the same pot(s) that supported national scientific ventures. Those who judged their interests to be thereby threatened worried about the 'drain' from national budgets and looked enviously at their French colleagues, whose international scientific collaborations were generally funded from their Foreign Office budget and who were therefore able to accommodate political or diplomatic considerations with little controversy.[48]

So global scientific bodies had to be handled with some care in the political arena. On the other hand, ICSU was a useful political tool for the scientists, providing a forum in which certain issues could be thrashed out. Most prominent in this context was ICSU's determined commitment to the free circulation of scientists, coupled with a policy of strict political neutrality. Free circulation required that no bona fide scientist should be excluded from an international scientific conference on grounds of 'race, religion, political philosophy, ethnic origin, citizenship, language or sex'.[49] This was strongly and consistently endorsed by the Royal Society (see Chapter 8). The principle was spelt out at the 1958 ICSU General Assembly, and developed further in later years.

The hazards of international politics to one side, however, ICSU tried to get on with its core task of advancing international science. Its

---

[47] CSP(66)2, copy at HWT C.66; OM/10(65); McAdam Clark to Frank Turnbull, 18 May 1965: TNA CAB 124/2983, and OM/8(66) (briefing meeting convened by Harrie Massey, 7 January 1966); Turnbull to McAdam Clark, 16 June 1965: TNA CAB 124/2983; OM/25(66) (minutes of a CSP meeting). To this list would be added in the 1980s the challenges, at a time of strong currency fluctuations, of paying subscriptions denominated in dollars or francs from a budget denominated in sterling.

[48] John Cockcroft, 'Scientific collaboration in Europe', *New scientist*, 24 January 1963, 170–2; OM/8(66).

[49] Frank Greenaway, *Science international*, Chapter 8; ICSU, *ICSU and the universality of science: 1957–2006 and beyond* (ICSU, 2006).

sustained sponsorship of research programmes demanding extensive collaboration, most obviously in geophysics and climate science, was pivotal in the development of those disciplines.[50] The Royal Society committed considerable resource to ensuring the success of these programmes, building on its experience of doing science globally during the 1957–8 International Geophysical Year. ICSU also became increasingly involved in capacity building, to the initial concern of the Society and the NAS.[51] The Society continued to give ICSU strong overall support, but during the 1980s it began to overhaul its approach to how it managed the UK's relations with ICSU and all the associated bodies. The web of sixty-two National Committees and Subcommittees established to deal with each element of ICSU had become unsustainable. The Society gradually chipped away at the superstructure, and in 1989 decided to fold all the National Committees into an overarching International Relations Committee, harking back – subconsciously, no doubt – to a feature of its prewar arrangements.[52] This major simplification was followed during the 1990s by steps wherever possible to outsource to appropriate learned societies the work of adhering to individual specialist unions, including sharing the subscription. The Society remained the UK adhering body for ICSU itself and for unions where there was no obvious learned society that could take on the role. It also led an initiative to bring the European national members together to share a common approach to ICSU matters where it was advantageous to do so.

By the end of the century, ICSU had about 120 national members, 30 union members and 20 associates. Over half the national members were national academies of science. The balance of rights and responsibilities between national members and union members was an intermittently touchy issue, not wholly resolved when ICSU dropped the word 'union' from its full name in 1998 and styled itself the International Council for Science. If there was a need for the world's national science academies to act collectively or have an umbrella body, that need could not easily be met by ICSU.

ICSU acted as principal scientific adviser to the 1992 UN Conference on Environment and Development in Rio de Janeiro. That conference, important in many ways, was criticised for giving population issues inadequate space on its agenda. The Royal Society Officers and their NAS

---

[50] ICSU, *ICSU and climate change* and *ICSU and polar research: 1957–2007 and beyond* (ICSU, 2006). See also interviews with Chris Argent, Arnold Burgen and Lorna Casselton.

[51] OM/36(83): notes on a meeting of RS and NAS Officers, May 1983.

[52] OM/49(80); OM/15(89), OM/66(89) and Royal Society, *Annual report for the year ended 31 August 1989*, 25–6.

counterparts discussed the matter at one of their periodic meetings in September 1991, and issued a joint statement on the interactions between population and sustainability. This attracted wide publicity. They then convened a conference in New Delhi in October 1993, together with the Indian National Science Academy and the Royal Swedish Academy of Sciences, with the Society providing the secretariat. Eighty-two national science academies were invited to participate and to sign a 'population statement' drafted by the four convenors, targeted at the 1994 UN Conference on Population and Development and at national governments. The statement pointed out that there could be little progress in dealing with global problems of resource consumption, socio-economic development and environmental protection unless there were serious efforts to deal also with the interrelated problem of population growth. It therefore identified a set of practical steps towards the goal of 'zero population growth within the lifetime of our children'. In the end, sixty-seven academies took part in the conference and sixty signed the statement.[53] The convenors were delighted.[54]

This experience of doing science globally went down well with the academies concerned. So well, in fact, that they decided to morph what had started as a one-off project into a permanent mechanism for global discussion of major scientific issues.[55] The InterAcademy Panel on International Issues, known as IAP, was launched in January 1995 as an extension of the New Delhi group, with the Society still providing the secretariat. IAP proved a useful vehicle for the Society to become more systematically involved in global policy issues with a scientific dimension. For example, sustainable consumption, an archetypal global issue, started to feature more prominently in the Society's thinking at this time, and the Society played a leading role alongside the NAS in IAP's influential week-long conference on the subject in Tokyo in May 2000. IAP also provided a means for globalising policy work that the Society had initially carried out in a UK context, for example on cloning and on ocean acidification.

---

[53] The Pontifical Academy was one of those that was sympathetic to the initiative as a whole but felt unable to sign the statement. Prompted by the Society's Foreign Secretary, Anne McLaren, it subsequently co-hosted a symposium 'Breastfeeding: science and society' with the Society and published a joint report on the subject.

[54] CM 10 October 1991, minute 39; CM 12 December 1991, minute 33; CM 9 April 1992, minute 26; CM 14 May 1992, minute 22; CM 30 November 1993, minute 11; Royal Society and National Academy of Sciences, *Population growth, resource consumption, and a sustainable world* (February 1992); Francis Graham-Smith, ed., *Population – the complex reality* (Royal Society, 1994). The collective statement was reprinted in summary on pp. 17–8 of the latter book. Also interview with Francis Graham-Smith.

[55] Interviews with Michael Atiyah, Brian Heap and Peter Lachmann.

The first attempt at establishing a global grouping of national science academies had led eventually to the creation of ICSU in 1931. In 1979, at a meeting hosted by the Australian Academy, the NAS Foreign Secretary Tom Malone resurrected the idea of a grouping only of academies as a counterbalance to what he perceived as ICSU's increasing vulnerability to being dragged into UNESCO politics. But the Royal Society President, Alex Todd, was not keen: he would go only so far as admitting a possible value to 'informal contacts between academies of reasonable and reasonably similar outlooks'. Ad hoc groups for specific purposes could work, but anything larger risked being ineffectual, and a permanent hand-picked group of academies meeting Todd's criterion of reasonableness would instantly offend those excluded from it.[56] However, the general optimism that followed the fall of the Iron Curtain and the collapse of apartheid, and the resounding success of the New Delhi conference, created an atmosphere in the mid 1990s where a long-term and inclusive grouping of the world's academies seemed both possible and purposeful.

IAP, comprising about 100 national academies of many different types and at many different stages of development, still had to work out exactly what it was for. It began by producing a series of statements, on the New Delhi model, about such topics as the future of cities, sustainability, human cloning, and mother and child health.[57] Flexibility was achieved by inviting each IAP member academy to sign the statements, which became formal IAP documents when two thirds had done so. They were then disseminated to UN bodies and other global organisations, and also regionally and nationally as appropriate. Alongside this, IAP undertook a series of capacity-building initiatives, sharing experience among its members in tackling various aspects of academy work. Both these strands of activity developed rapidly after the Italian Government offered a substantial long-term grant, enabling IAP to set up a much enlarged secretariat under the administrative umbrella of TWAS (originally called the Third World Academy of Sciences) in Trieste.

In 2000, the NAS instigated creation of the InterAcademy Council (IAC), with just fifteen national academies as members, to carry out more detailed studies of global policy issues by a process closely modelled on NAS practice. This initially caused some confusion in IAP ranks and much discussion about the relative missions of the two bodies and the relations between them. It was resolved by IAC being constitutionally tied into IAP in the sense of IAP electing the fifteen members on a rotating

---

[56] Alex Todd to Lloyd Evans (President of the Australian Academy of Science), 6 April 1979: TODD Acc 1021, box 33; OM/68(79) (meeting of RS and NAS Officers, June 1979); OM/89(79) (further correspondence between Todd and Evans, June/July 1979).
[57] The first 11 statements were gathered into a booklet *Statements 1993–2008* (IAP, 2008).

basis and identifying experts for project working groups. This helped IAP, which was still finding its feet; it also provided IAC with legitimacy as speaking for the global scientific community. The Royal Society then threw its weight behind the initiative. IAC was launched at the May 2000 IAP meeting, with warm endorsement from Kofi Annan, UN Secretary-General, who was seen as one of the main customers for its reports – an additional forum for the Society's global ambitions.

Communicating science globally

In its earliest days, the Royal Society's core activity was enabling scientists to communicate with each other. It did this by arranging meetings, by correspondence, and by launching the world's first successful scientific journal, whose frequent publication enabled those not able to get to the meetings to remain abreast of the latest developments. The *Philosophical Transactions* gave the Society an immediate presence throughout the scientific world.[58] It still does, alongside the other titles in the Society's now much expanded publishing operation.

Publishing is about business as well as about communication. The beginnings of the international research journal as a successful business model have been dated to just after the end of the Second World War.[59] For the Royal Society, they can be dated to the mid 1950s. Before the War, the Society had taken the view that, much as it cared about its journals, it was not too bothered about being a publishing business,[60] and in 1937 it had moved most of its publishing operation – particularly the two journals *Philosophical Transactions* and *Proceedings* (each in two parts, A and B) – to Cambridge University Press. By 1954, publishing was losing money at the rate of more than £5,000 p.a., or nearly 20 per cent of turnover. The editorial function had remained with the Society itself, but in 1954 the Society decided that it had to be more closely involved. It took the sales and distribution functions back in-house, outsourcing just the production side to CUP. It also instigated several other cost-saving measures, including a squeeze on Fellows' access to subsidised copies. Within two years the losses had been reversed. Annual sales of *Philosophical Transactions* and *Proceedings* rose rapidly from under 3,000 copies in 1954 to 5,500 in

---

[58] Its 1663 Charter gave the Society the very significant privileges of being allowed to publish, and to correspond with foreigners, on any matter to do with science. It made full use of this.

[59] Robert Campbell, 'Introduction', in Robert Campbell et al., eds., *Academic and professional publishing* (Chandos Publishing, 2012), 3.

[60] William Bragg (then PRS) to Frank Smith, 11 May 1938: MDA B3.1.

1965, then more gradually to a peak of 5,900 in 1974 before falling back to 5,100 by 1980.[61]

As the journals grew, from an annual output of about 250 papers in 1950 and 350 papers in 1960 to over 900 papers in 2000 and 1,800 papers in 2010, so the established ways of managing them became increasingly unsustainable. The practice was that the Physical and Biological Secretaries acted as editors of the A-side and B-side journals, respectively, alongside all their other responsibilities. By long custom, all papers were formally communicated through a Fellow. The Fellow was supposed to act as guarantor of quality, but in practice rarely could because of increasing specialisation, so that an additional refereeing process was necessary. Council acted as a Committee of Papers formally authorising publication of those that had survived the refereeing process. A modest reform in 1967 gave each Secretary a team of twelve associate editors to help manage the refereeing process.[62] The publishing operation drifted back into loss-making mode in 1971–2, as did the Society's overall finances. Major hikes in subscription prices, and various more incremental measures, followed as the pressure increased for publishing to generate surpluses. The 1971–2 deficit of £31,000 turned into a surplus of more than £100,000 in 1977–8, though worries about long-term profitability persisted.[63]

In 1983, the Secretaries convinced their Council colleagues of the need to appoint someone other than themselves as formal editors of the journals. Two were appointed, one for the two A-side (physical sciences) journals and one for the two B-side (biological sciences) journals. In 1988, each separate journal finally got its own dedicated editor.[64] Meanwhile, long-running debates continued about such generic matters as overall speed of publication and the case for a special 'rapid results' outlet, and the future of general, wide-ranging journals in an era of increasing specialisation. More domestic issues included encouraging Fellows to make more use of the Society's journals for their best work, the potential value of associate editors based outside the UK in stimulating overseas subscriptions and submissions of high-quality papers, and

---

[61] CM 1 April 1954, minute 11; CM 3 November 1955, minute 21; publication numbers from *Annual Reports*. On the Society's publications generally, see John S. Rowlinson and Norman H. Robinson, *The record*, Chapter 7.

[62] C/84(67). Refereeing had previously been undertaken via Sectional Committees.

[63] CM 22 March 1973, minutes 1, 5 and 6; C/130(73); CM 12 July 1973, minute 38(viii); CM 7 March 1974, minute 24(v); CM 10 July 1975, minute 32(ii); OM/40(79); OM/72 (80); C/155(84). Despite the worries, publishing generated a surplus of £2.1 M in 2012–3.

[64] CM 19 May 1983, minute 14; CM 14 July 1983, minute 14; CM 6 October 1988, minute 28.

how closely Council needed to be involved in day-to-day management. All that against a background of long-term decline in the volume of sales from their 1974 peak, and niggling concerns over whether the journals really were of the highest quality. And there was pressure for a more entrepreneurial approach to publishing. Another review of publishing policy beckoned.

The upshot of that 1987–8 review[65] was that Council relinquished its (nominal) control over acceptance of papers. Instead, it agreed a specific character and objectives for each individual journal (in some cases including provision for rapid publication of short papers), approved a series of changes in production processes to take advantage of new technology and effect cost savings, and established a Publications Management Committee to keep an eye on things. The hallowed requirement that all papers be submitted via a Fellow was dropped at the same time. But the pace of change continued to accelerate with the advent of electronic publishing, new business models and increased competition. The journals continued to generate surpluses, but there could be no certainty that they would continue so in future. Subscriptions to the four main journals dropped from 3,700 to 3,200 between 1990 and 1994 – about the market norm at that time – while submitted papers increased by about 50 per cent. The 1994–5 review of publishing[66] concluded that the Society should keep the publishing function in-house, and urged further experiment with scope and format. The Publications Management Committee was reconstituted and given greater editorial and commercial independence within broad guidelines established by Council. These developments went along with increasing professionalisation of publications staff, and indeed benefited greatly from their technical expertise. In 1996 the Officers explicitly affirmed that the Society's publications were central to its standing as a national academy of international repute.

There followed a marked expansion in the four core scientific journals, and the launch of several new ones: the monthly *Journal of the Royal Society Interface* in 2004; the bimonthly *Biology Letters* for rapid publication of short research articles in 2005; *Interface Focus* in 2011; the Society's first fully open-access, online-only journal, *Open Biology*, also in 2011; and *Royal Society Open Science*, with a new approach to peer review, in 2014.

One impact of all these developments was to give the Society's research journals a demonstrably greater global presence. Subscriptions had long been worldwide. Already during the 1970s, for example, only about 12

---

[65] C/192(88); CM 6 October 1988, minute 28. Also interview with John Enderby.
[66] C/31(95); CM 6 April 1995, minute 2.6; C/52(95); CM 18 May 1995.

per cent of sales were inside the UK, with the United States (36 per cent) and Japan (10 per cent) then the single largest foreign markets.[68] By 2013, the journals were being sold or distributed free of charge in 149 countries. In terms of subscriptions, the United States was then still the dominant customer, with 37 per cent of sales, followed by the UK (14 per cent), Germany (6 per cent), Canada (6 per cent) and Japan (5 per cent). But in terms of site licences – since journals were now available electronically – China dominated with 35 per cent of the total, followed by Brazil (22 per cent), Mexico (20 per cent), the United States (9 per cent) and the UK (3 per cent), these figures to some extent reflecting recently completed deals.

More striking, though, was the way in which authorship of papers in the Society's research journals became increasingly internationalised. As shown in Table 10.2, authors based in the UK accounted for 90 per cent of the total in 1950 but only 24 per cent in 2010. In 2010 authors from eighty-seven countries published papers in Royal Society journals.

Table 10.2 *Authors of papers in RS journals, by nationality of institution*[67]

	1950	1960	1970	1980	1990	2000	2010
UK	90%	82%	71%	55%	49%	38%	24%
USA	3%	6%	14%	21%	22%	23%	25%
Commonwealth	6%	8%	7%	4%	9%	9%	12%
Rest of the world	2%	4%	8%	19%	20%	30%	39%

The custom of holding weekly meetings in season for Fellows to discuss scientific developments goes back to the beginnings of the Society's existence. By the Second World War, 'season' had been narrowed to November–June, and a special statute allowed Council to choose to omit certain weeks within that period. In 1946, that still left twenty Thursday afternoons in which to 'consider, and discourse of philosophical experiments and observations . . . to view, and discourse upon, rarities of nature and art; and thereupon to consider, what may be deduced from them'.[69] The afternoons were given over occasionally to invited lectures

---

[67] All scientific journals published by the Royal Society in the year concerned. Data for 1950–70 by manual searching of the journals; data for 1980–2010 derived from *Web of Knowledge*. There may therefore be a discontinuity of detailed definition between 1970 and 1980. The 1987–8 review of publishing attributed the growth in papers of American origin up to that point to the brain drain and to the existence of page charges in some American journals: C/87(87), paragraph 3.4.

[68] C/130(81).

[69] Paragraphs 59–65 of the 1939 Statutes [the quoted extract is unchanged from the 1663 Statutes]; CM 20 July 1945, minute 7.

or to discussions of specific topics, but mainly they were devoted simply to the reading of papers – and the latter was often done badly, to the tedium of the (ever smaller) audience present. The main motive for attendance, apart from showing goodwill to the President who was expected to chair these occasions, was the sociable tea beforehand, which provided opportunity to gossip with friends and colleagues. The way forward was to increase the number of invited lectures and of focused discussion meetings, which were more trouble to organise but attracted quality audiences and produced coherent debates.[70]

That approach was eventually adopted formally by Council in 1961, and at Howard Florey's suggestion a committee was established to come up with good ideas for topics and speakers.[71] The committee was named for Robert Hooke, who had had to organise the weekly meetings three centuries previously. As Blackett foresaw,[72] the Society's move to the more spacious premises of Carlton House Terrace in 1967 gave the Hooke Committee greater scope. Discussion meetings of up to two full days duration became a staple part of the Society's scientific programme. Lectures, too, began to proliferate.[73] The reading of random papers on Thursday afternoons was phased out in 1971, and the requirement to hold weekly meetings at all was deleted from the Statutes in 1988 after 325 years on the books.

An important feature of the discussion meetings was that, after overcoming the Treasury's traditional reluctance, the Society was able to use its Parliamentary Grant to subsidise the costs of bringing in speakers from overseas.[74] This considerably enhanced the international reputation of the meetings, and allowed UK-based researchers to interact with the world leaders in their fields – the more so as there was no charge for attendance. In contrast to the old Thursday afternoon custom, the Society now actively tried to attract as wide a spectrum of the scientific community as possible to its meetings, including younger researchers. Successive Presidents encouraged the trend.[75] Publication in the

[70] C.G. Darwin, 'The "reading" of papers at meetings of the Royal Society', *Notes and records of the Royal Society of London*, 2 (1939), 25–7; A.V. Hill and Jack Egerton, memo, 'Ordinary meetings of the Society', Royal Society archives: Officers and Council, A.V. Hill, 1940–6.

[71] CM 13 July 1961, minute 19; CM 1 March 1962, minute 20; C/115(62); CM 12 July 1962, minute 25.

[72] P.M.S. Blackett, 'Anniversary Address, 1966', ix.

[73] Annual attendance at the scientific meetings rose from 2,000 in Burlington House days to over 5,000 by 1974. *Notes and records of the Royal Society of London*, 30 (1974), 11. See also OM/143(76) and OM/116(81) for summary statistics.

[74] OM/74(65), OM/93(65).

[75] Alan Hodgkin, 'Anniversary Address, 1974', 109; Alexander Todd, 'Anniversary Address, 1979', 373.

Society's journals[76] ensured a worldwide audience for the papers given at the discussion meetings.

By the mid 1990s, discussion meetings and scientific lectures, like publications, were deemed central to the Society's standing as an academy. Successive strategic plans affirmed their importance, stressing their value in promoting interactions among leading scientists from around the world. The early and innovative use of webcasting allowed much wider participation. From the unpromising beginnings of closed sessions for the reading of disconnected papers, meetings and lectures had become a significant vehicle for the Society to express its place in the global scientific community. And in 2010 the Society extended its capacity for such work when it opened Chicheley Hall, 55 miles north of London, as a venue for residential meetings.[77]

Expeditions

The Royal Society has a long history of sending scientific expeditions to distant parts of the world. Its active involvement with Captain Cook's three major voyages in the eighteenth century, and with the *Challenger* oceanographic expedition in the nineteenth, are highlights from earlier periods. At the beginning of the twentieth century it extended its influential engagement with Antarctic exploration.[78] After the Second World War, the habit was revived. The 1957–8 IGY involved the Society initially as the UK adhering body for ICSU, one of the Year's primary sponsors. It tested the Society's ability to conduct large-scale scientific research in logistically challenging locations across the world, in a multinational and politically delicate context. The success of the IGY rekindled the Society's enthusiasm for such matters. A string of further overseas ventures followed.

A Biological Expeditions Committee had been established in 1936, not with a remit to dream up possible expeditions but simply to advise the Society on applications for scientific investigation grants that had an expeditionary element. As preparations for the IGY were getting under way in the mid 1950s, the biological scientists got in on the fashion for expeditions by proposing to mark the centenary of *Origin of species* with a repeat of the *Beagle* voyage, but in the opposite direction. This would be

---

[76] By the mid 1980s, the discussion meetings were providing over half the material published in *Philosophical Transactions*: C/87(87).

[77] The acquisition and conversion of Chicheley Hall were driven strongly by the Executive Secretary Stephen Cox and the Finance Director Ian Cooper.

[78] G.E. Fogg, 'The Royal Society and the Antarctic', *Notes and records of the Royal Society of London*, 54 (2000), 85–98.

done in collaboration with the academies of Australia, New Zealand and Canada. Council established what eventually became known as the Southern Zone Research Committee to take it forward.[79] In the end, the *Beagle* proposal got no further than a substantial 'preliminary' land-based expedition to southern Chile in 1958–9, which the Society funded mostly from private sources. However, the SZR Committee's enthusiastic Chair, Carl Pantin, soon persuaded the Society to mount other expeditions.[80]

This was essentially opportunism. Key was the fact that, after the IGY and southern Chile, the Society had highly experienced staff in place to provide logistic and tactical support to expeditions. It also had ample capability to digest and disseminate the findings of expeditions through discussion meetings and publications. And it had the SZR Committee generating and advocating proposals for new ventures. Moreover, the Society had the close connections with the UK Government, with the Crown Agents for Overseas Governments and, through them, with the relevant authorities in other countries that were essential to sorting out the numerous administrative obstacles facing international expeditions.[81] Beyond these structural factors was the consideration that, following the 1963 Trend Report, promoting non-governmental international scientific cooperation was one of the few public functions that uncontroversially belonged in the Society's bailiwick. Provided the scientific rationale was strong, the finance available and the international politics not impossibly tricky, the Society in the 1960s proved very willing to be involved in scientific expeditions. It was something that it turned out to be good at, and it fitted with its international ambitions.

The expeditions were generally in the Southern Hemisphere.[82] Between 1960 and 1980, the Society was closely involved in at least nineteen scientific expeditions, dealing with tropical forests (Figure 10.3), volcanology, coral reefs and islands, botany and geology, and oceanography.[83] Each had its own mix of scientific and extra-scientific

[79] CM 19 April 1956, minute 39(i); CM 11 October 1956, minute 13.

[80] OM/36(59); OM 16 April 1959, minute 6(a); OM 21 May 1959, minute 2(b); interview with George Hemmen.

[81] The National Archives contain many government files relating to Royal Society expeditions in the postwar period.

[82] G.E. Fogg, 'The Royal Society and the South Seas', *Notes and records of the Royal Society of London*, 55 (2001), 81–103.

[83] George Hemmen, 'Royal Society expeditions in the second half of the twentieth century', *Notes and records of the Royal Society of London*, 64 (2010), S89–99; C/110(79); interview with George Hemmen; CM 5 March 1964, minute 25(vi). The expeditions were as follows: on tropical forests: North Borneo (1961 and 1964), the Solomon Islands (1965), Mato Grosso, central Brazil (1967–9), New Hebrides (1971); on volcanology: Tristan da Cunha (1962), Ascension Island (1964); on coral reefs and islands: Diego Garcia (1967),

Figure 10.3 Royal Society Solomon Islands expedition, July to December 1965. © George Hemmen

rationales, each generated research results published by the Royal Society, each reinforced the Society's position as a global player, and some gave rise to commemorative postage stamps.

The pattern changed in the 1980s. The SZR Committee recognised that scientific needs – and the prevailing culture – were moving towards a different way of doing things. Expeditions of a few months managed from London increasingly belonged to the past. But there was still a case for the Society to be involved in ecological and related research overseas, via long-term collaborative research programmes. Such programmes could have conservation, capacity-building and educational objectives, but had to be justified primarily on their contributions to scientific knowledge.[84] In that vein, the Society had already through the 1970s established and managed a permanent research station on the island of Aldabra in the Indian Ocean. This was handed over in 1980 to the Seychelles Islands

Aldabra (1967–80), Cook Islands (1969), Phoenix Islands (1973), Great Barrier Reef (1973 and 1975), Little Cayman (1975); on botany and geology: Southern Patagonia geological transect (1975–6), Southern Patagonia botanical transect (1975–80); and on oceanography: Indian Ocean (1960–5), Indian Ocean coelacanth expeditions (1969 and 1972).

[84] C/110(79); CM 14 June 1979, minute 34(v). The Tibet geotraverse of 1985 was the last major scientific expedition managed directly by the Society.

Foundation, which continues to run it.[85] In 1984 the Society built on another expedition, the 1964 North Borneo expedition, to establish a research station in the Danum Valley in Sabah, North Borneo, in close collaboration with the local government and scientific institutions, and remained involved for the following thirty years.[86] And in 1991 it was a founder member of an international consortium carrying out research at Lake Baikal in Siberia.[87]

The (Biological) Expeditions Committee, advising on grant applications, was closed in 1984 following the ending of the Scientific Investigations Grant (see Chapter 3). However, a small portion of the Scientific Investigations Grant was then used to set up the Overseas Field Research Grant scheme, overseen by a new committee, so that the Society could continue to support ventures organised by others.[88] This was merged into the renewed Research Grant Scheme in 1993. The expeditions unit as a distinct element within the staff structure closed around the same time. The SZR Committee, stimulus for most of the Society's postwar expeditions, was closed in 1987. The Society would find other ways of doing science globally.

---

[85] For background, see David Stoddart to David Thomas, 17 December 1979, plus attachments: D.C. Phillips papers, MS Eng.c.5476, L.26. On the geopolitical context, see Jennifer Goodare, *Representing science,* Chapter 4.
[86] CM 8 November 1984, minute 28.
[87] CM 23 May 1991, minute 35. The Society continued its support until 2009.
[88] OM/53(84).

11   Looking outward

The Royal Society is an elite institution. Yet in recent years it has acquired an increasing array of demotic interests. It is more democratic than Greenpeace. It is controlled by Fellows whose experience of every-day life in Britain is greater than that of *The Sun*'s controllers.[1]

Values

Why does the Royal Society still exist, its motto *Nullius in verba* unchanged since 1662[2] and its original Charter Book[3] still in use? The fact that science has become so important in human affairs is part of the answer, if not sufficient in itself. Nor is it simply that the Society has continued to find useful things to do, though clearly that, too, is prere-quisite. Underpinning all its busy-ness is the Society's aspiration to embody a set of values: the primacy of evidence, the importance of nurturing individual curiosity, the pursuit of outstanding achievement, the search for practical benefit, the duty to engage with society at large. None of these is straightforward. But each of them is vital to the continu-ing practice of science, and the Society's efforts to express these values as best it can through its culture and its various programmes of activity have given it enduring purpose.

Previous chapters have focused on those various programmes. That does not tell the whole story. The Society also answers to a basic human need – for recognition, encouragement, collegiality. Many Fellows of the Society, and many holders of research appointments or grants supported by the Society, have testified to the personal impact that association with

---

[1] From an editorial applauding the Royal Society's growing involvement in public life and comparing it favourably with a popular newspaper: 'Royal treasure', *Research fortnight* (12 May 2004), 2.
[2] See Preface.
[3] The Charter Book is a large vellum volume, to which new pages are periodically added. Each new Fellow signs the Book on admission, adding his or her name to those of all Fellows since the Society's beginnings. It can be an emotional occasion.

the Society has had for them. This creates strong momentum for the Society's continued existence. Conversely, and unavoidably, non-selection can cause real hurt for those who miss out by narrow margins. Yet it is striking that so many individuals who owe nothing to the Society are willing to give their time freely to help deliver its programmes. That, too, is key to continued existence.

A further key is independence, from government and from perceived vested interest. This is another complex attribute. As science became too important to be left to the scientists, it was inevitable that government agencies would move onto what the Royal Society had traditionally seen as its own turf, for example in giving grants or offering policy advice. It was the Society's very independence that was held to debar it from over-seeing significant sums of public money in the 1960s when the new research councils were being established. On the other hand, it was this same independence that made it effective in international diplomacy, albeit working very closely with the British Government.

Financial donations from non-government sources have been vital at numerous points in the Society's postwar history. This could be proble-matic to perceptions of independence when the Society was under hostile scrutiny for its views on such matters as energy or agriculture policy. Even donations from philanthropic sources could be a challenge if the donor's purpose did not match the Society's current priorities. In the end, a degree of pragmatism was called for: the Society had to negotiate with its funders, to juggle its various sources of income to best advantage, and to operate with due transparency in response to public interest in its affairs. This was as it should be for an organisation with the privileges of charitable status. But, ultimately, independence was a question not so much of formal structures as of the determination of the Society's leader-ship that its opinion and its favour could not be bought. Continued existence required the Society to insist at all times on making its own independent judgements about outstanding science and outstanding scientists, free from external interference.

Many organisations are involved in the promotion of science. Among them, the Royal Society is marked by its focus on the highest levels of actual and potential achievement in research, particularly but not only academic research. Its record in this has given it a central place in the networks of science. But during the postwar period, the Society increas-ingly recognised that the unique breadth of experience and talent to which it had access had to be harnessed to action, and for interests wider than itself. The Society could not easily have lasted so long if it existed just for its Fellows. The agitation over the presidential succession in 1945 was driven in part by the wish to see the Society more energetically engaged in public life and by the fear that, to the contrary, it might lapse into being a purely honorific body. And it was not only the leadership that needed to be outward looking. The Biological Secretary A.V. Hill lectured his Council colleagues in 1945:

The FRS should not be regarded as a decoration for services rendered, or a consolation prize in old age: but rather as admission to a working fellowship of men (and women) actively engaged in 'improving natural knowledge' and promoting the interests and use of science. It should involve duties and obligations as well as privilege and distinction.[4]

Hill might have been dismayed, ten years later, to hear the President Edgar Adrian rejoicing that 'the duties of the President are certainly less arduous than most people might suppose'.[5] Few subsequent Presidents would have echoed Adrian's relaxed assessment. From ordinary Fellows, too, the most common clamour was for more opportunity to contribute to the Society's work. The work ethos intensified over the period. For many it was indeed a 'working fellowship', and there was no shortage of useful things to do.

Continuity and change

Some matters were consistently high priority for the Royal Society throughout the postwar period, most obviously: the encouragement of the highest standards in science through election to the Fellowship and other forms of recognition; the promotion of scientific research by expert discussion and publication; and the defence of basic science. The Society still (in 2015) deals with all fields of natural science, despite the creation of separate national academies for engineering (in 1976, with considerable opposition from the Society) and for medical sciences (in 1998, with considerable help from the Society[6]). It still elects its Fellows from throughout the Commonwealth. It is still housed in prestigious premises. The silver gilt mace presented to the Society by King Charles II in 1663 is still paraded at Council meetings and other formal occasions.

So there was continuity. There was also change. A few things stopped. By the end of Howard Florey's pivotal presidency, the Society was no longer directly involved in running any of the UK's scientific institutions or facilities, though it was still consulted about certain appointments. In

---

[4] Paper dated 1 January 1945; copy at AE/1/6/1. Parentheses in the original: Hill was one of those campaigning for the election of the first female Fellows that year.

[5] Edgar Adrian, 'Anniversary Address, 1955', 156.

[6] Peter Lachmann, *First steps: a personal account of the formation of the Academy of Medical Sciences* (Academy of Medical Sciences, 2010); interview with Michael Atiyah.

1984 it ended the payment of grants to learned scientific societies to help them with the cost of publications, libraries and rates.[7] That year it also stopped the 130 year-old Scientific Investigations Grant scheme, though it then restarted it in 1989. It pulled out of IIASA in 1982, and left the European Science Foundation in 2006. The last scientific expedition organised by the Society was in the 1980s.

There were other significant changes, too, in the Society's priorities. In 1960, international activities were the Society's highest single priority in terms of annual expenditure, owing mainly to the cost of its work with ICSU. International activities remained at about 40 per cent of the rapidly growing total spend through the 1960s and 1970s, because of the Society's similarly rapidly growing involvement both in developing exchanges with far-flung parts of the world and in fostering the European dimension in science. In 1978 Council confirmed that this proportion was about right.[8] But the proportion fell soon afterwards as research appointments, and especially University Research Fellowships, took off, and by 2010 international activities accounted for just 12 per cent of total spend.

Within international, there were new priorities. From the late 1950s till near the end of the century, the promotion of international scientific exchanges in the context of mitigating diplomatic tension was a key focus. Scientific exchanges in the context of building a European scientific community became a major activity from 1967, and were complemented by a series of other European initiatives. In the 1990s and 2000s, capacity building became an increasingly important element of total international activity. Adherence to ICSU and related bodies, which had been a headline activity for the Society in 1960, took something of a back seat.

At the national level, the 1980s saw two long-established strands of work assume new importance and grow remarkably: research appointments, especially at the earlier career stages, and policy advice. The former had a big impact on the formation of future research leaders, and also served to put the Society in daily contact with a younger generation. Policy advice brought the Society increasingly to public attention as it increasingly addressed issues of general public concern as well as issues to do with the well-being of the Science Base. This experience reinforced the impetus coming from its education work and from its 1985 report on public understanding of science, that the Society should do a great deal more to interact directly with the public beyond the world of professional

[7] Total annual spend from the PGA on these grants at the time was £162,000.
[8] CM 13 July 1978, minute 41(v).

research science. Its response to that from 1985 onwards marked a further shift in priorities.

All of this required resources as well as good intentions. From the mid 1950s onwards, the Society was almost continuously trying to raise money from private philanthropy, from industry and from any other non-government source it could find (see Annex). Such money gave it independence, and also allowed it to experiment with new schemes for supporting science. Once it had established proof of concept with a given scheme, it could approach government much more effectively for long-term, substantial funding and grow that scheme to a useful scale. Success in securing private funds and success in escalating its Parliamentary Grant were mutually reinforcing. In recent decades the Society paid increasing attention also to its own income-generating activities such as publishing and letting conference facilities.

Resources meant staff as well as money. The Society's staff grew very considerably after the Second World War. For example, during David Martin's 1947–76 tenure as Executive Secretary, staff numbers grew from 30 to 100.[9] In later years, the growth was slower – 138 staff were employed in 2012 – but comparisons are not exact, because, as happened throughout the world of employment, certain staff functions (such as catering, mailing, reprographics and security) were outsourced, while others (most obviously secretarial) disappeared almost entirely, and new functions (science policy, press, IT, HR, development) were introduced. The annual salary bill rose from £60,000 in 1960 to £1 M in 1982 and £7.3 M in 2012. The staff became very largely graduate, reflecting the changing nature of the work being done, and in 2012 included 19 PhDs as against just two in 1960.[10] It all increased the Society's capacity to do things.

Culture

The move towards greater engagement with the outside world was par-alleled by a gradual cultural change towards greater inclusiveness in the conduct of the Society's internal affairs. A particular challenge was for the Officers to share more with Council and for Council as a whole to share more with the Fellowship – by providing earlier and fuller information, by consultation and, eventually, by opening up certain decision-making processes. Progress could be sluggish on this, despite periodic debates on ways of engaging Fellows more fully in the Society's work.

[9] Harrie Massey and Harold Thompson, 'David Christie Martin', 394.
[10] There are, of course, now more PhDs looking for work outside of scientific research.

For example the Officers in 1961 agreed that 'From time to time when there is important information about action taken by Council such information should be sent to Fellows.' Three years later, they wanted to formalise this by establishing a news bulletin for Fellows, but Council blocked it.[11] The bimonthly *RS News* was eventually launched to the Fellowship in 1980,[12] and quickly gained a much wider distribution, with an extra 2,000 copies going to scientific organisations, civil servants and the press. It was later reinvented, more than once.

Michael Atiyah pioneered postal ballots and contested elections (i.e. voting on a slate with more candidates than vacant places) for Council membership – though not for Officers – during his 1990–5 presidency.[13] But Fellows continued the centuries-old tradition of complaining that they had too little influence over Council activities, and Council members continued the analogous tradition of complaining about the power wielded by Officers. The Charter was updated in 2012, for the first time since the 1660s, primarily to enable ordinary Council members to serve longer terms (three years instead of one or two) and thus to contribute more effectively to corporate governance (see Annex). Numerous other innovations, such as establishment of a formal Nominations Committee, were introduced at the same time to enable Fellows to be more involved in the Society's decision-making. The Society recognised that a culture of heavily centralised control was not necessarily appropriate, and that making full use of the remarkable talent available in its Fellowship and its wider networks was more likely to ensure future vitality.

The Society became more open to non-Fellows. For example, the various schemes of research appointments and grants were opened to all fairly early in the period, though vacant research professorships were not publicly advertised until 1982. In 1988 it ceased to be necessary for papers by non-Fellows to be submitted formally via a Fellow for publication in the Society's journals. In 2001 holders of the Society's research fellowships were invited to make more use of the Society's Carlton House Terrace headquarters (Figure 11.1). And in 2002–3, those headquarters underwent a major refurbishment, which included the introduction of open-plan offices for staff and an upgrade of meeting facilities.[14] Numerous organisations subsequently hired those facilities for their own meetings and events, with the result that many individuals who had not previously encountered the Society visited its premises.

---

[11] OM 7–8 October 1961, minute 9; OM 9–10 October 1964, minute 3; CM 17 December 1964, minute 27. The reasoning behind Council's decision is not recorded.
[12] OM/152(76); interview with Peter Cooper.   [13] Interview with Michael Atiyah.
[14] Interview with Eric Ash.

Figure 11.1 Entrance to 6–9 Carlton House Terrace, home to the Royal Society since 1967. © The Royal Society

The 1964 proposal to establish a news bulletin for Fellows would also have created specialised staff capacity to handle press issues, but it came to nothing. An explicit Press Office function was eventually established, cautiously and on a small scale, in 1976 to promote public awareness of the Society's activities. But the Society was still lukewarm about the idea that it needed actively to promote public awareness of its own activities,[15] and it was not until 1995 that the Press Office began to be properly resourced. By the turn of the century it was in full swing, in recognition of the Society's need to defend its corner in a more competitive and controversial world. In 1996 the Society launched its first website – another key element in its interface with the wider world.

The Society became better at celebrating. For example, new Fellows admitted to the Society in the 1960s and 1970s had an outwardly very low-key experience, despite the personal significance of the occasion.[16] By the turn of the century, largely at the instigation of the Treasurer, Eric Ash,[17] admission involved a two-day seminar in which each new Fellow explained his or her research to the rest of that year's intake, a half-day introduction to the Society's multifarious activities and associated staff, a suitably dignified admission ceremony, and various social events in which spouses and other family members were included.

Trial by Select Committee

In February 2002 Ian Gibson, Chairman of the House of Commons Select Committee on Science and Technology, announced that his Committee was going to conduct an inquiry into the Royal Society, amid a swirl of comments about the Society being elitist, out of touch and biased against women. It was dressed up as an inquiry into government funding of scientific learned societies, but its primary target was evident from the outset. Journalists and publicists of various hues had a field day. The Society recognised the need to set out its stall fully and carefully, in both written and oral evidence, and invested a great deal of effort in doing so.

Despite all the posturing that accompanied the launch of the inquiry, the ensuing report was relatively muted and its conclusions, in its own words, 'broadly positive'.[18] The Society was said to 'achieve a great deal'

---

[15] OM/99(86); OM/59(88).

[16] Interviews with Geoffrey Allen, John Kingman and Liebe Klug.

[17] Interview with Bob May.

[18] House of Commons Select Committee on Science and Technology, *Government funding of the learned scientific societies* (fifth report of session 2001–2, HC 774–1), 5. Also interview with Julia Higgins.

with the funding it received from public sources. Its independence was not thought to be compromised by such funds. There was praise for the University Research Fellowships scheme, and encouragement for the Research Councils to imitate its key features. The Society's grant schemes merited continued support. Its international role was recognised. But its management of COPUS (see Chapter 6) came in for criticism, with the recommendation that COPUS be made independent of its three sponsoring bodies. The Society's relative lack of collaboration with the specialist learned societies was noticed. There were calls for greater transparency and inclusiveness.

The most contentious issue, though, was the under-representation of women in the Fellowship (see Annex for statistics). The Select Committee concluded that 'the present low level of female Fellows' did not represent any discrimination by the Society against women. It recognised that the roots of the issue went much deeper, to individual choices made at school and at early career stages, and it accepted the Society's arguments against instituting any form of positive discrimination to increase the proportion of women elected. Initiatives such as the Society's Dorothy Hodgkin fellowships (Chapter 3), it agreed, were more likely to produce long-term progress.

At the outset, newspapers had been encouraged to carry speculative statements that the Society was being investigated by the Equal Opportunities Commission. This was not in fact the case. However, there were a number of exchanges between the Society and the EOC, identifying initiatives already under way and further steps that might be taken to make it easier for more women to follow scientific careers. While it was recognised that there was then a long way to go, the EOC described itself as 'encouraged' by the Society's approach. The Society gradually increased its commitment to tackling diversity issues in science. For example, it established an active Equality and Diversity Network to monitor its own performance and to generate practical ideas, it supported national moves such as the Athena SWAN Charter promoting good institutional practice, and in 2011 it launched a four-year government-funded programme to investigate and mitigate barriers faced by particular groups to careers in science.[19] But there remained a long way to go.

The Select Committee inquiry did not revolutionise the Royal Society, but it did serve as a reminder that engaging with the wider public was not something that the Society could choose to do or not to do – it was going

---

[19] Georgina Ferry, 'The exception and the rule: women and the Royal Society 1945-2010', *Notes and records of the Royal Society of London*, 64 (2010), S163–72; Royal Society, *Submission to the Science and Technology Committee's inquiry on women in STEM careers* (September 2013); interview with Julia Higgins.

to be mandatory for the foreseeable future, and it would cover the Society's internal activities as well as scientific developments more generally. Over the previous decade or so the Society had been getting better at living its life publicly, and this would need to continue.

Putting talent to work

The history of the Royal Society in the nineteenth century centred on the struggle to focus the Society solely and unequivocally on high scientific achievement. By the end of the century that had been accomplished,[20] and it was further entrenched in the following decades. By 1945, there was no doubting that the Society's identity was rooted in being scientifically elite. But there was then a danger that this defining attribute could become an end in itself, and thus ultimately a liability as well as an asset. The first two postwar Presidents, Robert Robinson and Edgar Adrian, did little to change established attitudes on that score. But Cyril Hinshelwood, and particularly Howard Florey, understood that, instead of being suffocated by its own elitism, the Society had to become considerably more outward looking. Hinshelwood, Florey and their colleagues and successors set out to do something about it.

The fact that the Society did become more outward looking during the post-war period, in the sense of engaging more with the public, may be illustrated in cameo form by how it marked consecutive milestone anniversaries. The celebrations marking the Society's tercentenary in 1960 emphasised its position at the centre both of the global scientific community and of the British Establishment. An opening convocation in the Royal Albert Hall in the presence of the Queen, the Duke of Edinburgh and the King and Queen of Sweden,[21] an extensive supplement in *The Times* later published in book form, a special service in St Paul's Cathedral, and a series of BBC programmes on radio and television all portrayed the Society at the heart of the Establishment. Over 300 national academies and other scientific organisations from all parts of the globe sent formal tributes acclaiming the Society's pre-eminence in the scientific world, and many of them sent senior representatives to the convocation. There were scientific lectures and meetings, visits to laboratories and honorary degrees for the dignitaries. It was two and a half years in the planning, and lasted a little over a week.[22]

[20] Marie Boas Hall, *All scientists now*, 218.
[21] The King of Sweden was elected a Royal Fellow in 1959 and was formally admitted to the Society during the convocation.
[22] Harold Hartley, ed., *The tercentenary celebrations of the Royal Society of London* (Royal Society, 1961).

Figure 11.2 The Society's Foreign Secretary, Lorna Casselton, presenting a copy of Bill Bryson's celebratory volume *Seeing further* to the government's Foreign Secretary, David Miliband, at a meeting of the InterAcademy Panel, January 2010 (l to r): Howard Alper, David Miliband, Martin Rees, Chen Zhu, Lorna Casselton. © The Royal Society

The 350th celebrations in 2010 echoed the tercentenary in some respects. There was again a convocation of the Fellowship in the presence of the Queen and the Duke of Edinburgh, and other members of the royal family, during which Prince William, elected a Royal Fellow the previous year, was formally admitted to the Society.[23] There was extensive media coverage of the Society, a special issue of postage stamps, and a bestselling book edited by the popular writer Bill Bryson (Figure 11.2).[24] There was also a service at St Paul's Cathedral,[25] though on a much smaller scale than in 1960 and curtailed by a disruptive fire alarm. The Society was still at the heart of the Establishment.

---

[23] 'Convocation of the Fellowship of the Royal Society at the Royal Festival Hall, 23 June 2010', *Notes and records of the Royal Society of London*, 64 (2010), 217–27.

[24] Bill Bryson, ed., *Seeing further: the story of science and the Royal Society* (Harper Press, 2010).

[25] Rowan Williams, 'A sermon … on the occasion of the 350th anniversary', *Notes and records of the Royal Society of London*, 64 (2010), 213–5.

Figure 11.3 Martin Rees showing the Queen and Duke of Edinburgh some of the Society's treasures during the Society's 350th anniversary convocation at the Royal Festival Hall, 23 June 2010. © The Royal Society

But in other respects, 2010 was very different from 1960. Driven strongly by the Executive Secretary, Stephen Cox, and masterminded by a specially recruited Programme Director, Dominic Reid, it was over six years in the planning and was cast as a year-long programme in which boosting the public prominence of science and of the Society itself featured more strongly than events and initiatives aimed at directly advancing professional science.[26] The convocation was staged at the more accessible Royal Festival Hall, in conjunction with a scientific exhibition on the South Bank that ran for nine days and attracted over 50,000 visitors (Figure 11.3). In addition to activities in other major London cultural institutions, there were hundreds of events up and down the country, many organised by other bodies with Society encouragement, attracting over 300,000 visitors in areas where the Society normally had little presence. Both history and contemporary science were intensively covered. The year as a whole saw a 60 per cent increase in media coverage of the Society. It all signalled, according to one commentator, the

[26] Interviews with Lorna Casselton, David Read, Martin Rees and Dominic Reid.

Society's 'hunger to be seen as up to date, inclusive and important, not exclusive and aloof'.[27]

In the middle of all this, the science journal *Nature* commended the Royal Society for embodying 'the right kind of elitism'. Its track record was 'worthy of celebration. It stands today as a relatively successful model of what an independent national academy can achieve, having made itself both highly regarded in the corridors of power and prominent in public debates on major science-related issues.'[28] A more partial observer, Bill Bryson, commented: 'If we have an Earth worth living on a hundred years from now, the Royal Society will be one of the organisations our grandchildren will wish to thank.'[29] How does a private, elite body with relatively modest resources, committed to curiosity about the natural world, come to matter so much?

Part of the answer lies in the enduring importance of the values mentioned at the beginning of this chapter. The Society has, of course, fallen well short of fully epitomising each of these values at every point in its history, but to the extent that it is associated with them it is recognised as doing something significant. And doing something significant for a long time somehow adds to an organisation's aura.

Beyond that lies the Society's increasing realisation, from Hinshelwood and Florey onwards, that it needed to harness its concern with high scientific achievement to the widest possible approach to promoting science. In the immediate postwar years, a sense of entitlement and privilege could be discerned in some of the Society's dealings with the outside world. That could not survive the growing government involvement in science and the growing competition for control of science, let alone the growing public interest in science and wish for a say in the uses of science. The Society had to find niche areas, outside the business of electing Fellows, where its elite character gave it a competitive advantage, and it had to broaden its traditional range of target audiences as much as it could.

Being outward looking had several dimensions. One was to broaden the Society's remit beyond its own Fellows to embrace also the UK scientific community generally, including, albeit tentatively, those working outside the traditional Science Base. The most important generic thing the Society did here was to extend its role as patron of individual talent. It appointed outstanding individuals to posts where they could give free rein to their curiosity and carry out research on whatever they wanted and

[27] Colin Macilwain, 'In the best company', *Nature*, 465 (24 June 2010), 1002.

[28] Editorial, 'The right kind of elitism', *Nature*, 465 (24 June 2010), 986.

[29] Bill Bryson, ed., *Seeing further*, 13.

wherever they wanted, with personal support from senior Fellows and with minimal administrative and other distractions. Before the War, the Society used its private funds to establish the feasibility of such activity and to demonstrate its own suitability to manage it. From the early 1960s, when it first obtained public money for research appointments, the various schemes gradually gained critical mass and were opened increasingly to non-Fellows. In the 1970s, the focus of the research appointments began to move more towards the earlier career stages, and this accelerated from 1983 with the URF scheme, in which those appointed were typically in their early thirties. By 1981 research appointments already accounted for 22 per cent of the Society's £4.2 M Parliamentary Grant; in 2010, they were 79 per cent of a Grant then totalling £48.6 M.[30]

With a similar mix of private and public funding, the Society was also able to create opportunities for scientists to travel to conferences in other countries and to undertake collaborative research with overseas colleagues. Such schemes were initially targeted at Fellows, but, like research appointments, were fairly quickly opened up to wider participation. So, too, were the Society's various grant schemes. The Society also began to engage, or re-engage, with the applications of science and to grapple with what it could most effectively do to promote this complex area. All that made the Society more outward looking in the sense of engaging with a wide cohort of researchers beyond its own Fellowship. This was mutually beneficial as it markedly increased the number of active scientists with a stake in the Society.

Part of the international work was about promoting science by creating opportunity for British scientists to interact with colleagues from other countries. Part of it was the Society positioning itself to be able to lead policy-relevant initiatives when occasion arose, such as the 1993 New Delhi population conference and the 2005 G8 statements on climate change and on capacity building. And part of it was extending generic Royal Society influence. The Society had and has a global reputation, as those who travel outside the UK on the Society's behalf enthusiastically confirm.[31] It reinforced this by maintaining a high profile in international scientific fora, and by putting significant effort into nurturing its bilateral relations with key national academies and scientific organisations.

A further way in which the Society looked outward to promote science was in interacting more energetically with the policy-making process. The

---

[30] For financial details, see Annex. The 1986 corporate plan was already worrying that the scale of the Society's publicly funded research appointments might make it look like a government agency involved in managing science.

[31] Interviews with Lorna Casselton, Tony Epstein, Brian Heap, Julia Higgins and Stella Porter.

Royal Society does not wield formal power, in the sense that J.G. Crowther had speculated in 1944 might be appropriate for it to do. It cannot compel particular policy outcomes. Nor, on the whole, did it seek to have such power, at least since the reform of civil science following the Trend Report in the mid 1960s. It was, arguably, lucky then to have lost such managerial functions as it previously held, however disgruntled it felt at the time.[32] It focused instead on exercising influence.

The Society's influence derived from a number of factors. The most fundamental was its scientific authority, based on its reputation for discerning serious achievement in science. Safeguarding that reputation required taking constant pains with its selection processes and its official scientific pronouncements. A second factor was its independence. A third factor was its extensive networks and its ability to engage key individuals in its affairs. It had exceptional convening power, and made full use of it. The Society had some of the attributes of an insider, but it also operated outside the formal structures, its ties with government 'informal, discrete, ubiquitous'.[33] Particularly through the President and experienced Fellows in close touch with him, the Society was able to exercise influence through private exchanges with individual policy-makers and advisers, complementing the impact of its formal published contributions to policy. Such activity harnessed the Society's reputation for scientific excellence, but also depended strongly on the individuals involved.

The Society, then, could and did command attention when speaking about science. In earlier decades it responded to numerous requests, especially from government, for scientific advice. From the early 1960s it set out to interact more energetically and proactively with the policy-making process, setting parts of the agenda rather than simply responding to other people. Policy advice consistently featured strongly in internal reviews of strategic priorities from that time onwards. It expanded significantly after 1981, when it began to be supported by professional staff, and the Society gained a global reputation among national science academies for its effectiveness in this area. It was a niche that played to the Society's particular attributes, and it was a niche where the Society had demonstrable impact.

The Society looked outward to the public beyond the world of professional research science. Initially, this was through its interest in science and mathematics education, which developed from the mid 1950s into a significant combination of practical initiatives and inputs to education

---

[32] The Society later decided explicitly that it had no wish to emulate the massive scale of the policy advice work carried out by the National Research Council for the US Government: OM/36(83).

[33] See Chapter 1; also interviews with Brian Flowers, Brian Follett and Robin Nicholson.

policy. From the mid 1980s, the Society also explicitly embraced public understanding of science and public engagement with science as part of its core mission. It had previously been content to leave such matters to other organisations, but it came to see that its leadership role in science meant that it had to be directly involved. The anniversary celebrations in 2010 epitomised how far it had moved in this direction.

So during successive decades from 1960 the Society was increasingly committed to engaging with ever wider groups, as an integral part of its mission to promote science. In this it reflected the changing spirit of the times, and it recognised the need to seek talent from the broadest possible base. It also reflected social and political realities. Science was making ever greater demands on public funds and having ever greater impact on daily life, and the leaders of the scientific community had to respond to, and work with, the ever greater interest that science thereby evoked. The Society had to be unreservedly outward looking, and to harness its attributes to the public good, if it was to retain its significance. In responding to that challenge over the decades from 1960, it gained a new lease of life.

Annex: Running the Royal Society

This annex outlines how the Society was run in about 2012, and how its administration had evolved in the previous decades. It does not deal with the informal relationships and interactions that in practice play such a large part in daily decision-making and in shaping the overall feel of the place. Rather, it summarises the formal framework within which the Society's programme of activities is negotiated and implemented. That framework continues to evolve, of course, so this annex may not be wholly relevant to future arrangements, but it does cover the period of this book. It focuses on three aspects: governance, finance, and the process of electing Fellows.

Governance

The senior decision-making body in the Royal Society, as laid down in the original 1663 Charter, is its Council. That Charter also specified that the Council should comprise exactly twenty-one Fellows. They are the Society's Trustees under modern charity law. They held eleven formal meetings in 1960, and six in 2010.

Five Council members have specially designated roles: the President, Treasurer, Secretaries for the physical and biological sciences, and the Foreign Secretary. These five are known as Officers,[1] and like all Trustees they act in a voluntary capacity. There were two Secretaries from as early as 1663, Henry Oldenburg and John Wilkins being the original post holders. The custom of selecting one Secretary from the physical side and one from the biological side dates from 1827. The formal position of Foreign Secretary was envisaged under the 1663 Charter, but in practice it dates from 1719 and a legacy from a Fellow, Robert Keck, to support the costs of undertaking foreign

---

[1] Or, in some circumstances, as the 'President and Officers'. Occasional proposals to deal with the ever-increasing workload by creating a sixth Officer post have so far consistently come to nothing; for an early example, see OM 13 July 1961, minute 6(d).

correspondence. Philip Zollman was appointed in 1723 as the first person explicitly assigned to that role.[2]

The Officers have desks at the Society and in recent decades have typically given at least two days/week to Society activities. Their formal roles may be inferred from their job titles. Individual Officers also shoulder additional roles according to need and personal interest: in principle, every significant activity comes within the purview of one or another Officer. As well as participating in Council meetings and countless private discussions, the Officers meet together formally (i.e. with agendas and minutes), for example fifteen times in 1960 and seven times in 2010. These meetings had no explicit terms of reference until 1995, when, following Austin Bide's review of the Society's administration, the five Officers plus the Executive Secretary (the head of staff) were constituted as the Strategy Policy Board charged with 'developing and assessing the policies and objectives of the Society' and acting as an advisory body to Council. A Finance and General Purposes Committee was set up at the same time. In 2008 the two were rolled into a single entity called simply the 'Board' and comprising the Officers and Executive Secretary, which gradually came to take on also the role of guiding and overseeing implementation of Council's decisions.

Council members, including Officers, normally begin and end their terms of office on the Society's Anniversary Day (30 November or the nearest weekday if that falls at a weekend). Anniversary Day is also the occasion for the Society's AGM, and for the President to make a major speech (the Anniversary Address) aimed at both the Fellowship and the wider scientific community.

The practice that all Officers should serve a maximum of five years, with one retiring each year, was introduced by Michael Atiyah in 1992 and was written into the Supplemental Charter granted in 2012. No President had served more than five years since 1871, and Foreign Secretaries also usually served five years, but, before Atiyah's reform, the other Officers served variable periods of up to ten years. The reform was not simply a matter of arithmetical neatness, but was part of Atiyah's campaign to introduce greater democratisation in Royal Society practice and to broaden the field of possible candidates for these posts by making them less onerous.[3] Table A.1 lists the Officers and Executive Secretaries in post as at 1 January each year since 1945.

---

[2] The British Government did not create a unified post of Foreign Secretary until 1782, allowing the Royal Society to claim precedence in this particular matter.
[3] Interview with Michael Atiyah.

Table A.1 *Officers and Executive Secretaries in post, 1945–2014*

Year, at 1 January	President	Treasurer	Physical Secretary	Biological Secretary	Foreign Secretary	Executive Secretary[4]
1945	Henry Dale	Thomas Merton	Alfred Egerton	A.V. Hill	Henry Tizard	John Griffith Davies
1946	Robert Robinson			Edward Salisbury	A.V. Hill	
1947					Edgar Adrian	David Martin
1948						
1949			David Brunt			
1950						
1951	Edgar Adrian				Cyril Hinshelwood	
1952						
1953						
1954						
1955						
1956	Cyril Hinshelwood			Lindor Brown	Gerard Thornton	
1957		Bill Penney				
1958			Bill Hodge			
1959						
1960						
1961	Howard Florey	Alex Fleck			Patrick Linstead	
1962						
1963						

Year							
1964				Ashley Miles			
1965	Patrick Blackett					Tommy Thompson	
1966							
1967							
1968		Frederick Bawden					
1969			Harrie Massey				
1970	Alan Hodgkin			Bernard Katz			
1971							
1972						Kingsley Dunham	
1973		James Menter[5]	James Lighthill				
1974							
1975	Alex Todd						
1976		John Mason		David Phillips	Michael Stoker		
1977							Ronald Keay[6]
1978							
1979			Morris Sugden				
1980	Andrew Huxley						
1981					Arnold Burgen		
1982							
1983				David Smith[7]			
1984			Roger Elliott[8]				
1985	George Porter						Peter Warren[9]
1986		Robert Honeycombe			Tony Epstein		
1987							
1988				Brian Follett[10]			

Table A.1 (cont.)

Year, at 1 January	President	Treasurer	Physical Secretary	Biological Secretary	Foreign Secretary	Executive Secretary[4]
1989			Francis Graham-Smith			
1990	Michael Atiyah					
1991						
1992		John Horlock			Anne McLaren	
1993						
1994			John Rowlinson	Peter Lachmann[11]		
1995	Aaron Klug					
1996						
1997					Brian Heap	Stephen Cox[12]
1998		Eric Ash		Patrick Bateson		
1999			John Enderby			
2000	Bob May				Julia Higgins	
2001						
2002		David Wallace				
2003						
2004				David Read		
2005	Martin Rees		Martin Taylor			
2006		Peter Williams				
2007					Lorna Casselton	
2008			John Pethica	Jean Thomas		
2009						
2010						

2011   Paul Nurse
2012   Tony Cheetham   Martyn Poliakoff   Julie Maxton[13]
2013   John Skehel
2014   Alex Halliday
2015

---

[4] The head of the staff was termed Assistant Secretary from 1823 to 1962; then Executive Secretary; then, from 2011, Executive Director.

[5] Frederick Bawden died in office on 8 February 1972, and was succeeded by James Menter on 18 May.

[6] David Martin died in office on 16 December 1976. Ronald Keay, his deputy, immediately became Acting Executive Secretary, and was confirmed in post a month later.

[7] David Phillips resigned on 30 April 1983 to become Chairman of the ABRC, and was succeeded by David Smith on 1 May.

[8] Morris Sugden died in office on 3 January 1984, and was succeeded by Roger Elliott on 1 May.

[9] Peter Warren succeeded Ronald Keay on 20 May 1985.

[10] David Smith resigned on 31 July 1987 to become Principal of Edinburgh University, and was succeeded by Brian Follett on 1 August.

[11] Brian Follett resigned on 31 July 1993 to become Vice-Chancellor of Warwick University, and was succeeded by Peter Lachmann on 1 August.

[12] Stephen Cox succeeded Peter Warren on 1 July 1997. From 1985 to 1991 he had served as head of the Society's international activities.

[13] Julie Maxton succeeded Stephen Cox on 1 March 2011.

The 1663 Charter allowed the President to appoint an unspecified number of Vice-Presidents from among the members of Council. These posts have always been in the President's personal gift. They are often, but by no means always, allocated to the other Officers, and may include other Council members as well or instead. The duties are in the main those of hosting or chairing events or representing the Society externally, but may also involve taking responsibility for an activity such as developing relations with other organisations in an area of particular interest to the Society.

The 1663 Charter required that ten of the twenty-one Council members should retire each year, on Anniversary Day. Since the five Officers served multi-year terms, that meant inescapably that a certain number of Council members served only one year, much of which was spent getting up to speed on the breadth of Society business and dealing with the selection of individuals for election or other awards.[14] Coupled with the fact that the five Officers were very fully briefed, this led to the situation where many Council members felt that they could have little impact on Society strategy. It was a cause of constant complaint, which invariably foundered on the entrenched wisdom that it was too risky to change the Charter.[15] Eventually, however, Julie Maxton, newly appointed Executive Director who had been a practising barrister and Dean of Law at Auckland University, and then Registrar at Oxford University, steered the Society through the process of securing a Supplemental Charter in 2012. That Charter had the effect of enabling all ordinary Council members to serve terms of three years and Officers to continue with five-year terms, thus allowing all Council members to engage much more effectively with Society business. This may have considerable impact on the culture of the Society in future years.

An extensive network of committees reporting directly or indirectly to the Officers and Council enables the Society to draw on a wide range of expertise in carrying out its work. Committees are normally chaired by Fellows and include many non-Fellows among their members. The Society has always depended heavily on the willingness of many individuals to give freely of their time in this way. A Nominations Committee

---

[14] There is considerable private correspondence on this. See, for example, F.A.E. Crew to C.D. Darlington, 21 November 1944, describing his year on Council entirely in terms of those in his discipline whom he did or did not get elected or given medals, and warning Darlington (who was about to start his own one-year membership of Council): 'Power is concentrated in the hands of the Officers. You will spend six months in deciding who is to be elected and the next six months in deciding who shall be given medals.' Also similar correspondence between Darlington and Robert McCance, July 1954. C.D. Darlington papers, ms Darlington c.95.

[15] For example, OM 11–12 February 1984, minute 10.

was established in 2012 to provide a more transparent and more thorough process for identifying appropriate individuals for specific tasks within this structure.

The degree to which Officers seemed to dominate Council was one long-running source of internal dissatisfaction with the Society's governance. The degree to which Council itself seemed to have disproportionate power in such matters as determining its own successors was another. The controversies in 1935, 1945 and 1956 (see Chapter 1) were, among other things, about Fellows wanting more say in the running of the Society. Lack of democratic choice in the election of Council members was a particular sore point, surfacing again in, for example, 1970, 1981, 1984 and 1988.[16] Michael Atiyah came to the presidency in 1990 with a democratising agenda, and introduced competitive elections for ordinary members of Council. Competitive elections for Officers were regarded, then and later, as likely to deter the best candidates, though greater effort was put into consulting the Fellows before a decision on new Officers was reached.[17] The Society's 350th anniversary prompted a further bout of debate about how Fellows could participate more fully in running the Society. Council appointed a working group under Martin Taylor, which delivered a detailed report early in Paul Nurse's presidency with a series of practical proposals, including introduction of the Nominations Committee, a culture of more thorough consultation and changing the Charter to strengthen Council. These were all implemented over the following two years.

Only Council can formally speak for the Society. So, for example, a report drafted by a working group becomes an official Royal Society report through being formally approved by Council: until then it represents the views only of its authors. In practice, of course, authority to speak for the Society is delegated, explicitly and implicitly, to Officers (most obviously the President), chairs of working groups, certain senior staff, and other individuals as need arises. It would otherwise be impossible to engage in real time with the daily developments of public life. The corollary is that individuals prominently associated with the Society need to be clear when they are, and are not, speaking for the Society, and to be alert to the scope for being misinterpreted. Practice on this has evolved through experience.

[16] Correspondence between Philip Sheppard and David Phillips, December 1970–December 1971: D.C. Phillips papers, MS Eng c.5479; OM/24(81); OM 11–12 February 1984, minute 9; OM/107(88).

[17] Interviews with Michael Atiyah and Martin Rees.

Finance

The Royal Society has three income streams: donations from private sources (including Fellows' subscriptions), either to be used directly or to be invested and the income used; grants from public sources; and surpluses from commercial activities such as publishing and (more recently) the letting of conference facilities. In managing its finances in the postwar period, the Society had to ensure that it had sufficient in total for its overall needs, that it had sufficient flexibility to support its own initiatives without undue constraint, and that it was not so beholden to any one source as to threaten its independence.

Regular income from public sources (i.e. excluding exceptional items such as the IGY) overtook income from private sources in the 1950s. The main public source was the Parliamentary Grant-in-Aid, dating back to 1850 (see Chapter 3). Like other scientific bodies, the Society negotiated its PGA annually with the Treasury until a unified Science Budget was introduced in 1964 following the Trend Report; thereafter, it negotiated with DES and its successor bodies.[18] From time to time, additional grants were negotiated with other departments, notably the Foreign Office, for specific objectives. By unilateral action in 2010, the government replaced the PGA with an ordinary departmental grant, thus allowing it in principle to exercise greater scrutiny over how the money was used. Table A.2 analyses the growth of the Society's PGA since 1956.

The Society's private income was key to its independence and to its capacity to launch initiatives of its own choosing. Its postwar history was therefore punctuated with major campaigns to increase that income, alongside frequent smaller fundraising efforts for specific purposes. The 1960 tercentenary was an obvious opportunity for a campaign. Annual expenditure from private sources then stood at £85,000. The campaign target was set at £350,000, mainly from industry; this was duly achieved, and proved useful both for new initiatives and for plugging gaps that later appeared in the finances. The move to Carlton House Terrace in 1967 required £850,000 and a further appeal, which eventually was also successful. But the Society found itself moving into significant deficit in the early 1970s when seven-year covenants began to run out. This had been foreseen,[19] though not properly addressed. It precipitated long discussion

---

[18] Rothschild proposed in his 1971 report that the Society revert to negotiating its PGA direct with the Treasury, despite the Society telling him that it was content with the new arrangements. The government accepted the Society's preference. Rothschild to Hodgkin, 1 June 1971: Royal Society Committee on the Government Research and Development Study, 7/3/4/6; and meeting of the Cabinet Official Committee on Science and Technology, 21 April 1972: TNA T 224/2470.

[19] P.M.S. Blackett, 'Anniversary Address, 1966', x–xi.

Table A.2 *Income from Parliamentary Grant-in-Aid by purpose, 1956–2010 (£'000)*[20]

Year	Scientific investigations[21]	Research appointments[22]	International[23]	Other	Total	Total as % of Science Budget
1956–7	50	–	31	176[24]	257	n/a
1961–2	81	–	64	20	165	n/a
1966–7	152	90	189	125	556	0.9
1971–2	161	129	658	174	1,122	1.0
1976–7	331	296	963	447	2,037	0.9
1981–2	864	848	1,600	1,117	4,429	1.0
1986–7	–	2392	2,773	1,640	6,805	1.1
1991–2	2,174	6,769	4,576	2,216	15,735	1.7
1996–7	2,336	11,928	4,606	3,221	22,091	1.7
2001–2	2,042	15,828	4,559	3,243	25,672	1.5
2006–7	1,696	23,754	6,928	4,031	36,409	1.2
2010–1[25]	1,061	38,402	6,030	3,065	48,558	1.2

and a series of measures to contain costs and increase revenue (including doubling the Fellows' subscriptions).[26] A fresh appeal was launched in July 1974, which a year later had brought in £760,000.

Despite that, the Executive Secretary Ronald Keay minuted the President Alex Todd in 1980 about his concern that the PGA had climbed from 60 per cent of the Society's total expenditure in 1978–9 to nearly 70 per cent in 1980–1.[27] The rate of growth then slowed down, but the issue continued to trouble the Officers. Significant effort was put into raising private funds for specific new initiatives such as the Policy Studies Unit, fellowships and exchange schemes, and book prizes.[28] However, further increases in the proportion of publicly funded expenditure were inevitable as the URF scheme took off (Chapter 3), and by the mid 1990s PGA had reached 80 per cent of total spend. At that level the

---

[20] Data from published accounts for the year in question.
[21] The Scientific Investigations Grant up to 1984; Research Grants scheme from 1989.
[22] See Chapter 3.
[23] Mainly exchange fellowships, travel grants, international subscriptions and international research projects, but not expeditions.
[24] Including £142,000 for the IGY.
[25] For continuity. From FY 2011–12, the published accounts use different analytical categories.
[26] OM/15(73); OM 26 January 1973.
[27] Ronald Keay, 'RS staff management issues 1980–85': TODD Acc 1021, Box 33.
[28] OM/77(84); OM 12 July 1984, minute 3(k); CM 8 November 1984, minute 15; OM 18 April 1985, minute 3(a); OM 16 January 1986, minute 3(d).

Society's independence seemed to be in jeopardy, and in 1996 a new general campaign was launched to raise private funds. This produced £23 M in three years, thus easing the independence issue. But 94 per cent of the £23 M was earmarked for short- or medium-term projects, not all of them on the Society's original shopping list, so flexibility remained a problem.[29]

A permanent Development Office to support fundraising was established in 2001. Approaches to major foundations, to the Fellows themselves and to other donors raised nearly half the cost of a much-needed £12 M modernisation of the Society's premises in 2002–3. One consequence of this modernisation was a step change in the scope for generating income – and, in the process, raising the Society's visibility within the wider UK scientific community – by letting out the meeting rooms. The net surplus from such activity grew from £80,000 in 1999–2000 to £1.0 M in 2012–3. Over the same period, the surplus from publishing rose from £298,000 to £2.1 M. Such income could be used for any purpose and therefore provided much valued flexibility. The 350th anniversary, like the 300th, became the focus of a major fundraising campaign, this time producing about £100 M, helped by a completely unexpected £47 M legacy from the Australian lawyer Theo Murphy. All the campaigns also benefited substantially from the willingness of individual Fellows to support their own Society. By 2010, just 68 per cent of the Society's total annual expenditure was funded by the PGA.

Elections to the Fellowship

Every step of the process of electing Fellows[30] is subject to intense and continuous scrutiny. Council minutes are full of it. So, too, is Fellows' private correspondence – the merits of this or that individual, the justice of this or that tweak to the procedures, the desirability of this or that new objective. The debate tends to home in on the total number elected annually and the various categories of candidate.

Through to the mid nineteenth century, scientific eminence helped but was not a prerequisite for being elected. Quality control was weakened by the circumstance that Fellows could be elected at any time in the year and that there was no upper limit to the annual number elected. Between 1700 and 1850, numbers elected in any one year fluctuated from six (in 1783) to forty-seven (1834). Candidates were not systematically weighed against each other, and electing a scientifically weak candidate entailed no

[29] Royal Society, *Annual review 1998–99*, 24.
[30] This section focuses on Fellows: Foreign Members are discussed in Chapter 10.

obvious opportunity cost for scientifically stronger candidates. There were therefore few compelling incentives to make difficult choices. Up to 1847, scientifically distinguished Fellows averaged less than one third of the total Fellowship.[31]

In 1847, after extended debate, the rules were changed: to limit the number elected annually to fifteen, and to ensure that all fifteen were elected at the same time. That introduced competition and opportunity cost into the election process. It had the desired effect: already by 1860 professional scientists were in the majority within the overall Fellowship. Attempts to remove the limit of fifteen in 1875, and again in 1888, were successfully resisted.

One consequence was that the total number of Fellows declined, from over 750 in 1847 to under 450 by the end of the century. This was foreseen, and worried those concerned with the Society's finances. The 1847 revisions to the Statutes were therefore very strong on the need to pay subscriptions, which were then a far more important element of the Society's income than they are now. However, the 1847 revisions left the subscription at the 1823 level of £4, and left the admission fee at £10.[32] A second consequence of the new election rules was that Council increased its power in the Society's affairs by virtue of controlling the election slate: it produced the list of fifteen names that was put to the Fellowship for election. The quid pro quo was a modest degree of increased transparency in Council's activities.

The limit of fifteen annual elections stayed in place for over eighty years, until 1930.[33] It was then raised to seventeen. The rationale was that the growth of professional science meant candidates outside main-stream academic disciplines – for example field naturalists, explorers and those applying knowledge in engineering or medicine – were no longer getting a look in, whereas they had been strongly represented in the Royal Society before the days of specialisation. A second increase, to twenty in 1937, was intended to reduce the average age at election (which averaged forty-seven over the previous fifteen years), so that the Society could play a more effective role in public life. But the increase did not sufficiently solve the problem of an ageing Fellowship, and a further increase, to twenty-five, was agreed in 1945. This was prompted partly by the age issue, and partly by the growth of scientific activity not only in the UK but also, indeed particularly, in India and the British Dominions. The advent

---

[31] Henry Lyons, *The Royal Society 1660–1940*, 342.

[32] The admission fee was eventually raised to £20 in 1974 and £50 in 1981; it was abolished in 1984.

[33] The dates given here for changes in annual elections are the years in which the decisions were made rather than the years in which they were first implemented.

Table A.3 *Average age at election, 1940–2010*

Year of election	Average age at election
1940	44
1950	48
1960	48
1970	51
1980	53
1990	49
2000	55
2010	55

of female Fellows (the first two, Kathleen Lonsdale and Marjory Stephenson, were elected in 1945) further increased the pool of candidates.[34] Table A.3 shows, however, that the age issue could not be solved just by electing more Fellows.

The next increase, to thirty-two in 1964, was driven by the need to include more technologists and applied scientists in the Fellowship (see Chapter 4), though in the event only half the extra places were earmarked for them. A further increase, to forty in 1975, was a combination of promoting applied science and, more explicitly, responding to the growth in scientific activity since the Second World War. In 1998 an extra two places were added to facilitate equal nominal allocations to the physical and biological sciences, and two further places were added in 2003 to accommodate 'human sciences including scientific studies of ... epidemiology, economics, demography, human geography, biological anthropology and social behaviour' without opportunity cost elsewhere in the system. The most recent increase, to fifty-two in 2013, was primarily aimed at modifying the make-up of the Fellowship to include more general candidates, individuals with experience beyond mainstream academic research – rather like the 1930 increase.

Every proposal to increase the number of annual elections had to overcome the suggestion that it might dilute the quality of the brand. This could lead to acrimonious debate. For example, in 1979 Alex Todd tried to rescind the 1975 increase – which had been agreed before he became President – on the grounds that the standard of entry was edging 'below the proper level of excellence'. Council blocked that move,[35] but it was a long time before the next increase was proposed.

[34] By 2010, about 6 per cent of the total Fellowship, and 10 per cent of those elected in the previous decade, were female.
[35] CM 1 March 1979, minute 21; CM 14 June 1979, minute 8.

Whether any of these changes in numbers of annual elections had the intended effect depended on whether Fellows subsequently nominated candidates of the type being sought. On some occasions, for example in connection with the increases in 1964 and 2013, special groups were set up to identify potential candidates and catalyse nominations. Other measures were also taken. The 1847 reforms had included the requirement that six (rather than three, as previously) existing Fellows sign a candidate's proposal form. This constituted an extra layer of quality control. But it also made it harder for individuals outside the main disciplines and institutions to be nominated, and in 2001 the number of signatures was cut to two.

In 1960, the Society had 600 Fellows, of whom 13 (2.2 per cent) were women. In 2010, the Society had 1,356 Fellows. Of these, 72 (5.3 per cent) were women, but of the 220 Fellows elected in the five years to 2010, 9.5 per cent were women, suggesting that the overall female proportion would continue to grow. Since Fellows are elected for life, there is considerable inertia in such things. The 8,000th Fellow in the Society's history was elected in 1999.

Also in 1960, 46 per cent of all Fellows, and 53 per cent of UK-resident Fellows, were based in the 'golden triangle' of Oxford, Cambridge and London. By 2010, the proportion of UK-resident Fellows based in the golden triangle was virtually unchanged, at 52 per cent, though the proportion of the total Fellowship based in the golden triangle had dropped to 38 per cent, because of the greater number of Fellows now living outside the UK (see Chapter 10). Reducing the number of signatures required on a candidate's proposal form to two had not resulted in more candidates being elected from outside the golden triangle. Indeed, between 2004 and 2010, 63 per cent of newly elected Fellows resident in the UK were based in Oxford, Cambridge or London, possibly contrary to expectations. Whether these figures accurately represent the geographical distribution of scientific talent in the UK is a moot point. However, one of the challenges for the election process is to guard against any tendency for members of Sectional Committees (see below) inappropriately to favour candidates from their own institutions.

From 1917, the practice was introduced that, once nominated, a candidate for election to the Fellowship remained a candidate (unless elected, of course) for five years, after which the candidature could be immediately renewed. In 1963, this was changed to allow a candidature to run for seven years but, if unsuccessful, to require a three-year interval before a fresh nomination could be made. The intention was to constrain the total number of candidates, which had risen from eight per place to 12 per place over the previous decade, but the effect was modest and, as

Table A.4 *Candidates for election to the Fellowship*

Year	Places	Candidates	Candidates/place
1848	15	22	1.5
1900	15	90	6.0
1940	20	160	8.0
1970	32	360	11.3
1990	40	440	11.0
2010	44	645	14.7

Table A.4 shows, over the next fifty years there were typically ten to fifteen candidates for each place at any one time.[36]

Candidates are assessed by discipline-based Sectional Committees, which advise Council; Council then determines the final list on which the Fellowship votes at a formal meeting.[37] Discipline-based Sectional Committees were established in 1896 to provide expert scientific advice to Council about papers and medals.[38] There were initially six Sectional Committees. Their number and disciplinary coverage have since been subject to constant modification in response to the growth of science: one could, up to a point, track the evolution of science, or at least the Society's perception of it, by tracking the evolving pattern of Sectional Committees. In 1916 Council decided to expand the remit of the Sectional Committees by seeking their views on the suitability of candidates for election to the Fellowship. This role gradually increased in importance as the scale of the election process grew and Council needed more help: it is now the Committees' predominant function.

All Fellows, whether elected for mainstream research achievements or for wider contributions to science, are equal in the sense that there is only one category of (domestic) Fellowship: FRS. There have been intermittent proposals for other categories, none of which have prospered. For example, in 1936, R.A. Fisher and others proposed the creation of

[36] OM/59(62), C/141(62), C/92(72).
[37] The list always contains the same number of names as there are places. It is rare, but not completely unknown, for the formal election meeting to remove one of the names on the list.
[38] Joseph Lister, 'Anniversary Address', *Yearbook 1896–97*, 124. Sectional Committees had first been introduced in 1838, but two of them soon got caught up in accusations of partisan bias, and the Sectional Committees as a group were abolished: Marie Boas Hall, *All scientists now*, 68–70, 83–8, 126. On the inner working of Sectional Committees, see interviews with Patrick Bateson, Roger Elliott, John Enderby, John Kingman and David Smith.

Associates of the Royal Society. Their intention was to bring younger scientists into the Society's orbit. Council was sympathetic to the aim, but not to the proposed means, and hoped an increase in annual elections would do the trick instead. In 1962 Council rejected a further proposal for a category of associate membership, only for the idea to be floated again – partly for youth and partly to applaud achievement just below FRS standard – by Solly Zuckerman in 1963, by Rudolf Peierls and Nicholas Kurti in 1971, by John Ziman in 1976, and by Drummond Matthews in 1981.[39] When, from 2000 onwards, national academies in Germany, the Netherlands, Scotland, Sweden and elsewhere started setting up 'young academies', the Royal Society preferred to develop its informal network of URFs and other young scientists to fulfil the same function. It was important to avoid any chance of confusion with the species *Fellow of the Royal Society*.

One tangential route into the Fellowship has been a source of occasional controversy. The original 1663 Statutes allowed accelerated election for members of the Privy Council. This survived the 1847 reforms, such elections being deemed not to count towards the annual limit of fifteen. The privileged position of Privy Councillors was finally abolished in 1902, and replaced with a more wide-ranging Statute allowing the supernumerary election each year of one, or sometimes two, individuals who had 'rendered conspicuous service to the cause of science' or who were 'such that their election would be of signal benefit to the Society'. In 1916 Council, with minimal consultation, amended what was by then known as Statute 12 so as to restore the earlier position of Privy Councillors. This provoked David Bruce to organise a memorial asking for a rethink, which eventually secured the signatures of 241 Fellows – more than half the total Fellowship known to be in the UK at the time. The memorialists objected both to what they saw as the anachronistic deference shown to Privy Councillors[40] and to Council's high-handedness in pushing through the change. They won the argument, and the 1916 amendment was rescinded.

Between 1902 and 1945, thirty-one Fellows were elected under Statute 12. Between 1946 and 1982, a further twenty-three were elected. It was not always straightforward. Harold Wilson's nomination in 1969, for example, provoked significant disquiet among sectors of the Fellowship, essentially on political grounds, though there was no open revolt. But it was a different story in 1983. That year Council proposed two

[39] The idea has not gone away. See interviews with Sam Edwards and Brian Follett.
[40] 'It will be generally recognised', observed the memorialists, 'that the standing of Privy Councillors . . . has recently declined in the public estimation.'

individuals: the naturalist and broadcaster David Attenborough, whose candidature was entirely uncontroversial, and the Prime Minister Margaret Thatcher. Half the Prime Ministers since the end of the war had been elected under Statute 12 while still in office. But Mrs Thatcher was a divisive figure: some Fellows regarded her as the saviour of the nation, others the opposite. Her record on science was also ambivalent: she had studied chemistry and, up to a point, she championed basic research in Cabinet and protected the Science Budget, but she also imposed major cuts on university funding (see Chapters 3 and 5). Her candidature split the Fellowship, and the normally placid election meeting attracted an unprecedented turnout, marshalled by the key protagonists on each side.[41] In the end, she narrowly secured the required vote. The opposition had its revenge eighteen months later when Oxford University refused her an honorary degree.

The controversy provoked by Mrs Thatcher's election rumbled on for years. Over 70 per cent of the Fellowship responded to a request for comments in summer 1984 on whether and how to amend Statute 12. After several Special General Meetings and many Council debates, the upshot was to retain Statute 12 in something close to its original 1902 form, but to be explicit about avoiding elections that might be seen as support for a particular political party or as soliciting favours. The exceptional nature of what from 1995 was called Honorary Fellowship was underlined. The grouping together, from 2013, of General, Honorary and Royal candidates for the purposes of allocating places in the election process emphasised the Society's recognition of its need for worldly wisdom alongside scientific excellence.

[41] See interviews with Patrick Bateson, Arnold Burgen, Peter Cooper, Sam Edwards, Roger Elliott, Walter Hayman, Andrew Huxley, John Mason, Robin Nicholson, David Smith and Bill Stewart.

Sources

Archival sources

I have used material from the following archives, and am most grateful to the respective institutions and their staffs for help in locating items and permission to use them.

Bodleian Library, Oxford: C.D. Darlington papers, D.C. Phillips papers (which include many ABRC papers from his time as Chairman of ABRC)

British Association: Peter Briggs, *The BA at the end of the 20th century: a personal account of 22 years from 1980 to 2002*

Cambridge University Library: Gordon Sutherland papers (MS. Add.8353)

Churchill College, Cambridge: papers of Harold Hartley (prefaced HART); William Hawthorne (HATN) (by kind permission of the family); A.V. Hill (AVHL); Alex Todd (TODD)

ESRC/EPSRC Joint Information Services Unit: Council papers of the Science Research Council and Science and Engineering Research Council

HRH Duke of Edinburgh archive at Buckingham Palace: papers relating to the CEI

IET: John Coales papers (SC)

I Mech E: Christopher Hinton papers

Imperial War Museum: Henry Tizard papers

Royal Academy of Engineering: papers relating to the founding of the Fellowship of Engineering

The National Archives (TNA): series AB (UK Atomic Energy Authority), ADM (Admiralty), AT (Department of the Environment), BW (British Council), CAB (Cabinet Office), CO (Colonial Office), DSIR (Department of Scientific and Industrial Research), ED (Department of Education and Science), FD (Medical Research Council), FCO (Foreign and Commonwealth Office), FO (Foreign Office), OD (Overseas Development), POWE (Ministry of Power), PREM (Prime Minister's Office), T (Treasury), UGC (University Grants Committee)

Trinity College, Cambridge: Alan Hodgkin papers

All other archival references are to material held by the Royal Society. Some of this, identified by RMA ('Records Management Audit') numbers, is held off site as part of the Society's modern records and is catalogued at file level; the rest is held in the on-site archive and much of it is catalogued at item level. In particular, I have made extensive use of the following sources in the on-site archive:

> E.N. daC Andrade uncatalogued correspondence
> Patrick Blackett papers (references prefaced PB)
> George Lindor Brown papers (GLB)
> Henry Dale papers (HD)
> Alfred Egerton papers and diary (AE)
> Howard Florey papers (HF)
> Cyril Hinshelwood papers (CH)
> David Martin uncatalogued papers
> Robert Robinson papers (ROR)
> Tommy Thompson papers (HWT)
> Royal Society Council minutes (identified as CM plus the date of the meeting) and papers (identified as C/number(year)). Note that some of this material remains confidential at the time of writing
> Royal Society Officers' minutes (identified as OM plus the date of the meeting) and papers (identified as OM/number(year)). Note that some of this material remains confidential at the time of writing
> Modern Domestic Archives (MDA)
> Papers of various Royal Society committees, including those dealing with the Society's engagement with Alan Cottrell's engineering and technology committee (E&T), James Lighthill's industrial activities committee (IA), international relations (IR), policy on university funding (PUF), the Rothschild Inquiry (RD), scientific research in UK universities (STRUUK), Alex Fleck's technology committee (TC), and the Trend Inquiry (Trend)

Interviews

I conducted interviews with the following individuals, focused on their experiences of the Royal Society as an institution. Each interview was recorded and transcribed. I then edited the transcript for readability and invited the interviewee to amend it so as to ensure that it accurately reflected his or her view of the matter under discussion. The final transcripts are held at the Society's Centre for History of Science and may be consulted there.

> Sir Geoffrey Allen (FRS, 1976)
> Mr Chris Argent (Royal Society staff, 1961–91)
> Sir Eric Ash (FRS, 1977; Treasurer, 1997–2002)
> Sir John Ashworth (CSA, 1976–81)
> Sir Michael Atiyah (FRS, 1962; President, 1990–5)
> Sir Patrick Bateson (FRS, 1983; Biological Secretary, 1998–2003)
> Sir Colin Blakemore (FRS, 1992)
> Dr David Boak (Royal Society staff, 1999–2007)
> Sir Walter Bodmer (FRS, 1974)

Sir Arnold Burgen (FRS, 1964; Foreign Secretary, 1981–6)
Professor Lorna Casselton (FRS, 1999; Foreign Secretary, 2006–11)
Mr Peter Cooper (Royal Society staff, 1975–98)
Mr John Deverill (Royal Society staff, 1965–87)
Sir Sam Edwards (FRS, 1966)
Sir Roger Elliott (FRS, 1976; Physical Secretary, 1984–8)
Sir John Enderby (FRS, 1985; Physical Secretary, 1999–2004)
Sir Tony Epstein (FRS, 1979; Foreign Secretary, 1986–91)
Lord Flowers of Queen's Gate (FRS, 1961)
Sir Brian Follett (FRS, 1984; Biological Secretary, 1987–93)
Sir Hugh Ford (FRS, 1967)
Mr Terry Garrett (Royal Society staff, 1991–4)
Sir Francis Graham-Smith (FRS, 1970; Physical Secretary, 1988–94)
Professor Walter Hayman (FRS, 1956)
Sir Brian Heap (FRS, 1989; Foreign Secretary, 1996–2001)
Mr George Hemmen (Royal Society staff, 1955–85)
Dame Julia Higgins (FRS, 1995; Foreign Secretary, 2001–6)
Professor Robert Hinde (FRS, 1974)
Sir John Horlock (FRS, 1976; Treasurer, 1992–7)
HRH Duke of Edinburgh (FRS, 1951)
Sir Andrew Huxley (FRS, 1955; President, 1980–5)
Sir John Kingman (FRS, 1971)
Sir Aaron Klug (FRS, 1969; President, 1995–2000)
Lady Liebe Klug
Sir Ralph Kohn (FRS, 2006)
Sir Peter Lachmann (FRS, 1982; Biological Secretary, 1993–8)
Sir Bernard Lovell (FRS, 1955)
Sir John Mason (FRS, 1965; Treasurer, 1976–86)
Lord May of Oxford (FRS, 1979; CSA, 1995–2000; President, 2000–5)
Professor Noreen Murray (FRS, 1982)
Sir Robin Nicholson (FRS, 1978; CSA, 1981–5)
Lady Stella Porter
Professor Ken Pounds (FRS, 1981)
Sir David Read (FRS, 1990; Biological Secretary, 2003–8)
Lord Rees of Ludlow (FRS, 1979; President, 2005–10)
Mr Dominic Reid (Royal Society staff, 2007–10)
Sir John Rowlinson (FRS, 1970; Physical Secretary, 1994–9)
Lord Sainsbury of Turville (FRS, 2008; Science Minister, 1998–2006)
Sir David Smith (FRS, 1975; Biological Secretary, 1983–7)
Dr Fiona Steele (Fellowship of Engineering staff)
Sir William Stewart (FRS, 1977; CSA, 1990–5)
Sir Michael Stoker (FRS, 1968; Foreign Secretary, 1976–81)
Sir Peter Swinnerton-Dyer (FRS, 1967)
Sir Martin Taylor (FRS, 1996; Physical Secretary, 2004–9)
Sir John Meurig Thomas (FRS, 1977)
Lord Waldegrave of North Hill (Minister responsible for science, 1992–4)
Sir David Wallace (FRS, 1986; Treasurer, 2002–7)

Dr Peter Warren (Royal Society staff, 1977–97)
Ms Helen Williams (Department of Education and Science staff)

Published sources

ABRC, *Allocations of the Science Budget 1989–92* (13 December 1988, published February 1989).

ABRC, *The Science Budget: a forward look 1982* (October 1982).

ABRC, *Science Budget: allocations 1987–88* (3 December 1986, published February 1987).

ABRC, *A strategy for the Science Base* (May 1987).

ABRC, *A study of commissioned research* (Mason report, 1983).

ABRC, *Support of university scientific research* (the Merrison report: ABRC/UGC, 1982).

Academy of Medical Sciences, British Academy, Royal Academy of Engineering and Royal Society, *Fuelling prosperity* (April 2013).

Academy of Medical Sciences, Medical Research Council, Royal Society and Wellcome Trust, *The use of non-human primates in research* (December 2006).

Adam, David, 'British science champion quits post', *Nature*, 417 (6 June 2002), 577.

Adam, David, 'Suspicions intensify over elusive European Academy of Sciences', *Nature*, 419 (31 October 2002), 865.

Adrian, E.D. 'Address at the Anniversary Meeting, 30 November 1954', *Proceedings of the Royal Society of London. Series A, Mathematical and physical sciences*, 227 (1955), 279–87.

Adrian, E.D., 'Address at the Anniversary Meeting, 30 November 1955', *Proceedings of the Royal Society of London. Series A, Mathematical and physical sciences*, 234 (1956), 151–60.

Agar, Jon, '"It's springtime for science": renewing China-UK scientific relations in the 1970s', *Notes and records of the Royal Society of London*, 67 (2013), 7–24.

Agar, Jon, 'Thatcher, scientist', *Notes and records of the Royal Society of London*, 65 (2011), 215–32.

Agar, Jon, 'What difference did computers make?', *Social studies of science*, 36 (2006), 869.

Agar, Jon, 'What happened in the sixties?', *BJHS*, 41 (2008), 567–600.

Alter, Peter, 'The Royal Society and the International Association of Academies, 1897–1919', *Notes and records of the Royal Society of London*, 34 (1980), 241–64.

Arnott, Struther, Kibble, T.W.B. and Shallice, Tim, 'Maurice Hugh Frederick Wilkins', *Biographical memoirs of Fellows of the Royal Society*, 52 (2006), 455–78.

Atiyah, Michael, 'Address at the Anniversary Meeting, 29 November 1991', *Notes and records of the Royal Society of London*, 46 (1992), 155–69.

Atiyah, Michael, 'Address at the Anniversary Meeting, 30 November 1992', *Notes and records of the Royal Society of London*, 47 (1993), 109–18.

Balmer, Brian, Godwin, Matthew and Gregory, Jane, 'The Royal Society and the "brain drain": natural scientists meet social science', *Notes and records of the Royal Society of London*, 63 (2009), 339–53.

Bateson, Patrick, 'Genetically modified potatoes', *Lancet*, 354 (16 October 1999), 1382.

Beardsley, Tom, 'Human embryo experiments: societies urge a softer line', *Nature*, 302 (28 April 1983), 739.

Bevington, John C. and Gowenlock, Brian G., 'Sir Harry Work Melville', *Biographical memoirs of Fellows of the Royal Society*, 48 (2002), 291–308.

Blackett, P.M.S., 'Address at the Anniversary Meeting, 30 November 1966', *Proceedings of the Royal Society of London. Series A, Mathematical and physical sciences*, 296 (1967), v–xiv.

Blackett, P.M.S., 'Address at the Anniversary Meeting, 30 November 1968', *Proceedings of the Royal Society of London. Series A, Mathematical and physical sciences*, 308 (1968), v–xvii.

Blackett, P.M.S., 'Address at the Anniversary Meeting, 30 November 1970', *Proceedings of the Royal Society of London. Series A, Mathematical and physical sciences*, 321 (1971), 1–14.

Blackett, P.M.S., 'Wanted: a wand over Whitehall', *New Statesman* (11 September 1964), 346–50.

Blume, Stuart S., *Toward a political sociology of science* (Collier Macmillan Publishers, 1974).

Bodmer, Walter, 'Public understanding of science: the BA, the Royal Society and COPUS', *Notes and records of the Royal Society of London*, 64 (September 2010), S151–61.

Booth, Clive, 'A war of independence', *Times higher education supplement* (9 December 1983), 16.

Bowater, Laura and Yeoman, Kay, *Science communication* (Wiley-Blackwell, 2013).

Bown, William, 'A dissident view on life, the universe and democracy', *New scientist* (21 July 1990), 19.

Brimelow, Thomas, 'Anniversary dinner 1974', *Notes and records of the Royal Society of London*, 30 (1975), 5–14.

Brock, George, 'Freedom, but what else can we do for the Falklands?', *Times* (16 June 1982), 12.

Brumfiel, Geoff, 'Nanotechnology: a little knowledge … ', *Nature*, 424 (17 July 2003), 246–8.

Brumfiel, Geoff, 'UK scientists celebrate budget reprieve', *Nature*, 467 (27 October 2010), 1017.

Bryson, Bill, ed., *Seeing further: the story of science and the Royal Society* (Harper Press, 2010).

Bud, Robert, 'From applied microbiology to biotechnology: science, medicine and industrial renewal', *Notes and records of the Royal Society of London*, 64 (2010), S17–29.

Bud, Robert, *The uses of life: a history of biotechnology* (Cambridge University Press, 1993).

Burgen, Arnold, 'Academia Europaea: origin and early days', *European review*, 17 (2009), 469–75.

Burhop, Eric, 'The problem of Soviet scientists: a reply to John Ziman', *Nature*, 248 (12 April 1974), 542.

Butler, Declan, 'BSE researchers bemoan "ministry secrecy"', *Nature*, 383 (10 October 1996), 467–8.

Butler, Declan and Wadman, Meredith, 'Calls for cloning ban sell science short', *Nature*, 386 (6 March 1997), 8–9.

Buttimer, Anne, 'Academia Europaea: founders and founding visions', *European review*, 19 (2011), 153–253.

Cabinet Office, *Annual review of government funded R&D 1983* (HMSO, 1984).

Campbell, John, *Margaret Thatcher: the iron lady* (Jonathan Cape, 2003).

Campbell, Robert, 'Introduction', in Robert Campbell, et al., eds., *Academic and professional publishing* (Chandos Publishing, 2012), 1–14.

Cao, Cong, 'The Chinese Academy of Sciences: the election of scientists into the elite group', *Minerva*, 36 (1998), 323–46.

Clark, Ronald W., *Tizard* (Methuen, 1965).

Cockcroft, John, 'Scientific collaboration in Europe', *New scientist*, 24 January 1963, 170–2.

Collinge, John, 'Lessons of kuru research: background to recent studies with some personal reflections', *Philosophical transactions of the Royal Society of London. Series B, Biological sciences*, 363 (2008), 3689–96.

Collins, Peter, 'The British Association as public apologist for science, 1919–1946', in Roy MacLeod and Peter Collins, eds., *The parliament of science* (Science Reviews Ltd, 1981).

Collins, Peter, 'Editorial', *Notes and records of the Royal Society of London*, 64 (2010), S1–3.

Collins, Peter, 'Presidential politics: the controversial election of 1945', *Notes and records of the Royal Society of London*, 65 (2011), 325–42.

Collins, Peter, 'Realising our potential', *Science and public affairs* (Summer 1993), 5–7.

Collins, Peter, 'A role in running UK science?', *Notes and records of the Royal Society of London*, 64 (2010), S119–30.

Collins, Peter, 'A Royal Society for technology', *Notes and records of the Royal Society of London*, 64 (2010), S43–54.

Collins, P.M.D. and Bodmer, W.F., 'The public understanding of science', *Studies in Science Education*, 13 (1986), 96–104.

Collins, P.M.D., Hicks, D.M. and Wyatt, S., *Evaluation of national performance in basic research* (ABRC Science Policy Studies No1, 1986).

Collins, P.M.D., et al., 'Flows of researchers to and from the UK', *Nature*, 328 (1987), 27–8.

Connor, Steve, 'Science and sanctions', *Nature*, 320 (21 August 1986), 19–20.

COPUS, *To know science is to love it?* (Royal Society, 1998).

Cottrell, Alan, 'Edward Neville da Costa Andrade, 1887–1971', *Biographical memoirs of Fellows of the Royal Society*, 18 (1972), 1–20.

Cottrell, Alan, 'The rise and fall of science policy', *New scientist* (14 October 1976), 80–2.

Cozzens, Susan E., 'The discovery of growth: statistical glimpses of twentieth science', in John Krige and Dominique Pestre, eds., *Science in the twentieth century* (Harwood Academic Publishers, 1997), 127–42.

Crick, F.H.C. and Kendrew, J.C., et al., 'International conferences', *Nature*, 224 (4 October 1969), 93–4.

Crowther, J.G., 'The Royal Society', *New statesman and nation* (2 December 1944), 375.

Dale, Henry, 'Address at the Anniversary Meeting, 1 December 1941', *Proceedings of the Royal Society of London. Series A, Mathematical and physical sciences*, 179 (1942), 233–60.

Danckwerts, P.V., 'Science versus technology: the battle for brains', *Nature*, 200 (19 October 1963), 219–20.

Darmon, Gérard, 'European Science Foundation: towards a history', in John Krige and Luca Guzzetti, eds., *History of European scientific and technological cooperation* (European Communities, 1997), 381–402.

Darwin, C.G., 'The "reading" of papers at meetings of the Royal Society', *Notes and records of the Royal Society of London*, 2 (1939), 25–7.

Deverill, John J.P., 'Scientific exchange with the USSR under the agreement between the Royal Society and the USSR Academy of Sciences', in Craig Sinclair, ed., *The status of Soviet civil science* (Martinus Nijhoff Publishers, 1987), 255–77.

Dickson, David, 'The Leonardo project', *Times higher education supplement* (28 September 1984).

Dickson, David, 'US cuts back on official exchanges with USSR', *Nature*, 283 (7 February 1980), 513.

Doak Barnett, A., 'Exchanges in the process of "normalisation": an academic view', in Anne Keatley, ed., *Reflections on scholarly exchanges with the People's Republic of China, 1972–1976* (Committee on Scholarly Communication with the People's Republic of China, 1978), 45–50.

Doel, Ronald E., 'Scientists as policymakers, advisors, and intelligence agents: linking contemporary diplomatic history with the history of contemporary science', in Thomas Söderqvist, ed., *The historiography of contemporary science and technology* (Harwood Academic Publishers, 1997).

Edgerton, David, 'Science in the United Kingdom', in John Krige and Dominique Pestre, eds., *Science in the twentieth century* (Harwood Academic Publishers, 1997), 759–76.

Edgerton, David, 'The *White Heat* revisited: the British government and technology in the 1960s', *Twentieth century British history*, 7 (1996), 53–82.

European Science Foundation, *Voices of European science* (ESF, 1994).

Ewen, Stanley, 'Health risks of genetically modified foods', *Lancet*, 354 (21 August 1999), 684.

Fan, Fa-ti. 'Redrawing the map: science in twentieth-century China', *Isis*, 98 (2007), 524–38.

Feldberg, W.S., 'Henry Hallett Dale, 1875–1968', *Biographical memoirs of Fellows of the Royal Society*, 16 (1970), 76–174.

Ferry, Georgina, 'The exception and the rule: women and the Royal Society 1945–2010', *Notes and records of the Royal Society of London*, 64 (2010), S163–72.

Ferry, Georgina, 'Fifty years of EMBO', *Nature*, 511 (10 July 2014), 150–1.

Fishlock, David, 'A trickle not a flood', *Financial Times* (30 June 1987), 1 and 21.

Flather, Paul, 'The missing generation', *Times Higher Education Supplement* (24 September 1982), 8–9.

Florey, Howard, 'Address at the Anniversary Meeting, 30 November 1961', *Proceedings of the Royal Society of London. Series B, Biological sciences*, 155 (1962), 307–20.

Florey, Howard, 'Address at the Anniversary Meeting, 30 November 1963', *Proceedings of the Royal Society of London. Series B, Biological sciences*, 159 (1964), 393–404.

Florey, Howard, 'Address at the Anniversary Meeting, 30 November 1964', *Proceedings of the Royal Society of London. Series B, Biological sciences*, 161 (1965), 439–52.

Florey, Howard, 'Address at the Anniversary Meeting, 30 November 1965', *Proceedings of the Royal Society of London. Series B, Biological sciences*, 163 (1966), 425–34.

Flowers, Brian, *The ESF – the first twenty years!* (ESF, 20 November 1984).

Flynn, Laurie and Gillard, Michael, 'Pro-GM food scientist "threatened editor"', *Guardian* (1 November 1999).

Fogg, G.E., 'The Royal Society and the Antarctic', *Notes and records of the Royal Society of London*, 54 (2000), 85–98.

Fogg, G.E., 'The Royal Society and the South Seas', *Notes and records of the Royal Society of London*, 55 (2001), 81–103.

Foot, Rosemary, *Trading with the enemy: the USA and the China trade embargo* (Oxford University Press, 1997).

Gardner, John, 'The Gang of Four and Chinese science', *Bulletin of the atomic scientists*, 33 (September 1977), 24–30.

Gay, Hannah, *The Silwood circle* (Imperial College Press, 2013).

Godwin, Matthew, *The Skylark rocket: British space science and the European Space Research Organisation 1957–1972* (Beauchesne, 2007).

Godwin, Matthew, Bulmer, Jane and Balmer, Brian, 'The anatomy of the brain drain debate, 1950–1970s: witness seminar', *Contemporary British history*, 23 (2009), 35–60.

Goodare, Jennifer, *'Representing science in a divided world: the Royal Society and Cold War Britain'* (PhD thesis, University of Manchester, 2013).

Graham-Smith, Francis ed., *Population – the complex reality* (Royal Society, 1994).

Graham-Smith, Francis and Lovell, Bernard, 'Diversions of a radio telescope', *Notes and records of the Royal Society of London*, 62 (2008), 197–204.

Greenaway, Frank, *Science international: a history of the International Council of Scientific Unions* (CUP/ICSU, 1996).

Greenwood, J.W., 'The scientist-diplomat: a new hybrid role in foreign affairs', *Science forum*, 19 (1971), 14–8.

Gregory, Jane and Lock, Simon, 'The evolution of "public understanding of science": public engagement as a tool of science policy in the UK', *Sociology compass*, 2 (2008), 1252–65.

Gummett, Philip, *Scientists in Whitehall* (Manchester University Press, 1980).

Gummett, Philip J. and Price, Geoffrey L., 'An approach to the central planning of British science: the formation of the Advisory Council on Scientific Policy', *Minerva*, 15 (1977), 119–43.

Gummett, Philip and Williams, Roger, 'Assessing the Council for Scientific Policy', *Nature*, 240 (8 December 1972), 329–32.

Haberer, Joseph, 'Politicisation in science', *Science*, 178 (17 November 1972), 713–24.

Hailsham, Lord, *The door wherein I went* (Collins, 1975).

Hall, Marie Boas, *All scientists now: the Royal Society in the nineteenth century* (Cambridge University Press, 1984).

Harris, Henry, 'Howard Florey and the development of penicillin', *Notes and records of the Royal Society of London*, 53 (1999), 243–252.

Hartley, Harold, ed., *The tercentenary celebrations of the Royal Society of London* (Royal Society, 1961).

Hemmen, George, 'Royal Society expeditions in the second half of the twentieth century', *Notes and records of the Royal Society of London*, 64 (2010), S89–99.

Hennessy, Peter, Morrison, Susan and Townsend, Richard, *Routine punctuated by orgies: the Central Policy Review Staff, 1970–83* (Strathclyde papers on government and politics No 31, 1984).

Higgins, Julia, 'The Royal Society in Africa', unpublished paper given at the conference 'The Royal Society and science in the 20th century', 22–23 April 2010.

Hill, A.V., 'Age of election to the Royal Society', *Notes and records of the Royal Society of London*, 16 (1961), 151–3.

Hill, A.V., 'Cancelled visit of British men of science to the Academy of Science of the USSR', *Nature*, 155 (1945), 753.

Hill, A.V., *Memories and reflections* (unpublished, 1971/72; copies at Royal Society and Churchill College, Cambridge).

Hill, A.V., 'The needs of special subjects in the balanced development of science in the United Kingdom', *Notes and records of the Royal Society of London*, 4 (1946), 133–9.

Hinde, R.A. and Finney, J.L., 'Joseph Rotblat, 1908–2005', *Biographical memoirs of Fellows of the Royal Society*, 53 (2007), 309–26.

Hinshelwood, Cyril, 'The Tercentenary Address at the formal opening ceremony, 19 July 1960', *Notes and records of the Royal Society of London*, 16 (1961), 13–24; also at *Nature*, 187 (23 July 1960), 274–8.

Hinshelwood, Cyril, 'A visit to China', *New scientist*, 6 (5 November 1959), 858–60.

Hodgkin, Alan, 'Address at the Anniversary Meeting, 30 November 1971', *Proceedings of the Royal Society of London. Series B, Biological sciences*, 180 (1972), v–xx.

Hodgkin, Alan, 'Address at the Anniversary Meeting, 30 November 1972', *Proceedings of the Royal Society of London. Series B, Biological sciences*, 183 (1973), 1–19.

Hodgkin, Alan, 'Address at the Anniversary Meeting, 30 November 1973', *Proceedings of the Royal Society of London. Series B, Biological sciences*, 185 (1974), v–xx.

Hodgkin, Alan, 'Address at the Anniversary Meeting, 30 November 1974', *Proceedings of the Royal Society of London. Series B, Biological sciences*, 188 (1975), 103–19.

Hodgkin, Alan, 'Address at the Anniversary Meeting, 1 December 1975', *Proceedings of the Royal Society of London. Series B, Biological sciences*, 192 (1976), 371–91.

Hodgkin, Alan, *Chance and design: reminiscences of science in peace and war* (Cambridge University Press, 1992).

Hodgkin, Alan, 'Edgar Douglas Adrian, Baron Adrian of Cambridge', *Biographical memoirs of Fellows of the Royal Society*, 25 (1979), 1–73.

Home, R.W., 'The Royal Society and the Empire: the colonial and Commonwealth Fellowship. Part I: 1731–1847', *Notes and records of the Royal Society of London*, 56 (2002), 307–32.

Home, R.W., 'The Royal Society and the Empire: the colonial and Commonwealth Fellowship. Part II: after 1847', *Notes and records of the Royal Society of London*, 57 (2003), 47–84.

Horlock, John, *An open book* (The Memoir Club, 2006).

Horrocks, Sally M.,'The Royal Society, its Fellows and industrial R&D in the mid-century', *Notes and records of the Royal Society of London*, 64 (2010), S31–41.

Horton, Richard, 'Health risks of genetically modified foods', *Lancet*, 353 (29 May 1999), 1811.

House of Commons Select Committee on Science and Technology, *Government funding of the learned scientific societies* (fifth report of session 2001–2, HC 774-1 and HC 774-II).

House of Lords Select Committee on Science and Technology, *Science and government* (1981).

Huggins, William, *The Royal Society, or science in the state and in the school* (Methuen, 1906).

Hughes, Jeff, '"Divine right" or democracy? The Royal Society "revolt" of 1935', *Notes and records of the Royal Society of London*, 64 (2010), S101–17.

Hughes, Jeff, 'Introductory comments to final discussion session', *Notes and records of the Royal Society of London*, 64 (2010), S173–6.

Huxley, Andrew, 'Address at the Anniversary Meeting, 30 November 1981', *Proceedings of the Royal Society of London. Series B, Biological sciences*, 214 (1982), 137–52.

Huxley, Andrew, 'Address at the Anniversary Meeting, 30 November 1983', *Proceedings of the Royal Society of London. Series B, Biological sciences*, 220 (1984), 383–98.

Huxley, Andrew, 'Address at the Anniversary Meeting, 30 November 1984', *Proceedings of the Royal Society of London. Series B, Biological sciences*, 223 (1985), 403–16.

Huxley, Andrew, 'Address at the Anniversary Meeting, 30 November 1985', *Supplement to Royal Society News*, (December 1985), i–vii.

Huxley, Andrew, '*Nullius in verba*', *Nature*, 315 (23 May 1985), 272.

IAP, *Statements 1993–2008* (IAP, 2008).

ICSU, *ICSU and climate change: 1962–2006 and beyond* (ICSU, 2006).

ICSU, *ICSU and polar research: 1957–2007 and beyond* (ICSU, 2006).

ICSU, *ICSU and the universality of science: 1957–2006 and beyond* (ICSU, 2006).

Irvine, John, Martin, Ben, et al., 'Charting the decline in British science', *Nature*, 316 (15 August 1985), 587–90.

Irwin, Alan and Wynne, Brian, eds., *Misunderstanding science?* (Cambridge Unviersity Preess, 1996).

James, Frank and Quirke, Viviane, '*L'affaire Andrade*, or how not to modernise a traditional institution', in Frank A.J.L. James, *The common purposes of life: science and society at the Royal Institution of Great Britain* (Ashgate, 2002), 273–304.

James, Simon, 'The Central Policy Review Staff 1970–1983', *Political studies*, 34 (1986), 423–40.

Jellicoe, George, 'Lord Edward Arthur Alexander Shackleton', *Biographical memoirs of Fellows of the Royal Society*, 45 (1961), 486–505.

Jones, Richard, 'Introduction', in David Bennett and Richard Jennings, eds., *Successful science communication* (Cambridge University Press, 2011).

Jones, R.V. and Farren, W.S., 'Henry Thomas Tizard', *Biographical memoirs of Fellows of the Royal Society*, 7 (1961), 313–48.

Joravsky, David, *The Lysenko affair* (Harvard University Press, 1970).

Jubb, Michael, 'The scholarly ecosystem', in Robert Campbell, et al., eds., *Academic and professional publishing* (Chandos Publishing, 2012), 53–77.

Kaufman, Victor S., *Confronting communism: US and British policies towards China* (University of Missouri Press, 2001).

Kenward, Michael, 'Anglo-Soviet scientific exchanges frozen', *New scientist* (28 February 1980), 637.

Klug, Aaron, 'Address at the Anniversary Meeting, 29 November 1996', *Notes and records of the Royal Society of London*, 51 (1997), 121–31.

Klug, Aaron, 'Address at the Anniversary Meeting, 1 December 1997', *Notes and records of the Royal Society of London*, 52 (1998), 181–90.

Klug, Aaron, 'Address at the Anniversary Meeting, 30 November 1998', *Notes and records of the Royal Society of London*, 53 (1999), 157–67.

Klug, Aaron, 'Address at the Anniversary Meeting, 30 November 1999', *Notes and records of the Royal Society of London*, 54 (2000), 99–108.

Klug, Aaron, 'Address at the Anniversary Meeting, 30 November 2000', *Notes and records of the Royal Society of London*, 55 (2001), 165–77.

Krige, John, *American hegemony and the postwar reconstruction of science in Europe* (MIT Press, 2006).

Krige, John, 'The birth of EMBO and the difficult road to EMBL', *Studies in history and philosophy of biological and biomedical sciences*, 33 (2002), 547–64.

Krige, John, 'The Ford Foundation, European physics and the Cold War', *Historical studies in the physical and biological sciences*, 29 (1999), 333–61.

Krige, John, 'The politics of European scientific collaboration', in John Krige and Dominique Pestre, eds., *Science in the twentieth century* (Harwood Academic Publishers, 1997), 897–918.

Krige, John and Barth, Kai-Henrik, 'Introduction: science, technology and international affairs', *Osiris*, 21 (2006), 1–21.

Kubbinga, Henk, 'European Physical Society (1968–2008): the early years', *Europhysics news*, 39 (2008), 16–8.

Lachmann, Peter, *First steps: a personal account of the formation of the Academy of Medical Sciences* (Academy of Medical Sciences, 2010).

Layton, David, *Interpreters of science: a history of the Association for Science Education* (John Murray/ASE, 1984).

Lock, Simon, 'Deficits and dialogues: science communication and the public understanding of science in the UK', in David Bennett and Richard Jennings, eds., *Successful science communication* (Cambridge University Press, 2011).

Loder, Natasha, 'Royal Society: GM food hazard claim is "flawed"', *Nature*, 399 (20 May 1999), 188.

Lovell, Bernard, 'PMS Blackett', *Biographical memoirs of Fellows of the Royal Society*, 21 (1975), 1–115.

Lovell, Bernard, 'The Royal Society, the Royal Greenwich Observatory and the Astronomer Royal', *Notes and records of the Royal Society of London*, 48 (1994), 283–97.

Lyons, Henry, *The record of the Royal Society of London* (Royal Society, 1940).

Lyons, Henry, *The Royal Society 1660–1940. A history of its administration under its charters* (Cambridge University Press, 1944).

Macilwain, Colin, 'In the best company', *Nature*, 465 (24 June 2010), 1002–4.

MacLeod, Roy M., 'The Royal Society and the government grant: notes on the administration of scientific research, 1849–1914', *Historical Journal*, 14 (1971), 323–58.

MacLeod, Roy, 'The Royal Society and the Commonwealth: old friendships, new frontiers', *Notes and records of the Royal Society of London*, 64 (2010), S137–49.

Martin, Ben, Irvine, John, et al., 'The continuing decline of British science', *Nature*, 330 (12 November 1987), 123–6.

Martin, David, 'The Royal Society today', *Discovery*, 21 (7 July 1960), 292–302.

Masood, Ehsan, 'Cloning technique reveals legal loophole', *Nature*, 385 (27 February 1997), 757.

Massey, Harrie and Robins, M.O., *History of British space science* (Cambridge University Press, 1986).

Massey, Harrie and Thompson, Harold, 'David Christie Martin', *Biographical memoirs of Fellows of the Royal Society*, 24 (1978), 391–407.

Matthews, Robert, 'Now scientists must bridge the credibility gap', *Daily Telegraph* (26 October 2000).

May, Robert, 'Address at the Anniversary Meeting, 30 November 2001', *Notes and records of the Royal Society of London*, 56 (2002), 121–9.

May, Robert, 'Address at the Anniversary Meeting, 29 November 2002', *Notes and records of the Royal Society of London*, 57 (2003), 117–32.

May, Robert, 'Address at the Anniversary Meeting, 30 November 2004: Global problems and global science', *Notes and records of the Royal Society of London*, 59 (2005), 99–119.

May, Robert, 'Address at the Anniversary Meeting, 30 November 2005: threats to tomorrow's world', *Notes and records of the Royal Society of London*, 60 (2006), 109–30.

May, Robert, 'Under-informed, over here', *Guardian* (27 January 2005), 10.

May, Robert, *The use of scientific advice in policy making* (Office of Science and Technology, 1997).

McDonald, Alan, 'Scientific cooperation as a bridge across the Cold War divide: the case of the International Institute for Applied Systems Analysis', *Annals of the New York Academy of Sciences*, 866 (1998), 55–83.

McGucken, William, 'The Royal Society and the genesis of the Scientific Advisory Committee to Britain's War Cabinet, 1939–1940', *Notes and records of the Royal Society of London*, 33 (1978), 87–115.

McGucken, William, *Scientists, society and the state: the social relations of science movement in Great Britain, 1931–1947* (Ohio State University Press, 1984).

Melville, Harry, *The Department of Scientific and Industrial Research* (George Allen and Unwin, 1962).

Miller, David Philip, 'The usefulness of natural philosophy: the Royal Society and the culture of practical utility in the later eighteenth century', *British Journal for the History of Science*, 32 (1999), 185–201.

Morange, Michel, 'EMBO and EMBL', in John Krige and Luca Guzzetti, eds., *History of European scientific and technological cooperation* (European Communities, 1997), 77–92.

Morgan, Rose, *The genome revolution* (Greenwood Press, 2006).

Moszynski, Peter, 'Royal Society warns of risks from depleted uranium', *BMJ*, 326 (3 May 2003), 952.

National Academy of Engineering, *The National Academy of Engineering: the first ten years* (NAE, 1976).

van Noorden, Richard, 'Royal Society sets out case for investment in research', *Nature*, 464 (9 March 2010), 155.

Ogston, A.G., 'Harold Brewer Hartley', *Biographical memoirs of Fellows of the Royal Society*, 19 (1973), 348–73.

Olby, R.C., Cantor, G.N., Christie, J.R.R. and Hodge, M.J.S., *Companion to the history of modern science* (Routledge, 1990).

Oldham, Geoffrey, 'Chinese science and the Cultural Revolution', *Technology review*, 71 (October 1968), 23–9.

Oldroyd, David, *Earth, water, ice and fire* (Geological Society memoir 25, 2002).

Oriel, John, 'Too many learned societies?', *New scientist* (7 April 1960), 854–6.

van Oudenaren, John, *Détente in Europe: the Soviet Union and the West since 1953* (Duke University Press, 1969).

Palca, Joseph, 'South African exclusion causes academic schism', *Nature*, 319 (13 February 1986), 524.

Peyton, John, *Solly Zuckerman: a scientist out of the ordinary* (John Murray, 2001).

HRH The Prince Philip, 'Promoting engineering', *Ingenia*, 41 (December 2009), 12–6.

HRH The Prince Philip, 'Research and prediction. The inaugural Hartley Lecture, 21 May 1974', *Notes and records of the Royal Society of London*, 29 (1974), 11–27.

Phillips, David, 'A strategy for science in the UK', *Science and public policy*, 15 (1988), 3–12.

Phillips, Lord, et al., 'Lessons from the BSE Inquiry', *FST journal*, 17 (July 2001), 3–8.

Porter, George, 'Address at the Anniversary Meeting on 1 December 1986', *Science and public affairs*, 2 (1987), 3–10.

Porter, George, 'Address at the Anniversary Meeting on 30 November 1987', *Science and public affairs*, 3 (1988), 3–12.

Porter, George, 'Address at the Anniversary Meeting on 30 November 1990', *Science and public affairs*, 6 (1991), 3–14.

Pounds, Ken, 'The Royal Society's formative role in UK space research', *Notes and records of the Royal Society of London*, 64 (2010), S65–76.

van Praagh, G., 'Technology and the sixth-form boy', *Nature*, 199 (7 September 1963), 958.

Price, Derek de Solla, *Little science, big science* (Columbia University Press, 1963).

Price, Derek de Solla, 'The world network of scientific attachés', *Science forum*, 21 (1971), 34–5.

Raiffa, Howard, *History of IIASA*, talk given at IIASA, 23 September 1992 (transcript on IIASA website).

Rannestad, Andreas, *NATO and science: an account of the NATO Science Committee 1958–1972* (NATO Scientific Affairs Division, 1973).

Rees, Martin, 'Address at the Anniversary Meeting, 30 November 2006', *Notes and records of the Royal Society of London*, 61 (2007), 75–83.

Rees, Martin, 'Address at the Anniversary Meeting, 30 November 2010', *Notes and records of the Royal Society of London*, 65 (2011), 197–205.

Richardson, John, 'Exchanges in the process of "normalisation": US government perspective', in Anne Keatley, ed., *Reflections on scholarly exchanges with the People's Republic of China, 1972–1976* (Committee on Scholarly Communication with the People's Republic of China, 1978), 43–5.

Robinson, Robert, 'Address at the Anniversary Meeting, 30 November 1946', *Proceedings of the Royal Society of London. Series A, Mathematical and physical sciences*, 188 (1947), 143–60.

Robinson, Robert, 'Address at the Anniversary Meeting, 1 December 1947', *Proceedings of the Royal Society of London. Series A, Mathematical and physical sciences*, 192 (1947), v–xix.

Rogers-Hayden, Tee and Pidgeon, Nick, 'Moving engagement "upstream"? Nanotechnologies and the Royal Society and Royal Academy of Engineering's inquiry', *Public Understanding of Science*, 16 (2007), 345–64.

Rose, Hilary and Rose, Steven, 'Knowledge and power', *New scientist*, 42 (17 April 1969), 108–9.

Rowlinson, John S. and Robinson, Norman H., *The record of the Royal Society of London: supplement to the fourth edition for the years 1940–1989* (Royal Society, 1992).

Royal Society, *Academies of sciences in the constituent republics of the former Soviet Union: a current appraisal* (January 1992).

Royal Society, *Annual report* (bound into the *Yearbook* up to 1979, afterwards printed separately).

Royal Society, *BSE and CJD – the facts to date* (2 April 1996).

Royal Society, *Climate change controversies: a simple guide* (June 2007, updated December 2008).

Royal Society, *Climate change: what we know and what we need to know* (August 2002).

Royal Society, *Corporate plan: a strategy for the Royal Society 1986–1996* (February 1986).

Royal Society, *Corporate update and PES submission* (June 1995).

Royal Society, *Demographic trends and future university candidates* (April 1983).

Royal Society, *Disposal of radioactive wastes in deep repositories* (November 1994).

Royal Society, *The encouragement of scientific research in the United Kingdom* (June 1960, published 1961).

Royal Society, *Evidence to the Committee of Enquiry into the Organisation of Civil Science (Trend Committee)* (submitted 1963).

Royal Society, *The future of the Science Base* (September 1992).

Royal Society, *Genetically modified plants for food use* (September 1998).

Royal Society, *Genetically modified plants for food use and human health – an update* (February 2002).

Royal Society, *Grant-giving to universities for scientific research* (March 1961).

Royal Society, *The greenhouse effect: the scientific basis for policy* (July 1989).

Royal Society, *Guide to facts and fictions about climate change* (March 2005).

Royal Society, *The health hazards of depleted uranium munitions* (Part I, May 2001; Part II, March 2002).

Royal Society, *Infectious diseases in livestock* (July 2002).

Royal Society, *Into the new millennium: a corporate plan for 1997–2002* (May 1997).

Royal Society, *Knowledge, networks and nations: global scientific collaboration in the 21st century* (March 2011).

Royal Society, *Memorandum by the Council on the consultative document (Cmnd 4814) 'A framework for government research and development'* (February 1972).

Royal Society, *The needs of research in fundamental science after the war* (printed for private circulation, January 1945).

Royal Society, *The public understanding of science* (September 1985).

Royal Society, *Report of the ad hoc Biological Research Committee* (November 1961).

Royal Society, *Reaping the benefits* (October 2009).

Royal Society, *Review of data on possible toxicity of GM potatoes* (June 1999).

Royal Society, *The Royal Society Commonwealth Bursaries Scheme, 1954 to 1978* (1979).

Royal Society, *The Royal Society – the next 10 years* (March 1990).

Royal Society, *Science education 11–18 in England and Wales* (November 1982).

Royal Society, *Science in society: report* (July 2004).

Royal Society, *Science in society: the impact and legacy of the five year Kohn Foundation funded programme* (September 2006).

Royal Society, *The scientific century: securing our future prosperity* (March 2010).

Royal Society, *Submission to the Science and Technology Committee's inquiry on women in STEM careers* (September 2013).

Royal Society, *Support of geophysics in the United Kingdom* (June 1985).

Royal Society, *Vision for science and mathematics education* (June 2014).

Royal Society, *Yearbook* (published annually, starting 1897).

Royal Society and American Association for the Advancement of Science, *New frontiers in science diplomacy* (January 2010).

Royal Society and National Academy of Sciences, *Climate change: evidence and causes* (February 2014).

Royal Society and National Academy of Sciences, *Population growth, resource consumption, and a sustainable world* (February 1992).

Royal Society and Royal Academy of Engineering, *Nanoscience and nanotechnologies: opportunities and uncertainties* (July 2004).

Royal Society and Royal Academy of Engineering, '*Nanoscience and nanotechnologies: opportunities and uncertainties*': *two-year review of progress on government actions* (October 2006).

Royal Society and Royal Academy of Engineering, *Nuclear energy – the future climate* (June 1999).

Royal Society and Royal Academy of Engineering, *The role of the Renewables Directive in meeting Kyoto targets* (October 2000).

Royal Society et al., *Global response to climate change* (June 2005).

Royal Society et al., *The science of climate change* (May 2001).

Royal Society et al., *Transgenic plants and world agriculture* (July 2000).

Lord Sainsbury of Turville, *The race to the top: a review of Government's science and innovation policies* (HMSO, 2007).

Salomon, Jean-Jacques, *International scientific organisations* (OECD, 1965).

Salomon, Jean-Jacques, 'International scientific policy', *Minerva*, 2 (1964), 411–34.

Salomon, Jean-Jacques, *Science and politics* (Macmillan, 1973).

Schell, Orville, 'China's Andrei Sakharov', *Atlantic monthly* (May 1988), 35–52.

Schilling, Warner R., 'Science, technology, and foreign policy', *Journal of international affairs*, 13 (1959), 7–18.

Schroeder-Gudehus, Brigitte, 'Nationalism and internationalism', in R.C. Olby, G.N. Cantor, J.R.R. Christie and M.J.S. Hodge, eds., *Companion to the history of modern science* (Routledge, 1990), 909–19.

Schroeder-Gudehus, Brigitte, 'Science, technology and foreign policy', in Ina Spiegel-Rösing and Derek de Solla Price, eds., *Science, technology and society* (Sage Publications, 1977), 473–506.

SEPSU, *An analysis of the Royal Society Research Grant Scheme* (Royal Society, 1991).

SEPSU, *European collaboration in science and technology: pointers to the future for policy makers* (SEPSU Policy Study No 3, February 1989).

SEPSU, *Guide to European collaboration in science and technology* (December 1987; second edition, December 1990).

SEPSU, *Migration of scientists and engineers to and from the UK* (SEPSU Policy Study No 1: Royal Society, 1987).

SEPSU, *Two reports on University Research Fellowships* (Royal Society, 1993).

Sherrington, Charles, 'Address at the Anniversary Meeting, 30 November 1922', *Proceedings of the Royal Society of London. Series A, Mathematical and physical sciences*, 102 (1923), 373–88.

Sherrington, Charles, 'Address at the Anniversary Meeting, 30 November 1923', *Proceedings of the Royal Society of London. Series A, Mathematical and physical sciences*, 105 (1924), 1–16.

Sizer, John, 'A critical examination of the events leading up to the UGC's grant letters dated 1 July 1981', *Higher education*, 18 (1989), 639–79.

Smith, D.C., Collins, P.M.D., et al., 'National performance in basic research', *Nature*, 323 (1986), 681–4.

Smith, Kathlin, 'The role of scientists in normalising US-China relations: 1965–1979', *Annals of the New York Academy Science*, 866 (1998), 114–36.

Snowden, Christopher, 'Technological innovation in industry and the role of the Royal Society', *Notes and records of the Royal Society of London*, 64 (2010), S55–63.

Southwood, Richard, 'Surface Waters Acidification Programme', *Science and public affairs*, 5 (1990), 74–95.

Stanford Research Institute, *Possible non-military scientific developments and their potential impact on foreign policy problems of the United States* (Senate Committee on Foreign Relations, 1959).

Stevenson, Richard, 'Acid rain: swapping fishermen's tales', *Chemistry in Britain* (May 1990), 397.

Sunderlin, C.E., 'United States science offices abroad', *Nature*, 166 (15 July 1950), 87–8.

Suttmeier, Richard P., 'Scientific cooperation and conflict management in US-China relations from 1978 to the present', *Annals of the New York Academy of Sciences*, 866 (1998), 137–64.

Synge, R.L.M. and Williams, E.F., 'Albert Charles Chibnall', *Biographical memoirs of Fellows of the Royal Society*, 35 (1990), 57–96.

Thompson, Harold, 'Cyril Norman Hinshelwood', *Biographical memoirs of Fellows of the Royal Society*, 19 (1973), 375–431.

Thornton, H.G., 'A note on the visit to Russia of the Royal Society delegation in 1956', *Notes and records of the Royal Society of London*, 12 (1957), 230–6.

Thring, M.W., 'The efficient development of new ideas in industry', *Guardian*, 26 November 1963.

Todd, Alexander, 'Address at the Anniversary Meeting, 30 November 1976', *Proceedings of the Royal Society of London. Series A, Mathematical and physical sciences*, 352 (1977), 451–62.

Todd, Alexander, 'Address at the Anniversary Meeting, 30 November 1977', *Proceedings of the Royal Society of London. Series B, Biological sciences*, 200 (1978), v–xiv.

Todd, Alexander, 'Address at the Anniversary Meeting, 30 November 1978', *Proceedings of the Royal Society of London. Series B, Biological sciences*, 204 (1979), 1–14.

Todd, Alexander, 'Address at the Anniversary Meeting, 30 November 1979', *Proceedings of the Royal Society of London. Series B, Biological sciences*, 206 (1980), 369–80.

Todd, Alexander, 'Address at the Anniversary Meeting, 1 December 1980', *Proceedings of the Royal Society of London. Series B, Biological sciences*, 211 (1980), 1–13.

Todd, Alexander, *A time to remember: the autobiography of a chemist* (Cambridge University Press, 1983).

Todd, Alexander and Cornforth, J.W., 'Robert Robinson', *Biographical memoirs of Fellows of the Royal Society*, 22 (1976), 415–527.

Tooze, John, 'A brief history of EMBO', in *European Molecular Biology Organisation 1964–1989* (EMBO, 1989), 8–19.

Turney, Jon, 'Science suffering says FRS survey', *Times higher education supplement* (3 December 1982), 1.

Vig, Norman J., *Science and technology in British politics* (Pergamon Press, 1968).

Walgate, Robert, 'China wants 800,000 scientists by 1985', *Nature*, 274 (10 August 1978), 525.

Wang, Zuoyue, 'Science and the state in modern China', *Isis*, 98 (2007), 558–70.

Wang, Zuoyue, 'Transnational science during the Cold War: the case of Chinese/ American scientists', *Isis*, 101 (2010), 367–77.

Wang, Zuoyue, 'US-China scientific exchange: a case study of state-sponsored scientific internationalism during the Cold War and beyond', *Historical studies in the physical and biological sciences*, 30 (1999), 249–77.

Ward, Bob, 'The Royal Society and the debate on climate change', in M.W. Bauer and M. Bucchi, eds., *Journalism, science and society: science communication between news and public relations* (Routledge, 2007), 159–72.

Werskey, Gary, *The visible college* (Allen Lane, 1978).

Whitehead, J. Rennie, *Memoirs of a boffin* (published online, 1995).

Whyte, Neil and Gummett, Philip, 'Far beyond the bounds of science: the making of the United Kingdom's first space policy', *Minerva*, 35 (1997), 139–69.

Wilkie, Tom, *British science and politics since 1945* (Blackwell, 1991).

Williams, James H., 'Fang Lizhi's big bang: a physicist and the state in China', *Historical studies in the physical and biological sciences*, 30 (1999), 49–87.

Williams, Roger, 'Some political aspects of the Rothschild affair', *Science studies*, 3 (1973), 31–46.

Williams, Rowan, 'A sermon . . . on the occasion of the 350th anniversary', *Notes and records of the Royal Society of London*, 64 (2010), 213–5.

Williams, Ruth, 'Sir John Gurdon: godfather of cloning', *Journal of cell biology*, 181 (2008), 178–9.

Williams, Shirley, 'The responsibility of science', *Times Saturday review* (27 February 1971), 15.

Williams, Trevor I., *Howard Florey: penicillin and after* (Oxford University Press, 1984).

Williams, Trevor I., *Robert Robinson, chemist extraordinary* (Clarendon Press, 1990).

Wilmut, Ian, et al., 'Viable offspring derived from fetal and adult mammalian cells', *Nature*, 385 (27 February 1997), 810–3.

Wynne, Brian, 'Knowledges in context', *Science, technology and human values*, 16 (1991), 111–21.

Ziman, John, 'The problem of Soviet scientists', *Nature*, 246 (7 December 1973), 322–3.

Ziman, John, 'Public understanding of science', *Science, technology and human values*, 16 (1991), 99–105.

Command papers (all London: HMSO)

Annual report of the ACSP 1956/57 Cmnd 278 1957.

University development 1952/57 Cmnd 534 1958 [UGC quinquennial report].

Annual report of the ACSP 1959/60 Cmnd 1167 1960.

Annual report of the ACSP 1960/61 Cmnd 1592 1962 [includes case for what became the Trend Committee].

Annual report of the ACSP 1961/62 Cmnd 1920 1963.

Committee of Enquiry into the organisation of civil science Cmnd 2171 1963 [the Trend Report].

University development 1957/62 Cmnd 2267 1964 [UGC quinquennial report].

Annual report of the ACSP 1963/64 Cmnd 2538 1964 [valedictory report, with advice for its successors].

The brain drain Cmnd 3417 1967.

Report on science policy Cmnd 3007 1967 [report by the CSP].

University development 1962/67 Cmnd 3820 1968 [UGC quinquennial report].

A framework for government research and development Cmnd 4814 1971 [Green Paper: annexes include Rothschild and Dainton reports].

Report of the working group on scientific interchange Cmnd 4843 January 1972 [CSP review of the ESEP].

Framework for government research and development Cmnd 5046 July 1972 [response to Cmnd 4814].

University development 1967/72 Cmnd 5728 1974 [UGC quinquennial report].

Third report of the ABRC Cmnd 7467 1979.

Review of the framework for government research and development (Cmnd 5046) Cmnd 7499 March 1979.

Realising our potential: a strategy for science, engineering and technology Cm 2250 1993.

Index

'A' side (physical sciences)
  RS Secretaryship and publications,
    90, 252
  tradition of alternating Presidents, 13, 14
Aarhus conference, 1972, 219
ABRC (Advisory Board for the Research
    Councils)
  as successor to CSP, 64
  creation of, and the Dainton Report, 114
  differing perspectives of RS and, 79
  on European scientific collaboration, 221
  policy studies team within, 111
  preferred approaches over new research
    talent, 68
  reconstituted in 1990 and dissolved in
    1993, 120
  RS budget discussions, 1984, 71
  RTX proposal, 118
  A Strategy for the Science Base, report,
    1987, 72
  support for the research groups
    scheme, 64
  working group on genetic engineer-
    ing, 126
Academia Europaea, 226, 227, 229
Academia Sinica, 181
academic careers in the 1970s, 64, 65
academic research
  seen as underfunded in 1960, 34
  shift of RS concerns toward, 82
  spending on in 1955/56, 27
Académie des Sciences, 228
academies of science
  Europe-wide academies, 226
  See also national science academies.
Academy of Medical Sciences, 125,
    129, 262
accolade committee, CEI, 95
accolade functions
  globalisation of, 231
  in engineering, 97
  review of RS's, 106, 239

accommodation problems
  Andrade memorial and, 8
  Chicheley Hall, 256
  fundraising associated with, 284
  fundraising for new premises, 55
  in 1960, 27
  modernisation in 2002–3, 286
  move to Carlton House Terrace, 255
  opening of new premises, 1967, 213
  Robinson's ambitions, 23
  solution achieved under Florey, 24
  Treasury grant toward, 44
accountability, political. *See* public
    accountability
acid rain pollution, 140
ACME (Advisory Committee on
    Mathematics Education), 139
ACSP (Advisory Council on Scientific
    Policy)
  abolition and replacement by CSP, 30, 51
  and the *Encouragement* report, 37, 41
  Chairman's role in 1960, 28
  discussion of RS *Grant-giving* propo-
    sals, 44
  investigation into international scientific
    organisations, 246
  Overseas Scientific Relations committee,
    156, 160
  RS Officers as Chairmen and Deputy
    Chairmen, 22, 23, 29
  Trend Committee and, 47, 48
administrative review, 1995, 277
admission fees, Fellowship, 287
Adrian, Lord (Edgar) (FRS, 1923; PRS,
    1950–5)
  as presidential candidate in 1945, 14
  established attitudes unchanged by, 269
  on engagement with government, 23, 27
  on presidential duties, 262
  on recognition of scientific excellence, 33
  on Robinson's accommodation plans, 24
Advanced Fellowship scheme (SERC), 69

Printed in the United States
By Bookmasters